# PHYSICAL THERAPY PERSPECTIVES IN THE 21ST CENTURY – CHALLENGES AND POSSIBILITIES

Edited by **Josette Bettany-Saltikov**
and **Berta Paz-Lourido**

**Physical Therapy Perspectives in the 21st Century – Challenges and Possibilities**
Edited by Josette Bettany-Saltikov and Berta Paz-Lourido

**Published by InTech**

Janeza Trdine 9, 51000 Rijeka, Croatia



**Publishing Process Manager** Maja Jukic

**Technical Editor** Teodora Smiljanic

**Cover Designer** InTech Design Team

Additional hard copies can be obtained from orders@intechopen.com

Physical Therapy Perspectives in the 21st Century – Challenges and Possibilities, Edited by Josette Bettany-Saltikov and Berta Paz-Lourido
p. cm.
ISBN 978-953-51-0459-9

For EU product safety concerns:
IN TECH d.o.o., Prolaz Marije Krucifikse Kozulić 3, 51000 Rijeka, Croatia,
info@intechopen.com or visit our website at intechopen.com.

# Meet the editors

Dr Josette Bettany-Saltikov qualified as a Physical Therapist in Malta in 1986 and has a Master's degree in Orthopaedic and Sports Physical Therapy from the State University of New York at Buffalo (USA).Her PhD from Kings College, University of London, investigated the Topographical, kinesiological and Psychological factors in patients with scoliosis. Josette has over 25 years experience of conducting research studies in back shape, posture and spinal deformities and over 17 years experience of teaching research methods to Physical Therapists and other allied health professionals. She has published and presented widely at both national and international conferences and is both a manuscript reviewer and on the editorial board of a number of International Journals.

Dr Berta Paz Lourido qualified as Physical Therapist in Spain in 1995. She is also Bachelor in Pedagogy, Post-graduated Degree in Manual Therapy, Post-graduated Degree in Cooperation and International Development. Director of the on–line postgraduate degree for physiotherapists: "Physiotherapy in Primary Health Care and Community Care", 2005-2008. Currently involved in 3 research projects in the fields of Primary Health Care, education in physiotherapy in the Bologna process and the quality of life in people with disabilities. Past projects developed were related to quality of education in physiotherapy, internationalisation, physiotherapy in home care and cooperation for development projects. She has widely contributed with scientific presentations in international conferences regarding physiotherapy, public health and education, as well as published her research in international journals and books.

# Contents

# Preface

We are very excited to introduce the first InTech book dedicated to advancements in the field of physical therapy. Indeed in the last few decades physical therapists have made great strides in advancing both the research knowledge base within physical therapy whilst at the same time making significant improvements in clinical practice.

In our society today physical therapists face continuous challenges related to the advancement of physical therapy practice, medicine and technology in addition to the effects of the global economic downturn. An ongoing challenge for the profession is the need to underpin physical therapy practice with the best available evidence. For the past five decades, physical therapy has been a burgeoning science as evidenced by the numerous national and international physical therapy organisations to be found worldwide.

The aim of this book is not to cover all areas of research and practice in physical therapy. There are many emerging contexts that require the intervention of physical therapists and it is not possible to include all of them within a limited space. We consider that this book highlights a number of key aspects related to the way that physical therapy is currently contributing to better overall health of citizens everywhere who are suffering from numerous diverse medical conditions. Because of this, different topics and research approaches will be found within this book, ranging from very specific interventions to broader concepts within physical therapy practice that consider the social determinants of health.

As this book will demonstrate, many researchers throughout the world are contributing greatly to our understanding of the use of a range of diverse physical therapy approaches for a wide array of medical conditions. More recently, increasing and innovative contributions have been made by physical therapists worldwide towards this important enterprise. They are also providing strong evidence to support the benefits of exercise to both avoid and treat acute and chronic diseases.

This book contains new information on physical therapy research and clinical approaches that are being undertaken into numerous medical conditions; biomechanical and musculoskeletal conditions as well as the effects of psychological factors, body awareness and relaxation techniques; specific and specialist exercises for the treatment of scoliosis and spinal deformities in infants and adolescents; new thermal agents are being introduced and different types of physical therapy interventions are being introduced for the elderly both in the home and clinical setting. Additionally research into physical therapy interventions for patients with respiratory, cardiovascular disorders and stroke is being undertaken and new concepts of wheelchair design are being implemented.

Our hope is that this book will become an important compendium and resource for physical therapists who treat a wide array of clinical conditions. Additionally, we hope these reviews will act to stimulate researchers throughout the world to continue this important work and solve persistent clinical questions posed by physical therapy researchers and clinicians worldwide.

**Dr. Josette Bettany-Saltikov**
Teesside University, Middlesbrough
UK

**Dr. Berta Paz-Lourido**
University of the Balearic Islands
Spain

# Part 1

# Physical Therapy for Scoliosis and Spinal Deformities in Infants, Adolescents and Adults

1

# Physical Therapy for Adolescents with Idiopathic Scoliosis

Josette Bettany-Saltikov[1] et al.*
[1]*Teesside University, Middlesbrough,*
[1,2]*UK*

## 1. Introduction

Scoliosis is a three-dimensional deformity of the spine. In its most common form, idiopathic scoliosis (70% to 80% of cases), the causes are unknown (Rowe 2003). AIS is discovered at 10 years of age or older, and is defined as a curve of at least 10°, measured on a standing radiograph using the Cobb technique (Parent et al, 2005). While the prevalence of AIS is around 3% in the general population, almost 10% of those diagnosed with AIS will require some form of treatment; usually observation or scoliosis-specific exercises (SSE) for mild curves, braces for moderate curves and spinal surgery for severe curves (Cobb angle >50$^0$). Up to 0.1% of the population is at risk of requiring surgery (Lonstein, 2006). A severe form of AIS is more commonly found in females. Typically, AIS does not cause any health problems during growth (except for extreme cases). However, the resulting surface deformity frequently has a negative impact on adolescents` body-image and self-esteem that can give rise to quality of life (QoL) issues and in worst cases, psychological disturbances (Maclean et al, 1989). Adolescent patients are generally treated in an attempt to halt the progressive nature of the deformity. No treatments succeed in full correction to a normal spine, and even reduction of the deformity is difficult (Danielson and Nachemson, 2001). If scoliosis surpasses a critical threshold, usually considered to be 30° Cobb, at the end of growth, the risk of health and social problems in adulthood increases significantly (Negrini, 2005). Problems include reduced quality of life, disability, pain, increased cosmetic deformity, functional limitations, sometimes

* Tim Cook[2], Manuel Rigo[3], Jean Claude De Mauroy[4], Michele Romano[5], Stefano Negrini[5], Jacek Durmala[6], Ana del Campo[2], Christine Colliard[2], Andrejz M'hango[7] and Marianna Bialek[7]
[1]*Teesside University, Middlesbrough, UK*
[2]*SpineCorporation, UK, Spain*
[3]*Elena Salva Institute, Barcelona, Spain*
[4]*Clinique du Parc – Lyon, France*
[5]*Italian Spine Scientific Institute, Milan, Italy*
[6]*Medical University of Silesia, Katowice, Poland*
[7]*Fits Institute, Poland*

pulmonary problems, and progression during adulthood (Weinstein et al, 2003). Because of this, management of scoliosis also includes the prevention of secondary problems associated with the deformity.

### 1.1 Current consensus and clinical practices

The level of evidence in the conservative management of AIS is not high, whatever treatment is considered. Treatments applied in this field include surgery (fusion), bracing and/or exercises. In the past electrical stimulation has also been used but without significant results; other treatments not recommended by the current guidelines (Weiss et al, 2006) include manipulations and insoles. The existing evidence concerning these treatments, which is classified according to the Oxford Centre for Evidence Based Medicine (Philips et al, 2001), can be summarized as follows: scoliosis-specific exercises (SSE) can be recommended as a first step in the treatment of AIS to avoid and/or limit curve progression (grade A); bracing is recommended when SSE`s are unable to prevent progression (grade B); and surgical fusion is the unavoidable treatment when AIS is either causing symptoms (rarely), conservative treatment has failed or a well-informed patient requests fusion (grade C recommendation) (Weinstein et al,2008).

Considering the evidence, the treatment approaches adopted by various orthopaedic surgeons and physicians specialised in the field of conservative management of scoliosis are divided indicating a lack of clinical equipoise across the professions and different countries. In general these approaches can grossly be split into two opposing groups: the first group regard the exercises as useless, while the second group use them and advocate their efficacy (Negrini et al, 2005). Similarly, bracing has been abandoned by some (Dolan and Weinstein, 2007) while others support its use on the basis of the existing weak evidence about efficacy; fusion is generally considered to be necessary when AIS either exceeds a certain degree, previous treatments have failed or AIS causes symptoms, but indications vary widely according to the preference or not of the treating physician/surgeon for conservative management (Dolan and Weinstein, 2007). These two conflicting approaches seem to prevail in two different regions of the world: while in the US and UK, the wait and see strategy prevails, in various parts of continental Europe, Eastern and Southern Europe conservative treatment (SSE`s and bracing) is considered to be of benefit to the patient and used routinely by the large majority of scoliosis physicians and surgeons.

A possible reason for the negative beliefs towards SSE within the clinical community in the United Kingdom is the lack of knowledge within the physical therapy community and associated clinical specialists. These pathological condition-specific exercises are not taught at either undergraduate or post-graduate level within the physiotherapy curriculum in the UK. Most clinicians (both physiotherapists and surgeons) in the UK normally do not appreciate the difference between SSE and general physiotherapy. Scoliosis-specific exercises consist of individually adapted exercises that are taught to patients in a centre that is totally dedicated to scoliosis treatment. The patients learn an exercise protocol that is personalized according to medical and physiotherapeutic evaluations. Usual generalised physiotherapy (GPT), on the other hand, is more generic, usually consisting of low-impact stretching and strengthening activities like yoga, pilates or tai chi (taiji), but can include many different exercise protocols according to the preferences of the therapist. The understanding within the generalised AIS treating community in the UK and USA may be based on the effectiveness of generalised physiotherapy which has not been shown to be effective.

### 1.2 Quantity and quality of the research to date and their limitations

Recent systematic reviews (Negrini et al, 2009) have shown the possible effects of SSE`s on scoliosis primarily in terms of Cobb angle, based on controlled studies, which were mainly observational and partly prospective. A Cochrane Review (Romano et al, 2009) (co-authored by 3 of the current authors: Bettany-Saltikov, Negrini and Romano) on the effectiveness of scoliosis-specific exercises for patients with idiopathic scoliosis (currently being peer-reviewed) found that, despite a comprehensive search of published and unpublished literature, only two studies met the stringent Cochrane methodological criteria. Of these only one was a randomised controlled trial; this trial compared a protocol of exercises, electrostimulation, traction and postural training (Wan et al, 2005) to a protocol of electrostimulation, traction and postural training. This study provided very low quality evidence in favour of SSE`s versus the same protocol without exercises. More recently, a prospective controlled cohort study comparing the SEAS exercises versus usual physiotherapy (Negrini et al, 2008b), also provided very low quality evidence in favour of SEAS exercises. The outcome most frequently used across previous studies was the Cobb angle; only Negrini's study considered the more patient-centred outcome of brace avoidance as a main outcome.

Further, another systematic review that also included observational trials was conducted by Negrini et al in 2008 as an update to a previous review conducted in 2003. This review was included in the DARE Cochrane Database (Negrini et al, 2003c). 19 studies were retrieved, including one RCT and eight controlled studies; 12 studies were prospective. In total the 19 papers included considered 1654 patients and 688 controls in all. The highest-quality study (RCT) compared two groups of 40 patients, showing an improvement of curvature in all patients in the intervention group after six months. Apart from one old study (conducted in 1979 and of very low methodological quality using general physiotherapy, not SSE), all studies confirmed the efficacy of scoliosis-specific exercises in reducing the progression rate (mainly in early puberty) and/or improving the Cobb angles (around the end of growth). SSE`s were also shown to be effective at reducing brace prescription. Although the authors of this review concluded that the current evidence on exercises for AIS is of level 1b, the only RCT reported within the review had a number of serious methodological issues. This raises the need for a well conducted RCT.

The aims of Scoliosis-Specific Exercises considered in various research protocols to date include: limiting or halting scoliosis progression, improving physical functioning and reducing scoliosis patients` disability and avoiding more invasive methods of treatment such as bracing. In the worst patients (fused, or elderly in a flexed posture) pulmonary rehabilitation has also been considered. If scoliosis does progress beyond a certain critical threshold (generally considered to be 30 degrees), bracing is generally considered (by physicians or surgeons who normally believe in bracing) to be the subsequent step of treatment with the aim of avoiding surgery. Nevertheless, it has been shown that braces have psychological consequences on adolescents during a crucial pubertal period of spinal growth when relationships with the opposite gender are generally initiated and body self-image and self-esteem develops (Falstrom et al,1986). Surgery has also been shown to have a significant psychological impact, as well as causing considerable functional limitations due to the fusion of the spine (Hawes 2006b). Hence, there is a promising role for therapeutic scoliosis-specific exercises, that do not have any unwanted psychological consequences.

The International Scientific Society On Scoliosis Orthopaedic and Rehabilitation Treatment (SOSORT) has proposed and supports the use of SSEs and gives indications for their use (Weiss et al, 2006). Furthermore and most importantly, scoliosis-specific exercises based on specific auto-correction and stabilization are also supported by a recent consensus of specialists in the field of the conservative management of scoliosis (Weiss et al, 2006). Numerous scoliosis-specific exercise approaches to the treatment of mild to moderate scoliosis are available. The following SSE approaches that will be discussed in this chapter include; The Scientific Exercise Approach to Scoliosis (SEAS), written by M.Romano and S. Negrini; The Barcelona Scoliosis Physical Therapy approach (BSPTS), written by Dr. Manuel Rigo; The Lyon approach, written by Dr. Jean Claude De Mauroy; The Functional Individual Therapy for Scoliosis (FITS) approach, written by Andrejz Mhango and Marianna Bialek; The DoboMed approach, written by Prof. Jacek Durmala; and finally the SpineCor approach, written by Ana del Campo and Dr. C Coillard. Each approach will now be discussed in turn.

## 2. SEAS approach (Italy)

### 2.1 Introduction

SEAS is an acronym for "Scientific Exercise Approach to Scoliosis". The name indicates that this approach is based on scientific principles, which is a very important feature of this treatment approach. The continuous improvements and developments to the original method results from the constant introduction of new knowledge derived from the scientific literature.

### 2.2 History of the SEAS method

The SEAS method originates from the Lyon approach where a number of the basic characteristics to the approach had already been developed. This includes: improving the patient's awareness of their deformity, autonomous correction by the patient, the use of exercises to stimulate a balance reaction, as well as the performance of in-brace scoliosis specific exercises using the brace as a training tool (Romano et al, 2008).

### 2.3 Principles of the SEAS method

The difficulty with treating patients with idiopathic scoliosis is the impossibility of working directly on the cause of the deformity, which is still unclear. Each type of treatment whether surgery, bracing, or scoliosis-specific exercise (SSE), is aimed at minimizing the effects of the symptoms of the disease. In the SEAS approach the two main treatment objectives are active self-correction as well as the improvement of spinal stability. The self-correction component can be defined as the search for the best possible alignment within three dimensional spatial planes, that are obtained autonomously by the patient.

These are some assumptions the form the basis of the SEAS approach:

Conservative treatment of scoliosis has the aim of preventing the progressive deformation of the vertebrae, caused by the constant asymmetric pressure on them.

The self-correction obtained by the active movement of the patient lasts for the duration of this movement.

Even using very demanding treatment approaches that involve performing exercises for several hours a day it is not possible to maintain the correct position after the exercise sessions finishes.

The purpose of the SEAS exercises is therefore to find a strategy that helps the patient search for the position of self-correction, as they move throughout the day during their usual activities of daily living. In our concept, this can only be done by developing a specific reflex neuromotor spinal reaction, that when performing different destabilizing everyday activities drive the spine toward corrections instead of postural collapse.

For this reason, an essential aspect of self-correction, structured according to the SEAS approach, is that this movement has to be performed in a local `direct' manner: 'direct' means a self-correction performed by the patient focusing only on moving the spine, without any external aids (supports, specific body positions...) or movements of other body parts (limbs, head...).

According to the SEAS approach the execution of an "indirect" self-correction movement does not achieve the aim on which this concept is based, i.e. moving from the "search of the best passive alignment" to the "functional stimulation of the alignment reflex". In fact, neurophysiologically, an active self-performed movement can be integrated into motor behaviours ("alignment reflex") better then passive ones; moreover, as we will see below, an active self-correction can be "challenged" in many very different situations (exercises) simulating real everyday life ("functional stimulation") which is better than specific static positions requiring specific supports.

## 2.4 Description of the SEAS approach

During the execution of an "active" self-correction we can see:

Appreciable improvement of the aesthetic component of the trunk.
Improvement of the plumbline and the weight distribution (also of the peripheral joints).
Improvement of the postural component.

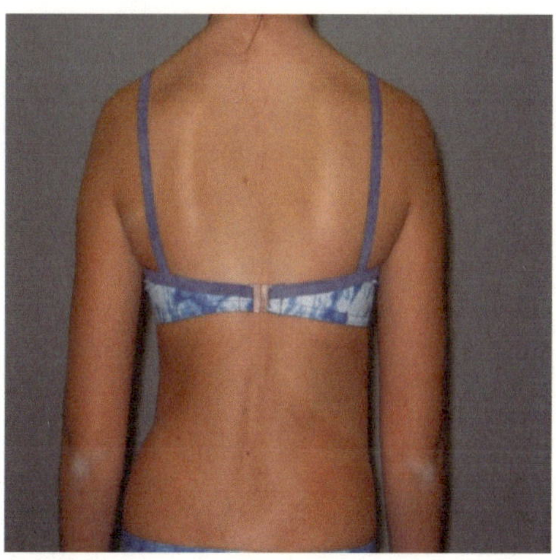

Fig. 1. Before Active Self Correction.

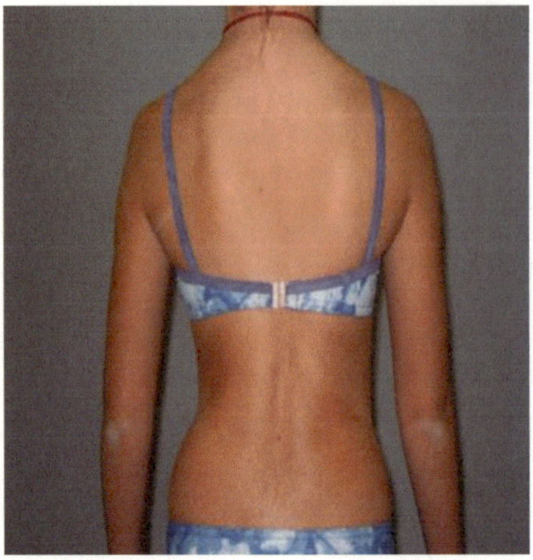

Fig. 2. After Active Self Correction.

The modifications are not only postural but also measurable on X-Ray. SEAS exercises focus on the three primary principles, listed below in relation to their importance.

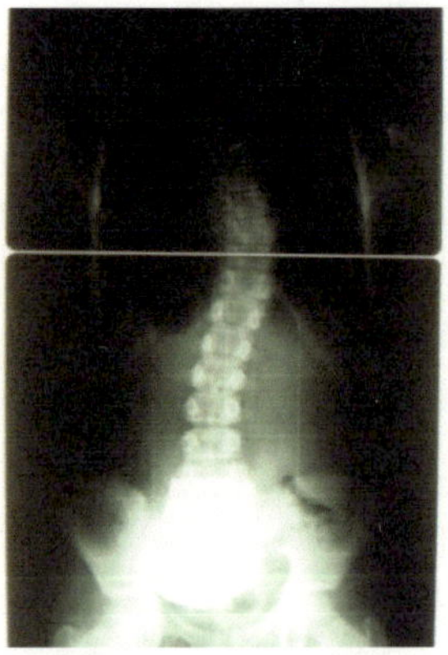

Fig. 3. Before Active Self Correction.

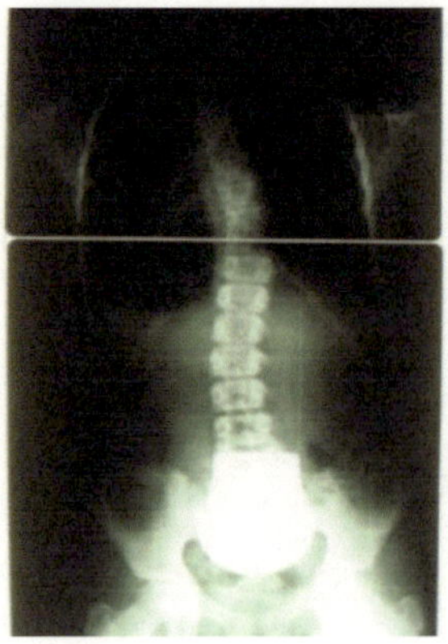

Fig. 4. After Active Self Correction.

**Principle 1. The SSE exercises use an element of "distraction" for training the maintenance of self-correction**

If we define the term "scoliosis specific exercise" as a movement performed by a patient in order to counteract the pathology, in scoliosis, the specific exercises are structured to have a direct corrective effect on the curve. In the majority of corrective methods that are based on exercises, self-correction is already integrated into the movement performed by the patient. In the SEAS approach, these 2 elements (self-correction and exercise) are integrated but performed in succession. **Self-correction**, especially when performed in various directions, is the real movement against the misalignment. The "exercise" is added to the self-correction in order to train the response in the correct position in different situations of postural stress. Therefore the exercise is only an element of the complex activity performed by the patient in order to counteract the curve.

**Principle 2. The purpose of the SSE exercise is to improve the primary 'target function', i.e. the stability of the spine**

The **target functions** are those elements, which need to be improved with the treatment. The main function is the stability of the spine. The progression of an unstable scoliosis is always towards an aggravation. The scoliotic spine can hence be described as a structure whose constitutional elements are no more able to maintain the physiological alignment. The asymmetrical distribution of loads and the progressive deformation of the vertebrae increases the difficulty of the spine to preserve stability. For this reason one of the primary aims of the exercises in this approach, is the stimulation of the musculature with the greatest potential for stabilization. This aim is achieved by asking the patient to perform various actions which tend to destabilize the spine, for example, imbalance exercises, exercises with addition of loads and exercises with dynamic components; the patient must be able to maintain active self-correction and spine stability in spite of these destabilizations, training in this way the target function.

**Principle 3. The aim of the exercise is to improve the deficit found during the initial assessment (strength, muscular retractions, motor coordination ...)**

The SEAS approach requires an accurate assessment of the patient prior to treatment. In addition to the normal measurements of the specific pathological assessment, (Cobb angle, Bunnell, plumbline, sagittal posture and aesthetic parameters), the patient performs a number of tests in order for the physiotherapist to evaluate the physical condition of the patient. These include testing the strength and elasticity of the muscle groups that mainly influence the posture of the pelvis and spine, and testing neuromotor development (balance, proprioception, hand-eye co-ordination etc...). The purpose is to obtain a reliable overall assessment of the patient`s abilities. The outcome of this assessment will be used to determine the choice of exercises that may improve the deficits discovered during the assessment, and to identify those exercises which are most suitable for the improvement of the stability of the spine. For example, if during the assessment, difficulties of balance are discovered, just those exercises which contain important elements of balance will be used in order to improve the stabilization of the spine.

In a word the fundamental principle on which the SEAS approach is based, is the word "control". The patient is asked to always verify, during the performance of the exercise, the correct maintenance of the selected self-correction. In order to facilitate this `control` the patient uses a series of standard questions that they ask themselves during treatment.

**The four questions**

**Question 1**: Whilst performing SEAS exercises, patients are always told to start from where the spine is in a position of basic support. This means that before performing any exercise the patient always needs to check that they are not in a position of uncontrolled relaxation. For this reason, the first question is: **Is my spine supported?** At this stage, the patient does not yet perform the correction. Regardless of the position (sitting, standing etc..) from which the patient starts performing the exercise, he is requested only to maintain the basic control of the spine, being aware that he is not relaxed. Once he/she is aware of this precondition the patient performs the self-correction chosen by the therapist on the basis of their curvature, imaging, aesthetics, and posture.

**Question 2:** To verify that they have successfully performed the self-correction, the patient asks themselves the second question: **Is my body now more symmetrical than before?** The first test is visual **(I see that my body is more symmetrical than before!**) because the patient performs the exercises in front of the mirror. But then over time (usually months) they will be more and more linked to somatosensory perceptions **(I feel** that my body is more symmetrical than before!) because the exercises will be carried out without the help of the mirror. The patient consequently will gradually, over time, perform exercises that will aim to make it increasingly more difficult to maintain the self-correction. The question that they will now ask themselves at this point will be:

**Question 3: Whilst doing the exercise am I able to maintain the correction?** Based on the reply of the patient the therapist will be able (beyond the simple observation) to understand whether the difficulty of the exercise is appropriate to the patient's ability to maintain self-correction during the execution of the exercise. In fact, if the patient replies "no" to this question the therapist will know that the patient should perform an exercise that is less difficult.

**Question 4:** The patient performs the exercise for about 10 seconds then slowly relaxes, returning from the self-correction to his/her normal position. The question that they ask themselves now will be: **Am I able to see that my body returns back to the original position it was in before performing the self-correction?** That means that the patient must be able to observe that the position of the trunk has changed from the position of correction to its usual position. This is probably the most important `control' performed by the patient to verify if the exercise has been carried out properly. The relaxation phase is the phase where movements of the body mass occur as a result of an elastic recoil and not as a consequence of active movement. If the patient answers "no" to this question the self-correction has been lost during the execution of the exercise and the exercise performed has lost its corrective specificity, becoming a mere gymnastic exercise. After this the patient is required to improve their perform of the exercise or, if this is not possible and the patient finds the exercise too difficult to maintain, then the main aim which is self-correction will need to be changed to a less demanding one.

**Application protocol of the SEAS exercises**

The typical application of the SEAS protocol consists of the teaching of an individual exercise program that is preceded by patient assessment and followed by family counselling. The SEAS approach attaches great importance to counselling because the patient's family is considered to be an important member of the therapeutic team, without whose great support it is very difficult to obtain the compliance necessary to achieve the final result. The average session with a physiotherapist lasts approximately 1.5 hours and the sessions are performed every 3 months (3-4 times a year) by a skilled therapist. The patient is provided with a copy of the exercises on an appropriate format (DVD, USB Flash Drive) so that the patient can complete the program either at home or at a nearby gym with the help of a therapist, personal trainer or family member. The program includes 2-3 sessions a week lasting about 45 min, or a 15 min daily session, according to the preferences of the patient and the family.

**Strength of the SEAS approach**

One of the most important elements that distinguish this approach from other treatment methods is the absolute attention and consideration of the patient`s individual characteristics for the design of the treatment program. The three-dimensional self-correction is not only defined by the pattern of scoliotic deviation but also by the patient's ability to perform the exercises. Initially the direction of correction will be chosen from the simplest ones suited to the specific case and that the patient is able to follow correctly. Gradually, as the ability to execute and to control the correction becomes more refined, self-correction will become more complex up to the best execution. Even the choice of exercises follow the same path, whereby the level of difficulty should always take into account the patient's ability. The difficulty of the exercise should increase in parallel with the increase of the patient's ability. Another typical feature of the SEAS protocol is that the choice of exercise takes into consideration the eventual different stages of the treatment. For example, a patient with a prescription for bracing will make an initial treatment phase in which the objectives of the exercises will be focused on improving the mobility and the plasticity of the trunk and the spine to allow the brace to achieve the best possible corrective effect. Subsequently the exercises will change, and may try to take advantage of the forces of the brace as a useful tool to obtain a better modelling of the trunk. The ISICO clinic has a very useful tool that can be used to help develop an appropriate exercise program. This free software is available online at www.scoliosismanager.it

## 2.5 Research results

Over several years a set of essential outcomes of the SEAS approach, applied at different stages of treatment, have been published in articles that can be found on PubMed. SEAS exercises are effective in slowing down the progression of the curves and in reducing the rate of prescription of braces. The results of patient groups treated with SEAS exercises have been compared in several studies to patient groups that have performed other types of exercise (Negrini et al, 2006b). Even a worst-case analysis study has shown a statistically significant difference in treatment failure among patients treated with SEAS exercises compared to patients who have followed different exercise programs (Negrini et al, 2008a), with a particular emphasis on the possibility of reducing the rate for the need for surgery (Negrini et al, 2008b). As already mentioned patients who wear a brace have an initial program of intense mobilization to increase the corrective action of the brace. The group of patients treated with the SEAS protocol obtained better results than the control group who carried out various exercises (Negrini et al, 2006a). Other important effects of the SEAS exercises are their effectiveness in reducing the loss of correction due to the gradual weaning of the brace (Zaina et al, 2009) and the ability to control the progression of curves in adult patients (Negrini et al, 2008c).

## 3. Barcelona Scoliosis Physical Therapy School (BSPTS)

### 3.1 Introduction

The 'Barcelona Scoliosis Physical Therapy School' - BSPTS - is the physiotherapy method integrated into the rehabilitation approach used in the 'Elena Salvá Institut' in Barcelona, Spain. BSPTS is a physiotherapy method that can be defined as a therapy plan of cognitive, sensory-motor and kinaesthetic training which teaches the patient to improve her/his scoliosis posture and soft tissue imbalance, utilizing the assumption that scoliosis posture and soft tissue imbalance promote curve progression. The BSPTS recognises the importance of a multidisciplinary team approach to treating scoliosis which should include a medical doctor, physiotherapist, orthotist and psychologist.

### 3.2 History of the BSPTS method

The BSPTS was founded by Elena Salvá PT (1926-2007) circa 1968. It follows the Schroth principles, adapted from the original intensive-in-patient rehabilitation protocol used at the Katharina Schroth Clinic (Bad Sobernheim, Germany) to an equally intensive but out-patient regimen. The original Schroth principles were described by Katharina Schroth (Lehnert-Schroth, 2007). She opened her first clinic in Meissen in 1921. Katharina Schroth and her daughter Christa Lehnert-Schroth, also a physiotherapist, emigrated later to the former West Germany and opened a second clinic in Bad Sobernheim in 1961.

### 3.3 Principles of the BSPTS method

A paper describing the method from K.Schroth has recently been published by her grandson Hans Rudolf Weiss (Weiss, 2011). The primary principles can be briefly summarized in three points:

1. In idiopathic scoliosis, muscle imbalance is secondary to the deformity and its progression. Muscle imbalance can only be corrected by reaching the best possible three-dimensional 'self correction' before isometric muscle tension is used to stabilize this position.
2. Repetition of the best corrected posture, with the help of propioceptive and exteroceptive stimulation as well as visual control (mirror), is a useful mechanism to achieve a 'corrective body schema' to substitute for the 'scoliosis body schema'.
3. After the best possible 3D correction is achieved, specific breathing mechanics can be introduced to increase the corrective effect, whilst at the same time re-shaping the deformed trunk (this system was initially called 'orthopaedic breathing').

Secondary principles describe how to correct the postural component of the scoliosis in 3D according to a previously diagnosed curve pattern. Primary and secondary Schroth principles have been well described in numerous papers and books (Weiss, 2011).

The BSPTS, after closely following all of the newly acquired information on the three-dimensional nature of idiopathic scoliosis, has slightly but significantly modified the secondary principles. However, the primary principles are unchanged. In this section, the secondary principles about specific 3D correction according to the BSPTS are described.

For any physiotherapist interested in scoliosis treatment it is essential to understand the three-dimensional nature of idiopathic scoliosis. We have recently defined idiopathic scoliosis as a multi-factorial three-dimensional deformity of the spine and the trunk, which appears and can progress during any of the rapid periods of growth in apparently healthy children. The aetiology is unknown but some models explain how idiopathic scoliosis develops and progresses (Burwell, 2009). The basic scoliosis deformity is classically described as a lateral deviation of the spine associated with axial rotation, however, it can be better defined as a combination of torsional regions separated by junctional zones, each region formed by a variable number of vertebrae in extension (or lordosis) deviated and rotated to the same side (Dubousset 1992). The scoliotic spine suffers a mechanical torsion, divided into inter-vertebral torsion and intra-vertebral torsion, the first being predominant close to the junctions and the second at the apical region. On the other hand, torsion produces a translation of the apical region. In a typical right thoracic scoliosis, translation follows an arch in a clockwise direction (Perdriolle & Vidal 1985). The vertebrae at the apical region present a typical morphological deformity: 1) Lateral wedging, from the lateral translation; 2) Dorsal wedging, from the ventral translation. Dorsal wedging is related to a relative anterior spinal overgrowth (RASO), due to eccentric growth caused secondarily by torsion.

From the physiotherapy perspective it is necessary to differentiate the structural component and the non-structural component. The structural component is related to bone deformity and does not occur in a specific moment but over time as a result of asymmetrical loading, according to the 'Vicious Cycle model' (Stokes et al, 2006). The non-structural component involves a purely postural component which can be reduced by self-correction in a moment, as well as a soft tissue component. The asymmetry of the soft tissues is partially fixed, but changable by a) exercises to increase flexibility and b) manual therapy intervention in a relatively short time of intensive physiotherapy. There is a relatively accepted dogma in scoliosis physiotherapy stating that idiopathic scoliosis is a flat back deformity or lordotic deformity and consequently there is a need to use exercises that increase kyohosis in the thoracic spine. Some schools recommend re-kyphosant exercises, which were generally performed by promoting trunk flexion in a single plane, the saggital plane. However, the geometry of the spine is not always hypokyphotic/hypolordotic, but highly variable (Legaye & Orban 1995), and the structural flat back is closely related to the torsional element.

### 3.4 Description of the BSPTS method

The BSPTS recommends looking at every patient to recognize the abnormal geometry of the spine and then, by using a general principle of correction called 'detorsion', try to attain the best correction until halted by the structural component. After attaining the best correction of the non-structural component, some degree of 3D deformity related to the structural component will still be noted. This may be minor or major depending on the severity of the treated scoliosis. At this point, the therapist can ask the patient to gently intensify the corrective forces. However, the patient must always use the combination of forces in the three planes rather than correcting in an isolated plane, as would be the case when using bending exercises in any single plane (frontal or sagittal), or by derotating one section of the body against an adjacent one. From the educational point of view, obviously the corrective forces and movements need to be decomposed plane by plane. The physiotherapist will train the patient to become familiar with all of the corrective forces and movements using 'plane by plane' singular, gentle corrections, whilst looking, from the very beginning, for a synchronic combination of all corrective forces and movements. The principles of three-dimensional correction according to the BSPTS can be briefly described as follow:

1. Self-elongation from a stable corrected pelvis. Self-elongation has a de-torsional correction effect by itself. Lateral deviation and rotation will decrease whilst the sagittal geometry of the spine will become flattened making the existing structural flat back more obvious. No matter what geometry is observed initially, after correcting lateral deviation and rotation by self-elongation, the geometry of the scoliotic spine will be flatter since the scoliotic spine is structurally flat, generally speaking. Afterwards the physiotherapist will teach the patient to fight against the structural flattening of the spine in order to re-create a more physiological kypho-lordotic profile. Self-elongation is an essential principle used to apply tension to all of the elastic structures that have to be stretched by de-collapsing all the concavities in the frontal and sagittal planes.
2. The combination of self-elongation with a corrective tension in any part of the ventral or dorsal trunk, consecutively from caudal to cranial is a correction principle called 'straightening'. The second correction principle is called 'asymmetrical sagittal straightening' which means that one hemi-body will be corrected in opposition with the other. Throughout 'asymmetrical sagittal straightening' the patient creates a 'pair of forces' for derotation at each desired plane or level. Thus, an additional derotation effect can be achieved by combining the first two corrective principles; 'self-elongation' and 'asymmetrical sagittal straightening'.
3. The third corrective principle is related to the frontal plane. Improvement of the correction in the frontal plane is then demanded from gentle movements and/or tensions performed in the frontal plane following the classical schema of blocks from Lehnert-Schroth. The schema of blocks differentiates two basic functional types called 'three' and 'four' curve patterns (Lehnert-Schroth 2007). The Schroth classification has been further re-defined to improve its reliability, by adding new radiological criteria (Rigo et al 2010). The BSPTS uses this classification to define specific corrections according to the observed functional curve pattern.
4. Breathing mechanics are then used to increase the de-torsional effect by creating an internal pair of forces for derotation, called Schroth rotational breathing. According to the BSPTS, asking the patient to increase, during inspiration, the sagittal diameter of the rib cage is a putative mechanism to fight against the thoracic structural flat back under the assumption that thoracic structural flat back reduces such a diameter. Deep abdominal muscle activation during correction will be used as a mechanism to prevent lumbar kyphosis, by maintaining a constant activation of the 'thoraco-lumbar fascia'. This is probably the part of the technique that requires more assistance from the physiotherapist.
5. Stabilization is achieved by muscle activation maintaining the best possible correction in expiration. Muscles in the concavities will be activated throughout eccentric isometric tension whilst the convex muscles will activate in concentrically (Figure 1).

The full description of the specific correction for each functional spine pattern as well as the description of any of the recommended exercises is beyond the objectives of this short brief. The principles of the BSPTS have previously been described in other publications (Rigo et al, 2008). Nevertheless, no matter how in-depth such a description can be made in a written publication, this will only be a supportive tool for therapists who have already been instructed during an extensive training course by the BSPTS. This is clearly a limitation and a weak point of this methodology. There is a relatively long learning curve for physiotherapists even after the successful completion of a theoretical-practical course, before clinical results can be achieved. Thus, physiotherapy for scoliosis should be considered a sub-specialty for devoted physiotherapists.

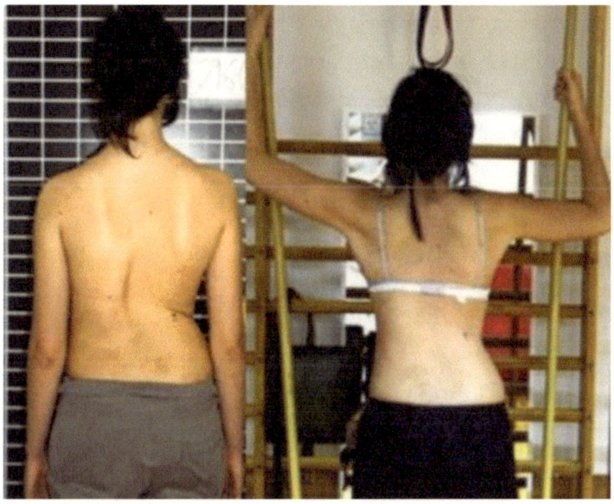

Fig. 5. A mature girl with a severe right thoracic/left thoracolumbar scoliosis (4 curves) performing an exercise in upright position with two poles and a resistant element on the left hip. She does maximum muscle activation by pushing down with the two poles, only after reaching the best 3D postural correction.

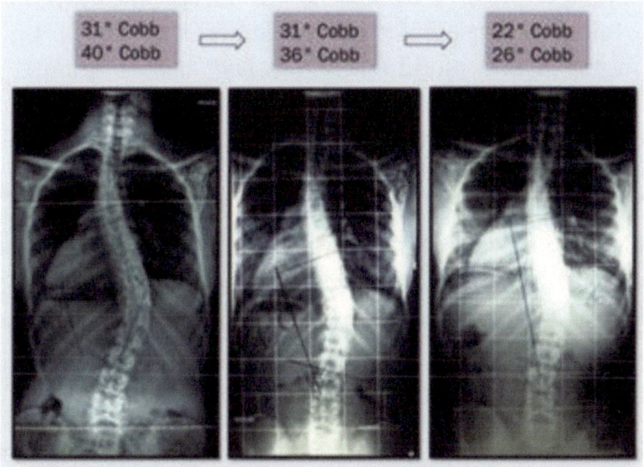


Fig. 6. The radiological evolution of a different girl with a right low-long thoracic curve combined with upper structural thoracic curve. The girl was almost mature when she started her exercise program (left X-ray). The Cobb angle was significantly reduced one year later (right), with an intermediate improvement (middle). Although all the Xrays were taken in an independent place, and the girl was instructed not to correct, some signs in the last X-ray - left ribs and shoulder - suggest that she was actively correcting herself. Thus this improvement should be interpreted with caution.

### 3.5 Research results

Results of the classical Schroth method have been widely published and summarized in several papers. Improvement of breathing function, pain, back asymmetry, posture, muscular imbalance and the Cobb angle in the short-term have been shown (Weiss, 2011). Specificity of the exercises has been shown by the BSPTS (Jelacic et al, 2011). However, no research-based paper has shown that specific exercises following Schroth principles, or any other physiotherapy principle, can prevent curve progression in rapidly progressive scoliosis during the acceleration phase of the pubertal spurt, before the peak of growth and maintained through until reaching maturation. Thus, nobody can claim to halt curve progression in those cases, using the name of Schroth or the name of BSPTS or of any Schroth variant.

## 4. The DoboMed physical therapy approach (Poland)

### 4.1 Introduction

The Dobosiewicz's Method (DoboMed) is a conservative management approach for idiopathic scoliosis that addresses both the trunk deformity as well as respiratory function impairment. The DoboMed approach has incorporated both Klapp`s position for kyphotization of the thoracic spine as well as Lehnert-Schroth's approach for active asymmetrical breathing into its method.

### 4.2 History of the Dobomed method

The method was developed in 1979 by Prof. Dobosiewicz and has been used routinely in Poland since 1982. Initially it was tentatively trialed on an outpatient group and has since been continually improved and modified. It was later used (regularly since 2000) in the Department of Rehabilitation of the Medical University of Katowice, Poland, as an intensive in-patient rehabilitation approach for patients with scoliosis. From the start, this approach was used either as a sole physical therapy method or combined with bracing (Cheneau brace). Prof. Krystyna Dobosiewicz died in two thousand and seven. She was both a physiotherapist and a physician. Prof. Dobosiewicz was also very familiar with the Klapp's and Lehnert-Schroth's methods. It was from these beginnings that she started to create her own approach to the treatment of scoliosis.

### 4.3 Principles of the DoboMed method

The basic aim of this method is to prevent progression and/or decrease the curvature of scoliosis. The second aim is to improve respiratory function. Small, moderate and large curves (idiopathic scoliosis) can all be treated with DoboMed, however the effectiveness of the therapy depends on the curve flexibility and the patient's compliance. Active cooperation is the basic requirement for using DoboMed, therefore DoboMed is not recommended for small children.

### 4.4 Description of the Dobomed method

This Dobomed approach is a biodynamic method of 3-dimensional auto-correction of idiopathic scoliosis based on the pathomechanics of idiopathic scoliosis. The basic technique of active three-dimensional correction involves mobilization of the primary curve towards curve correction, with special emphasis on `kyphotization' of the thoracic spine and/or `lordotization' of the lumbar spine. This mobilization is performed in closed kinetic chains and developed upon a symmetrically positioned pelvis and shoulder girdle. The pelvis and shoulders are positioned first and kept stable for the duration of the exercise and during the inspiration and expiration phases (Fig.7). For example, in primary thoracic curves the desired movement of the thoracic vertebrae is towards kyphosis, which is a backward displacement and axial derotation towards neutral. This is obtained partly by active movement of the thorax on a stable pelvis and shoulder girdle and partly by active 'asymmetrical' breathing. The frontal plane correction occurs automatically as the sagittal and axial planes are being corrected. Lateral flexion of the spine is not required in thoracic curves. Symmetrical positioning of the pelvis and the shoulder girdle is something that is unique to the DoboMed method.

The method consists of three parts. The first part to the approach is the main corrective technique and comprises of an active three-dimensional self-correction of the spine and rib-cage. The 3-D self-correction in forward bending is an original component to the DoboMed method. This basic technique is one of the key components of the DoboMed approach. The preparatory and final phases can be altered. Both basic and short exercises as well as combined exercises can be used in this phase. In the preparatory phase, non-specific physiotherapy may be used as a warm-up prior to the spine-specific exercises for each session. This position is thought to facilitate active correction between two symmetrical and stable zones, and helps to consolidate the correct postural habit beyond the therapeutic session. The exercises are designed in closed kinematic chains in order to enhance their effectiveness. This is obtained by the fixation of the pelvis and the shoulder girdle with the upper and lower limbs. At the beginning of the session, after warming up, exercises in low positions are performed (Fig.7). These positions free the back muscles from the influence of gravity. Probably because of that the largest correction of scoliosis is observed in these low positions. Maximum active kyphotization of the thoracic spine and lordotization of the lumbar spine with simultaneous 3D correction of the spinal deformation is performed in between performing the exercises in low positions (Fig 10, 11).

Next, active 3-dimensional auto-correction exercises are performed in upright positions with the spine positioned vertically with gravity fully affecting the back muscles. All exercise positions require strict symmetrical arrangement by fixation of the pelvis and the shoulder girdle with the upper and lower limbs during all phases of the respiratory cycle (Fig.7). The course of action focuses on the vicinity of the apical vertebra. On the concave side of the curvature a strong local pressure is applied (Fig 12), and on the convex side a subtle facilitation is applied (Fig 12). The correction and facilitation are 'phase-locked' with the particular phases of the respiratory cycle. A strong local pressure is applied on the concave side during inspiration, and a subtle facilitation is applied on the convex side during expiration. During expiration the achieved correction or hypercorrection is stabilized by an isometric contraction (Fig. 10-12).It is also important that when patients try and fix each position, the patient should have already corrected their pelvis position in 3 planes (frontal, sagittal and horizontal) (Fig.7).

As DoboMed may be considered as a difficult method, frequent checking of the patient by the physiotherapist is required. In our practice the best results were achieved by daily systematic exercises actively supervised by the parents, who were previously trained in the approach during the initial in-patient rehabilitation period.

The following are the principal distinctive features of the DoboMed approach:

1. Symmetrical positions for exercising;
2. Asymmetrical active movements to accomplish 3D scoliosis correction;
3. Thoracic spine mobilization to increase thoracic flexion;
4. Transverse plane derotation;
5. Specific treatment emphasis is focused on the area of the curve apex;
6. Concave rib mobilization to expand and derotate the ribs;
7. External facilitation;
8. Respiration-directed movements of the thorax and spine to improve respiratory function;
9. 3D displacement of vertebrae to obtain 3D scoliosis correction.

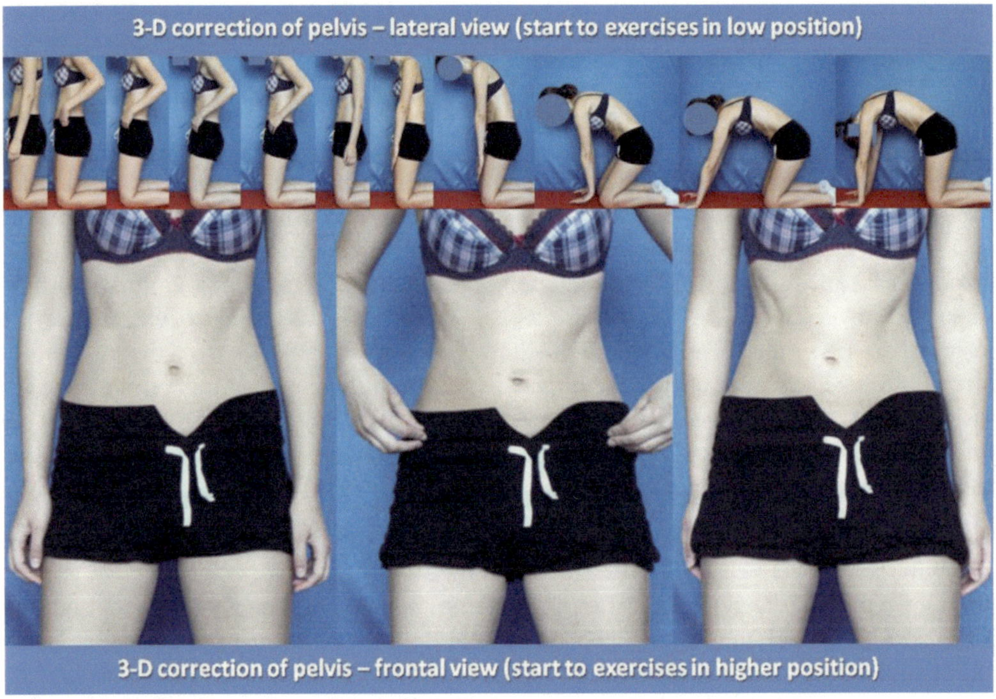


Fig. 7. 3D correction of the pelvis in the frontal and lateral view.

### 4.5 Research results

Results of therapy based on the DoboMed approach have been found to have an inhibitory effect on curve progression in idiopathic scoliosis. The radiological results were assessed on the basis of retrospective and prospective studies. They demonstrated that in most cases stabilization of curves in children with progressive idiopathic scoliosis treated occurred when using DoboMed (fig 8-9). The best effects of the treatment were observed in single scoliosis curves. The improvement of respiratory function, assessed by spirometry values (vital capacity, forced expiratory volume in one second), were also noted. The general exercise efficiency evaluated using ergospirometry was observed to increase significantly during physiotherapy with the DoboMed approach.

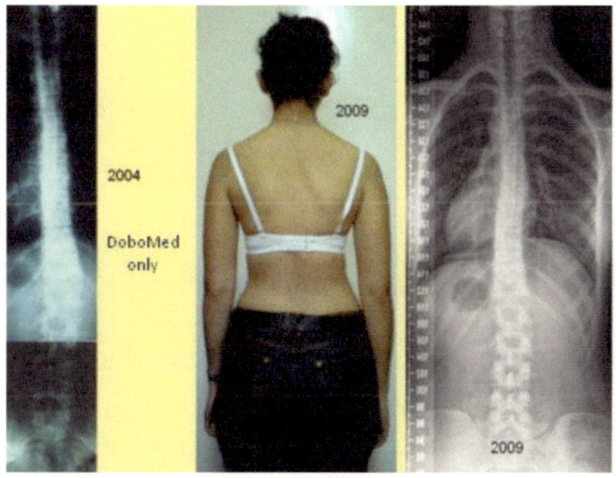


Fig. 8. Dobomed Exercise Case 1.

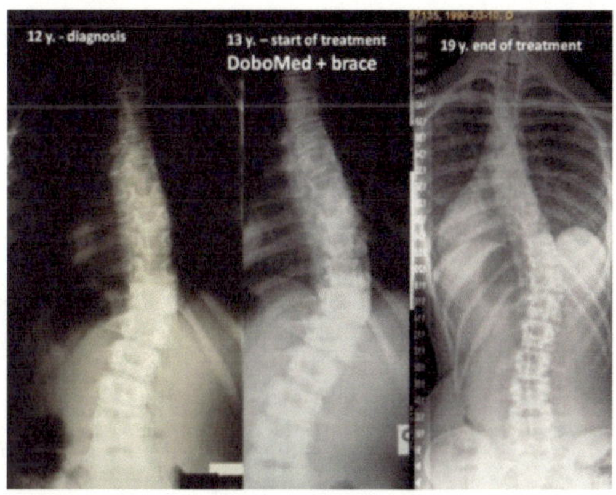


Fig. 9. Dobomed Exercise Case 2.

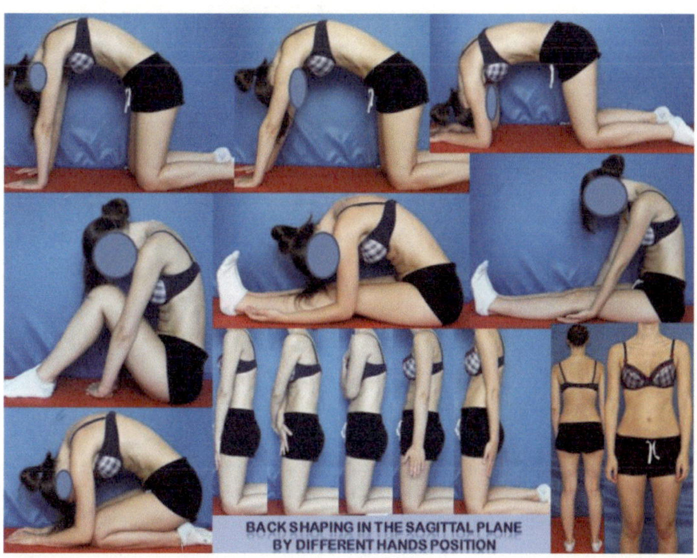


Fig. 10. "Shaping" the back by the use of different hand positions in the sagittal plane.

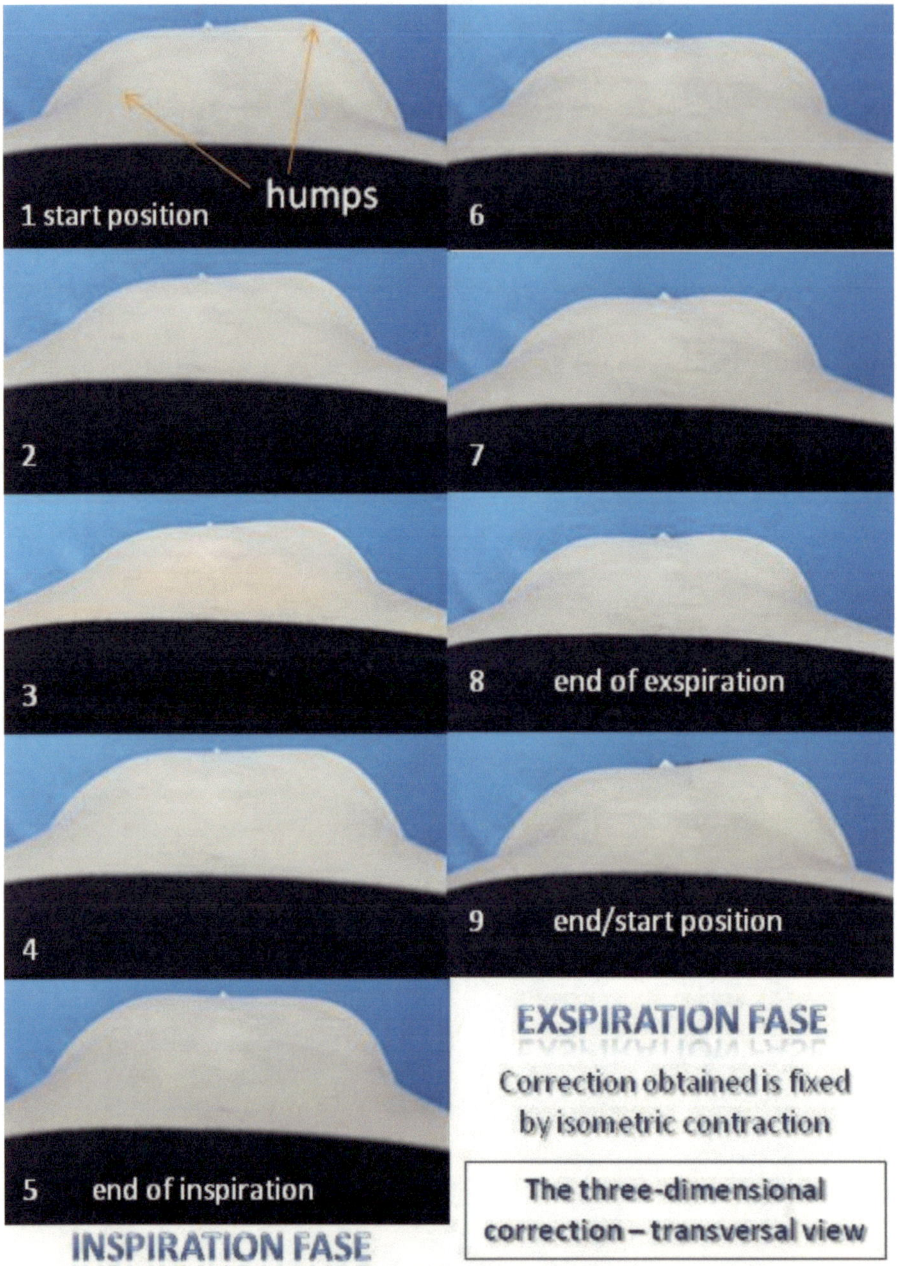


Fig. 11. The DoboMed may be considered to be a relatively difficult method; therefore, frequent checking of the patient to ensure they are practicing the exercise correctly is mandatory. The best effectiveness was achieved by daily exercises that were actively supervised by the parents, who had previously been educated in the DoboMed approach during the initial inpatient rehabilitation period.

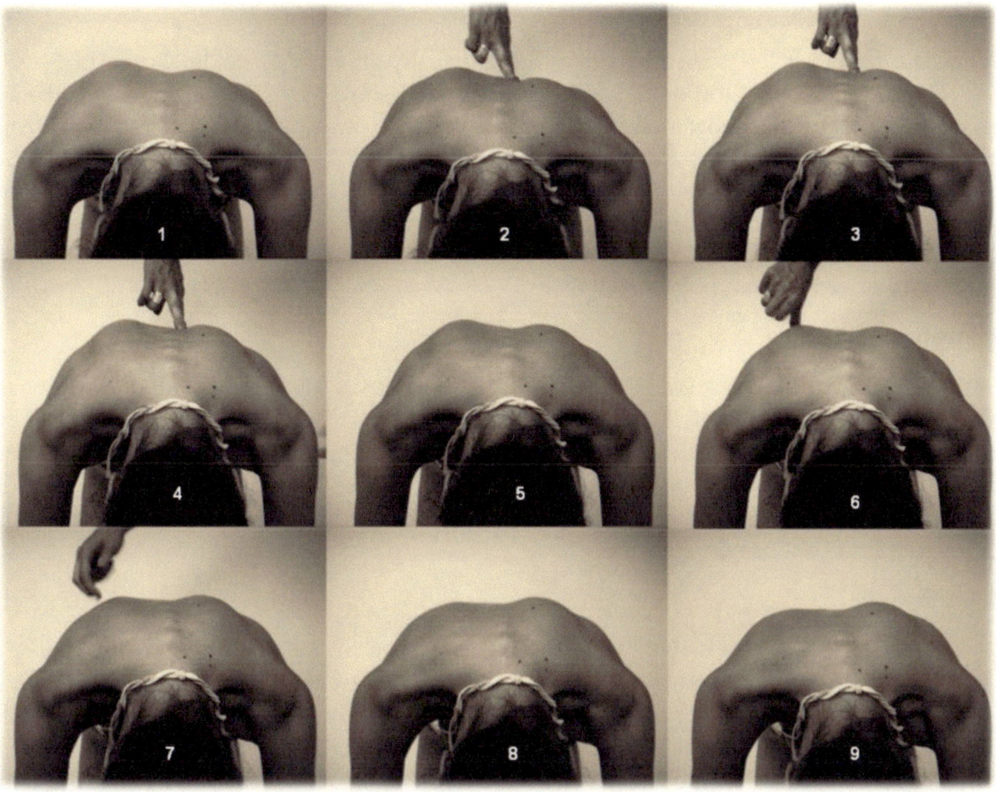


Fig. 12. The DoboMed approach to trunk derotation.

## 5. The Lyon approach (France)

### 5.1 Introduction

Physiotherapy is an integral part of the Lyon management for scoliosis together with the production of plaster cast before the Lyon brace.

### 5.2 History of the Lyon method

The Lyon method was developed 60 years ago by Pierre Stagnara together with the Lyon brace (Stagnara 1978 & Dubousset 1996). It has been tested by many physiotherapists including Georges Mollon (1986) and Paul Ducongé (2002). The Lyon method is not intended to provide the physiotherapist with an original technique and specific exercises, but rather it is intended to be a way of approaching and understanding scoliosis.

### 5.3 Principles of the Lyon method

**Stage I: Lyon approach to Assessment**

The Lyon approach considers that there are three elements that are important to guide therapy:

**1. The patient's age**

When the patient is less than 10 years old, it is usually an Early Onset Scoliosis (EOS). We know that there is a maturation delay in balance and posture. Physical therapy will be based on exercises favouring coordination and balance in a fun way, such as ball games. During puberty, growth occurs at the bone level and is inhibited by the paravertebral and myofascial structures. The readjustment of myofascial tension occurs globally by symmetrical range of motion exercises. The hypotonia of the musculature is also a characteristic of this period. We must insist on the maintenance of muscle tone over time by endurance exercises in aerobic metabolism.

**2. Postural Imbalance**

Multiple postural defects may be associated with scoliosis. The problem is that it is very difficult to determine what is a defect or compensation. If in doubt, examine the child walking; if the head is projected at the support polygon of the feet, there is no need to act on the segmental pelvic or scapular girdle correction. Similarly, if one leg is shorter on the side of the rib hump, it is likely that compensation by heel lift will increase the rib hump and in this case we will not use the shoe lift.

**3. Cobb angle**

- Below 25 °, it is impossible to predict the evolution of scoliosis and it behaves like a chaotic system described by de Mauroy (2008). It is essential to explain to parents that physical therapy cannot prevent the scoliotic earthquake, but scoliosis-specific exercises can produce an earthquake-resistant construction that will limit the damage.
- Above 25 °, the linear mechanism with a vicious circle of evolution has been described by Stokes: the asymmetric pressure on the vertebral body and disc modifies the growth of the vertebral body with accentuation of the deformity. The distance of the apical vertebral body from the line of gravity further increases the asymmetry of pressure. This pressure needs to be reduced by exercising, sitting position and sport.

**Stage II: Awareness of trunk deformity**

As the child has never seen his back, we must therefore make them aware of the corrective possibilities while also avoiding a dramatizing of the situation. Scoliosis is not a disease, but is an adaptation of the spine. This correction will be made by the orthopaedic mirror or using a camcorder and a screen. The feedback from the squared mirror is very useful during exercises of auto-elongation or walking with a small sand bag on the head. We must stimulate the perception of horizontality and verticality.

At this stage, we must also check and change the sitting position. It's the writing position that will focus our efforts. The feet are placed behind the chair; the buttocks in front of the seat, the anterior thorax just touch the front edge of the table and both forearms lie on the horizontal work plan. This position reduces stress on the disks and prevents excessive flexion of the spine. (fig 13)

**Stage III: What to do: Example exercises**

Gymnastic exercises **avoiding spinal extension** are the basis of the Lyon method of physiotherapy. They must be simple in order to be repeated every day at home. The exercises are symmetrical, mainly made in the supine position; indeed, the radiological scoliotic angle is higher in the standing position than in the lying down position. The standing exercise called the "grand porter" consists of walking on tiptoes with a sandbag on the head. The child tries to stretch the spine along the vertebral axis while controlling balance by looking at the feedback mirror (fig 14). During therapy the exercises will be gradually adapted by changing various parameters such as pace, intensity, duration.

Fig. 13. Usual sitting position for writing and using the computer.

Fig. 14. "Grand porter".

**Stage IV: What not to do and why?**

Much more important and rarely described are exercises and postures to avoid in scoliosis.

1. The functional plane is the sagittal spine. In this plane the extreme magnitudes should be avoided. By extension the flat back is accentuated, and by flexion, the disc pressure is increased. Furthermore, especially beyond a 25 ° degree rotation, the apical vertebra tends to further increase the existing rotation due to the paraspinal muscles lever arm and the position of the Instantaneous Centre of Rotation. Spinal muscles of the concave and convex side become agonist exerting an auto deforming action. (fig 15)
2. The deep breathing favours the rotation of the apical vertebra, therefore, avoid shortness of breath.

3. The modelling of the rib hump in the sagittal plane favours a flat back. For a true derotation in the prone position, place a cushion under the opposite anterior chondrocostal prominence and press the inner slope of the rib hump. (fig 16)
4. Strengthening the superficial body building type of muscles is useless, we must work on endurance on the deep paraspinal muscles.

**Stage V: Sport or only physiotherapy**

Many physiotherapists ask their patients to stop practicing sport. The Lyon method allows the pursuit of sport in so far as the contraindications of step 4 are met. So it's more how to practice the sport rather than the sport itself that matters. Our best results were obtained in the Lyon brace in the group of children who performed more than 5 hours of sport a week. The best sports for scoliosis are combat sports; the worst, rock climbing, as it promotes too much extension of the spine. We must adapt the sport to the age of the child. Before puberty, sports of balance and coordination are preferred. Swimming is excellent during the pubertal growth phase. After the age of 15 years the axial impact sports promotes calcium fixation to the bone and improve bone mass.

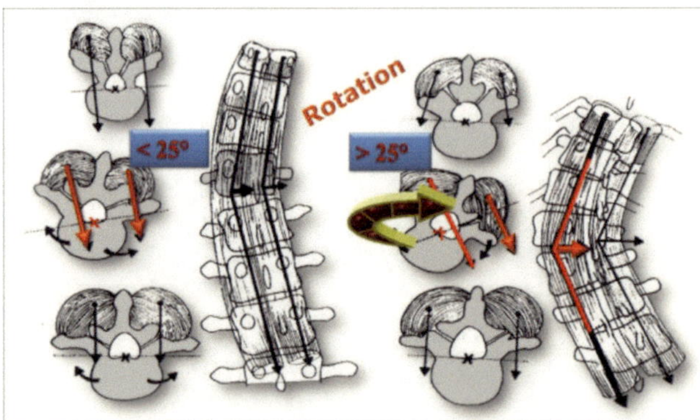


Fig. 15. Increasing rotation during trunk flexion for a scoliosis of more than 25°.

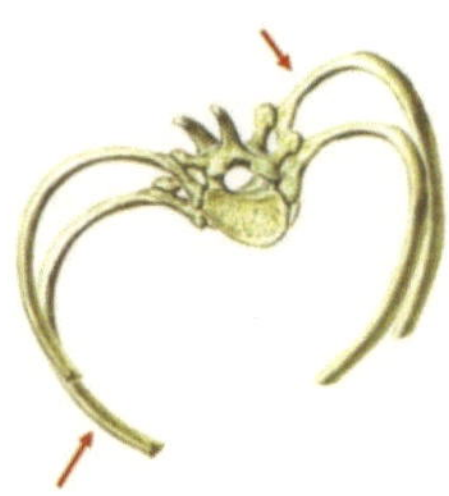

Fig. 16. Biomechanical basis of apical vertebral derotation.

## 5.4 Description of the Lyon method

When scoliosis is evolving, i.e. moves from a chaotic phase to a linear phase, the Lyon method combines physiotherapy and the Lyon brace. The Lyon brace is always preceded by a plaster cast that allows a real lengthening of the concavity beyond simple mobilization (de Mauroy 2011). Physiotherapy is greatly facilitated by the plaster cast. This is a true three-dimensional correction. During the time in a plaster cast (one to four months), physiotherapy is intensified with at least two sessions a week supervised by the physiotherapist. The work plan includes:

- Breathing control with use of expiratory reserve volume because the pelvis is fully stabilized.
- 3D mobilization of the spine (fig 17)
- Mobilization of the ilio-lumbar angle (lumbar scoliosis)
- Therapeutic patient education (food control to avoid cast syndrome, skin care ...)
- Sitting position check

Fig. 17. Auto-3D correction of scoliosis with Lyon plaster cast.

The advantage of the plaster cast for scoliosis under 30 degrees is that the brace is worn only during the night. Physical therapy will continue at least once a week. When the scoliosis curve exceeds 30°, the brace must be worn during part of the day. The physiotherapist will perform physical therapy with or without brace. (fig 18)

Fig. 18. Group physiotherapy in Lyon brace.

### 5.5 Research results

Psychologically physiotherapy in a group is better because the child feels less alone in their treatment. Unfortunately, the Lyon physiotherapy method for scoliosis does not follow a universal standardised protocol, but has to adapt to each child and develop during growth. It is therefore very difficult to quantify results in terms of angular correction for scoliosis, considered to be but it is essential when the Lyon brace is prescribed.

## 6. FITS (Poland)

### 6.1 Introduction

Functional Individual Therapy of Scoliosis (FITS) treatment is based upon the inclusion of many elements selected from a variety of other therapeutic approaches that have been adapted to form a different treatment concept.

### 6.2 History of the FITS method

In 2003 Bialek and M'hango decided to create their own program of therapy to improve postural problems and scoliosis. Many new ideas were developed during rehabilitation camps since the camps hosted many scientific studies. This was done in cooperation with Dr Wieslaw Chwala of the Department of Biokinetics at AWF Cracow, who was carrying out EMG examination and 3D gait analysis using the Vicon system. Moreover, Spirometry and Moire projection examinations were also conducted. As a result of these studies a new program for diagnosing and treating postural problems and scoliosis was created, which the authors started implementing in their practice by the end of 2003. Having observed positive effects of their work with scoliosis M. Bialek and A. M'hango decided to draw up a Physical Therapy course program of scoliosis diagnosis and therapy for physiotherapists and physicians. This resulted in the 'Functional Individual Therapy of Scoliosis' (FITS) concept being created, and in 2004 the first course based on the FITS concept was held. It received PTF (Polish Society of Physiotherapy) accreditation in 2005 (www.fits.pl). In 2006 cooperation with the orthopedist Dr Tomasz Kotwicki from Poznan began, which triggered further evolution of the FITS concept. On average three new groups, totaling about 70 trainees, attend our course and over a hundred children go to FITS rehabilitation camps every year. To date almost 450 trainees have completed our course and 700 children have been to FITS rehabilitation camps.

### 6.3 Principles of the FITS method

FITS is a method of diagnosis and therapy for idiopathic scoliosis. It may be used as a separate system for scoliosis correction, a supportive therapy to bracing, preparation of children for surgery, or for the correction of the shoulder and pelvic girdles after surgical intervention. Different techniques have been used from a variety of therapeutic methods in search of the most effective means of clinical and structural scoliosis correction. These techniques had proven themselves to be significantly effective in a short period of time. They were then modified and systematized, which was the starting point for the FITS concept.

FITS is a complex, asymmetrical and individual therapy, which can be used in a child of any age regardless of the Cobb angle. It requires the child to take an active part in the therapy process, which is guided by an experienced and specialist therapist. FITS therapy is conducted in an out-patient clinic or as an in-patient by means of a 1 or 2 week course of treatment. It is recommended that the therapy is delivered by a physiotherapist together with an orthopedist and a psychologist.

The main objectives of the concept can be listed as follows:

To make a child aware of the existing deformation of their spine and trunk, and to indicate the direction of scoliosis correction.
To release myofascial structures which limit the three-plane corrective movement.
To increase thoracic kyphosis through myofascial release and articular mobilisation.
To teach correct foot loading; which will improve the position of the pelvis and realign the scoliosis.
To improve the lumbo-pelvis stabilization.
To release the myofascial tension between the apex of the scoliosis and the iliac crest (which limits spine shift towards scoliosis correction).
To teach the correct shift of the spine in the frontal plane in order to correct the primary curve whilst stabilizing, or maintaining in correction, the secondary curve.
To facilitate the right three-plane corrective breathing in functional positions (breathing with concavities).
To indicate the right patterns of scoliosis correction and any secondary trunk deformation related to curvature (asymmetry of head position, asymmetry of the lines of shoulders, shoulder blades, waist triangles and pelvis).
To teach balance exercises and improvement of neuro-muscular coordination with scoliosis correction.
To teach correct pelvis weight bearing in a sedentary position and to correct other spine segments in gait and activities of daily living (Bialek M.,M'hango A., 2010).

### 6.4 Description of the FITS method

There are three stages of the FITS concept:

**Stage 1: Examination of the patient**

Examination of the patient using a classical clinical assessment and a FITS specific assessment. Classical assessment includes: history, course of treatment, X-ray analysis and examination of a patient in three different planes.

FITS specific clinical assessment involves examination of the distance from the plumb line to the following: anal cleft, the apex of the primary and secondary curve, the edge of the scapula as well as checking the position of both scapulas. What has also been included is the observation of the type and location of compensation, position of the pelvis and measurement of ATR using the Bunnell's scoliometer. In addition, the assessment of lower extremity alignment in a standing position and gait is also conducted. Furthermore, the length of muscles in the lower limbs, pelvic girdle, shoulder girdle and trunk are checked for any indications of asymmetry. Finally the assessment of the possibility for scoliosis correction in standing and sitting positions is performed (Bialek M.,M'hango A., 2010).

**Stage 2: Preparation for 3D correction: detection and elimination of the myofascial restrictions which limit a three-plane corrective movement by using different techniques of myofascial relaxation**

The authors of this concept emphasize the importance of a child's awareness of the type of scoliosis and the trunk deformation caused by scoliosis. We analyzed X-ray pictures of the child, the three-dimensional position of scoliosis on a model of the spine, and we show the direction of correction. In our opinion, making a child a partner and not the subject of therapy significantly increases motivation to exercise and improves the effects of therapy at the same time (Bialek M.,M'hango A;2008, 2010). Many years of work experience with scoliosis has shown significant myofascial limitations in the area of many muscle chains (Myers, 2009). The limitations are particularly visible whilst attempting to perform passive corrective movement in functional positions. In order to specify the direction of therapy and effectiveness of therapeutic procedures we use an examination of corrective movement as a test during each session. Corrective movement at the beginning of therapy can be done only in one plane - shift, rotation, or flexion/extension. In further stages of therapy three-dimensional corrective movements also need to be included (Bialek M.,M'hango A., 2010). When a patient has been fully examined we move on to relaxation of structures restricting correction by using techniques such as; contract-relax technique, myofascial release (Chaitow, 1996, 2002), trigger point therapy (Travell & Simons, 1992), joint mobilization (Bialek, 1997; Colot & Verheyen, 2002; Kaltenborn, 1998; Lewit, 2001) and neuromobilisation. These techniques are often used in the area of myofascial bands according to Myers; SBL (superficial back line), DFL (deep front line), LL (lateral muscle line), SL (spiral muscle line), SFL (superficial front line) (Myers, 2009).

**Stage 3: Three-dimensional correction: building and stabilisation of new corrective posture patterns in functional positions**

We start building and stabilising new corrective patterns of posture in functional positions by teaching correct foot loading with the use of sensory motor balance training according to Greenman (Greenman, 2003). By observing children with scoliosis we notice unsettled stabilization of the lower part of trunk, especially during everyday activities (Lee, 2001; Richardson et al., 2004). Stabilization exercises of the lower part of the trunk have to be done before starting the teaching of corrective patterns of the upper part of trunk and shoulder girdle.

Facilitation of the right three-plane corrective breathing should be done after diaphragm release and restoration of the best possible joint mobility in the thoracic spine and thorax. The aim of our therapy is the facilitation of breathing movements towards the concavities which can be achieved by a 'derotative' breathing exercise (Dobosiewicz et al., 2005). The effectiveness of these exercises can be improved by adding elongation of the scoliosis concavity by using upper and lower limb patterns. In every case, attention should be paid to the correct position in the sagittal plane. Teaching patterns that correct scoliosis, and other trunk deformations that are associated with curvatures, is done in open and closed kinematic chains with the use of TheraBand (Ellenbecker & Davies, 2001). Each 'limb pattern' consists of a correction in the sagittal, frontal and transverse planes. The choice of each element of every corrective pattern depends on Cobb angle, size and direction of trunk rotation, position of the spine in the sagittal plane, and location of functional compensation

(Bialek, 2001). In scoliosis treatment it is most desirable to correct the primary structural curve in three planes. However, in many cases this might be too difficult, due to a considerable degree of curvature or insufficient 'correctivity' of curvature. In these cases we believe that creating and developing functional compensatory curves above and below the primary curve is beneficial

By functional compensation we mean creating a rotation which is directed in the opposite direction to the rotation found in the primary scoliosis, not creating a structural curvature. The compensation relates only to soft tissues and not the structures seen in X-ray photos. This counter-rotation should not exceed 3-4 degrees ATR.

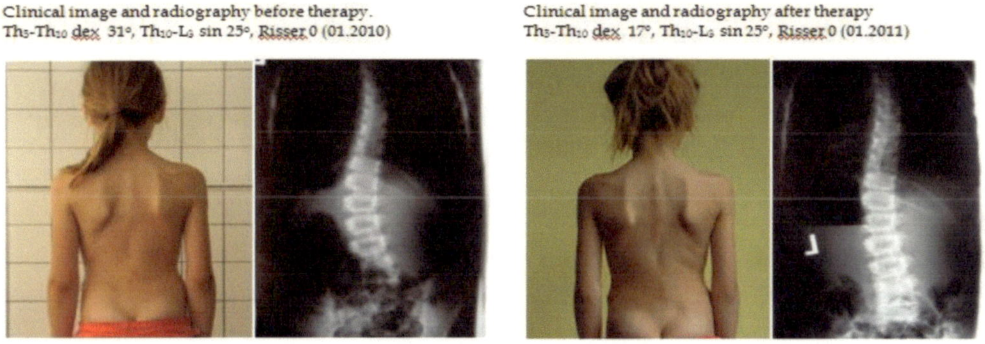


Fig. 19. Before and after therapy pictures of Marta, our 10-year-old patient

Once the functional compensation has been achieved, our continued work focuses on decreasing the primary curvature by continuation of only elongation and shift. In a scoliosis with a small Cobb angle, which is more likely to be corrected, we can achieve a balanced posture and a good clinical effect by creating a functional compensation (Bialek et al., 2007). It is very important to be able to establish at which level the compensation is most desirable and where the compensation should be increased, decreased, or where there should be none. The type and size of functional compensation can be evaluated by measuring trunk rotation using Bunnell's scoliometer (Bunnell, 1984).

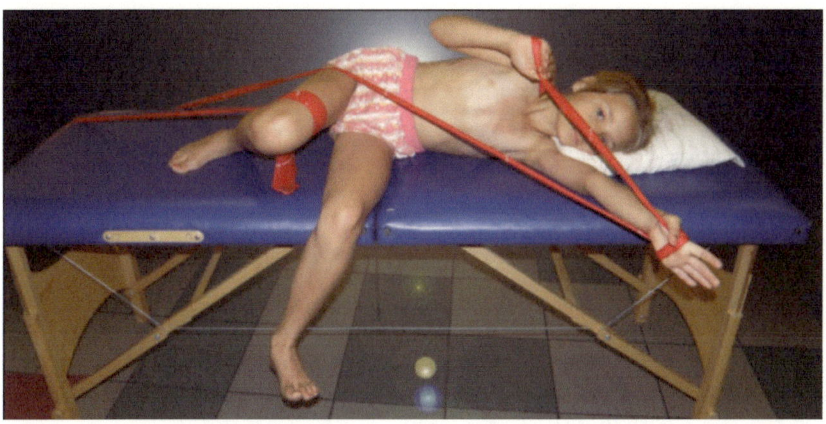

Fig. 20. Exercise in pattern correction.

**Strengths of the FITS method**

Individual approach to children.
Awakening patients' awareness, making them partners not subjects.
Examination of spinal flexibility and of active and passive correctional movement capability.
Removal of any myofascial limitations which restrict correct movement before educating the child of the correct patterns and proper postural reeducation for their specific curvature.
A comprehensive approach to patients and 3D scoliosis correction.
Teaching daily functional activities in scoliosis autocorrection.
Selection of movement patterns according to the change of the patient's clinical state

**Limitations of the FITS method**

The method is complex and requires utmost accuracy.
Strict cooperation of therapists and parents in the therapeutic process of a child is required.
Long duration of the therapy, until completion of bone maturity.

### 6.5 Research results

**Study design.** Retrospective study according to SOSORT and SRS criteria.

**Materials**

Three-hundred and seventy four patients received conservative treatment for idiopathic scoliosis according to the FITS concept between 2005 and 2010. One-hundred and fifteen of them were included in this study because they were conducted according to the SRS criteria of the initial Cobb angle and Risser sign. Patients were analyzed separately in two groups:

**Group A** - 68 girls and 10 boys, who received the FITS therapy without bracing. The group comprised children older than 10, with a Cobb angle between 10° and 25° and a Risser sign between 0 and 2. The children were classified into two subgroups:

Subgroup A1 - single thoracic (Th) or thoracolumbar (Th/L) or lumbar curve (L) (52 children).
Subgroup A2 - double scoliosis - thoracic (Th) and thoracolumbar (Th/L) or lumbar (L) curves (26 children).

**Group B** - 34 girls and 3 boys who received the FITS therapy combined with bracing. The group was comprised of children older than 10, with a Cobb angle between 26° and 40° and a Risser sign between 0 and 2. The children were classified into two subgroups:

Subgroup B1 - single thoracic (Th) or thoracolumbar (Th/L), (5 children).
Subgroup B2- double scoliosis - thoracic (Th) and thoracolumbar (Th/L) or lumbar (L) curves (32 children).

**Method**

The Cobb angle and Risser sign were analyzed at the initial stage and at the 2.8 year follow-up. The percentage of patients who improved (defined as a decrease of Cobb angle of more than 5°), were stable (± 5°), or progressed (increase of Cobb angle of more than 5°) was calculated. The clinical assessment comprised: the initial and follow-up Angle of Trunk Rotation (ATR), the plumb line imbalance, the scapulas level, and the distance from the apex of the primary curve to the plumb line.

| Subgroup | Improvement | Stabilization | Progression |
|---|---|---|---|
| A1 | 50.0% | 46,2% | 3,8% |
| A2 | 50.0% | 30.8% | 19.2% |
| B1 | 20.0% | 80.0% | 0.0% |
| B2 | 28.1% | 46.9% | 25.0% |

Table 1. Percentage values of scoliosis improvement, stabilization and progression

| Subgroup | N pts in each group | N pts with progression | % progression in either curve | N pts with progression in Th curve | % progression in Th curve | N pts with progression in Th/L curve | % progression in TH/L curve | N pts with progression on both curves | % progression in both curves |
|---|---|---|---|---|---|---|---|---|---|
| A1 | 52 | 2 | 3,8% | 0 | 0% | 2 | 3,8% | - | - |
| A2 | 26 | 5 | 19,2% | 4 | 15,4% | 4 | 15,4% | 3 | 11,5% |
| B1 | 5 | 0 | 0 | 0 | 0% | 0 | 0% | - | - |
| B2 | 32 | 8 | 25% | 6 | 18.7% | 3 | 9,4% | 1 | 3,1% |

Table 2. Percentage of progression of more than 5° divided into Th and Th/L or L curve.

**Results**

In Group A1 in single structural scoliosis, 50.0% of patients improved, 46.2% were stable and 3.8% progressed. In Group A2 in double scoliosis, 50.0% of patients improved, 30.8% were stable and 19.2% progressed. In Group B1 in single scoliosis, 20.0% of patients improved, 80.0% were stable and no patient progressed. In Group B2 in double scoliosis, 28.1% of patients improved, 46.9% were stable and 25.0% progressed (Table 1).

Progression of scoliosis more than 5° was highest in the thoracic scoliosis in group B2 (18.7%). This was slightly less in Th and Th/L or L component of double scoliosis in group A2 (15.4%). Progression in both curves was greater in group A2 (11.5%) than in group B2 (3.1%)(Table 2).

## 7. The SpineCor method

### 7.1 Introduction

The SpineCor physiotherapy method is a postural re-education method that involves a combination of three-dimensional Corrective Movements and global muscle rebalance exercises.

### 7.2 History of the SpineCor method

In the early 1990's a group of researchers from Sainte-Justine University Hospital (Montreal, Canada) led by Professor Charles H. Rivard and Dr. Christine Coillard began investigating the etiology and pathogenesis of Idiopathic Scoliosis. The purpose of this research was to gain a better understanding of the causes of scoliosis and the factors that determine its progression, which in turn would lead to more effective treatment. Whilst a full understanding of the etiopathogenesis of Idiopathic Scoliosis may still be many years away, this research provided the group with the relevant information to better manage the scoliosis progression with conservative treatment. The research concluded that idiopathic scoliosis should be described as a **neuro-musculo-skeletal pathology** since there is clear evidence of abnormality in all three areas in advanced cases. It was found that the initiating factor of Idiopathic Scoliosis is genetic, however the pathogenesis involves a sequence of critical events resulting in scoliosis which is defined not only as a 3-dimensional deformation of the spine, but also as a pathology involving a specific postural disorganization and a neuromuscular dysfunction (Coillard et al, 2001).

### 7.3 Principles of the SpineCor method

The understanding of scoliosis as a neuro-musculo-skeletal pathology made Dr. Coillard and the team of researchers at Sainte-Justine Hospital realise that in order to treat scoliosis effectively it was necessary to develop a new therapeutic approach which treats the muscular, neurological and osseous elements of scoliosis at the same time. The result was the SpineCor 3-dimensional approach to scoliosis which is based upon the use of dynamic forces by the application of a curve specific **Corrective Movement.**

The Corrective Movement provides a dynamic opening of the scoliosis curve with the ultimate objective of decreasing the spinal deformation. This is possible by the alternation of compression and distraction forces on the vertebral growth plates, whilst at the same time obtaining progressive correction of the postural disorganization and the muscular asymmetry. Over time this will result in neuromuscular integration of the new movement strategy. The integration of the Corrective Movement into the patient's brain provides sustainable long term improvement of curves and permanent stable changes to patients` postures.

The SpineCor dynamic brace was developed as a tool which is used to maintain the Corrective Movement the necessary length of time to get its integration into the patient. The SpineCor brace may be considered as a postural re-education tool rather than a brace. It provides low impact rehabilitation exercises 10's of thousands of times per day that have a positive effect on the spinal abnormal loading, as well as providing neuromuscular re-education with the final objective of neuro-muscular integration.

After years of positive outcomes with SpineCor Brace treatment the SpineCor research team thought that specific exercises that supported the Corrective Movement Principle would also have a positive effect on the treatment of Idiopathic Scoliosis. The SpineCor Physiotherapy Method was then developed, initially designed to complement the SpineCor bracing, however in some circumstances it can be used as a 'stand alone' treatment. The SpineCor physiotherapy program involves curve specific postural re-education and muscular re-balance exercises that comply with the Corrective Movement Principle.

A good understanding of the Corrective Movement Principle is a fundamental requirement to learning and correctly applying the SpineCor Physiotherapy Method. The Corrective Movements have been designed based upon the regional kinematics of the spine. In scoliosis the spinal deformation, the postural changes and the muscular imbalance affect the normal movements of the spine. The research carried out at Sainte Justine Hospital found that in scoliosis patients the movement patterns are maintained but the movement amplitude is unbalanced in opposite directions. As movement patterns in scoliosis patients are maintained and the spinal deformation and the postural disorganization are linked, the Corrective Movements should be performed in the same plane of maximum movement amplitude of a specific spinal segment but in the opposite direction correcting the specific postural disorganization. In simple terms, the Corrective Movements were designed to overcorrect the specific postural disorganization. The postural overcorrection will open the scoliosis curve to decrease the spine deformation and at the same time allow for progressive correction of the muscular asymmetry. Individual or combined regional Corrective Movements of the spine create the overall Corrective Movement Position. A single curve with its apex at T12 or L1 will have a different Corrective Movement than a curve with its apex at L2 or L3 simply because the kinematics in these two regions is different. This makes each Corrective Movement specific for each type of scoliosis curve.

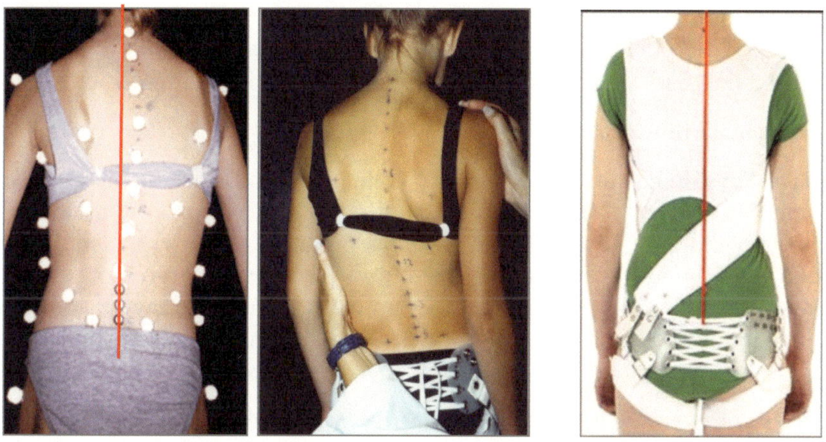

Fig. 21. Corrective Movement and fitting of the SpineCor brace for a Right Thoracic Type 1 curve. The moderate tension in the elastic bands allows the repetition and amplification of the Corrective Movement.

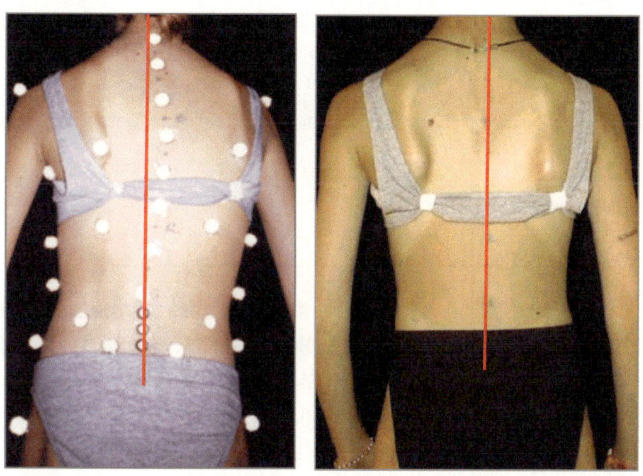

Fig. 22. The same patient is shown here before (left) and after treatment (right) with the SpineCor brace. The Corrective Movement has been integrated providing permanent stable changes to the patient's posture.

### 7.4 Description of the SpineCor method

The main goal of the SpineCor physiotherapy exercises is to provide further repetition and amplification of the Corrective Movements in order to help the consolidation of the progressive curve reduction and to reinforce neuro-muscular integration and postural reorganization. When physiotherapy is used in conjunction with SpineCor bracing it complements the action of the brace. This differentiates this method from the exercises programs associated with rigid braces, where exercises are frequently performed in order to compensate for the collateral effects of the brace. As the SpineCor brace has no side effects, the brace and physiotherapy work with the same treatment principles to reach the same objectives. On the occasions that the curve has a Cobb angle that is too small for bracing (under 20 degrees), physiotherapy may be used on its own. On these occasions the goals would be to gain muscular balance, improve posture, reduce pain, and prepare the patient for bracing in those cases where it seems certain that this will be required.

**Postural Re-education Exercises**

The SpineCor Physiotherapy Method program addresses postural re-education exercises as the first step. No scoliosis patient feels crooked despite the fact that they have abnormal posture and movements, since scoliosis develops gradually and the motor control centers have time to adapt to the information received from the peripheral receptors and accept that information as being valid. It is essential then to teach patients an awareness of their own postural disorganization and how to mobilize the different postural levels (shoulders, thorax, pelvis), before patients are taught their specific Corrective Movement position.

Postural re-education exercises involve an **active self-correction** where patients learn to reproduce the Corrective Movement position without external aids. The position of the entire body is controlled and patients are taught to reproduce each of the different components of the Corrective Movement in the 3 planes of deformation (fig 23). Once patients are capable of controlling their posture and maintaining the Corrective Movement position in the 3 planes of the space, they are taught to amplify this position working with the breathing cycles. To help the neuromuscular integration of the correction the program also includes exercises to develop balance reactions on exercise balls and unstable surfaces. Patients are also taught to keep the Corrective Movement position while walking and performing daily living activities.

**Muscular Re-balance Exercises**

The SpineCor Method program addresses muscular rebalance exercises in the second step.

These exercises are curve specific and they are always performed from the Corrective Movement position learnt in the previous stage. This position by itself automatically provides a 3-dimensional correction of the postural disorganisation as well as the correction of the muscular imbalance. Muscular re-balance exercises seek to obtain a progressive correction of the muscular imbalance through **global asymmetrical muscular work**. From the Corrective Movement Position patients are taught specific isolated movements against resistance looking for postural overcorrection of each component of the Corrective Movement independently (fig 23). The position of the segments of the body that do not take part in the exercise is facilitated for a better postural control.

### 7.5 Research results

The effectiveness of the Corrective Movement principle in the treatment of Idiopathic Scoliosis through its application by the SpineCor brace has been demonstrated by the treatment results within studies published by the Sainte-Justine research team (Coillard et al, 2003, 2007, 2008, Vachon et al, 2006), and by independent centres from different countries (Potackzek et al, 2008; Szwed et al, 2009; Vera, 2009; Tsakiri et al, 2009).

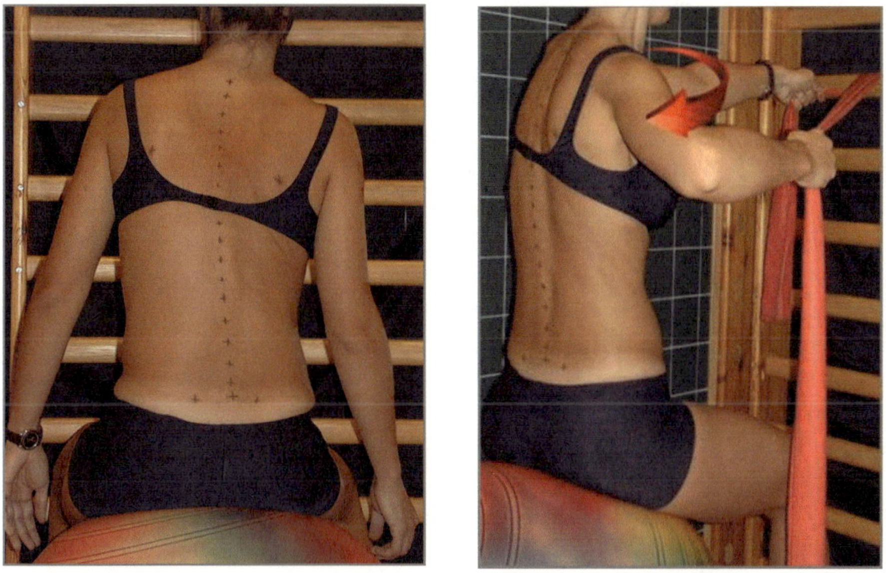

Fig. 23. Corrective Movement position achieved by the patient (left) and global asymmetrical muscular work on latissimus dorsi muscles (right) for a Right Thoracic Type 1 curve.

Our experience has shown that the reinforcement of the Corrective Movement principle by means of a scoliosis class specific Physical Therapy exercise program has a positive effect on its neuromuscular integration (del Campo, 2010). However at this present time there have been no comparison studies where the effectiveness of the SpineCor Physiotherapy Program has been evaluated against the Natural History of Idiopathic Scoliosis in pre-bracing groups of patients, or where the effectiveness of the SpineCor Physiotherapy Program on its own has been evaluated against the combined use of the SpineCor Brace and the Physiotherapy program. These studies will take place in the future when a significant number of patient results for both treatment groups in each case are collected.

## 8. Overall conclusions

In conclusion this chapter has reviewed six different scoliosis-specific exercise approaches to the treatment of adolescent idiopathic scoliosis that are currently used within different European countries. While the approaches differ, the overall aims are the same whatever method is used, which is to realign the spine, rib-cage and pelvis to the `normal` anatomical postures. Furthermore although research advances have been made to demonstrate the effectiveness of some of these approaches many more high quality research studies such as randomised controlled studies are urgently needed, as well as studies comparing the effectiveness of the different exercise approaches.

## 9. References

Bialek M.,M'hango A. (2008). FITS Concept Functional Individual Therapy of Scoliosis,In: *The Conservative Scoliosis Treatment,* Grivas T., IOS Press,250-261, ISBN 978-1-58603-842-7, Amsterdam

Bialek M.,M'hango A. (2010). Complex diagnostics and therapy of spinal curvatures and scoliosis according to FITS Concept (Functional Individual Therapy of Scoliosis), In: *Dysfunctions of the Locomotor System The Role of Osteopathy and Manual Therapy in the Treatment of Soft Tissues,* Majchrzycki, M.; Warzecha, D.; Kocur, P., Poznan University of Medical Sciences,96-109, ISBN 978-83-7597-107-1, Pozna

Burwell RG, Aujla RK, Grevitt MP, et al. Pathogenesis of adolescent idiopathic scoliosis in girls - a double neuro-osseous theory involving disharmony between two nervous systems, somatic and autonomic expressed in the spine and trunk. Possible dependency on sympathetic nervous system and hormones with implications for medical therapy. Scoliosis 2009 4:24 (31 October 2009)

Chaitow, L. (2002). *Positional Release Techniques* (second edition), Churchill Livingstone, ISBN 0 443 07081 4, London

Chaitow, L. (2006). *Muscle Energy Techniques* (third edition), Churchill Livingstone, ISBN 978-0443101144, London

Coillard, C. & Rivard, C. (2001). Etiology of Idiopathic Scoliosis: an unsynchronized growth or why a system can turn chaotic. *Résonances Européenes du Rachis.* Volume 9, N° 29, (2001), pp. 1123-1139

Coillard, C., Leroux, M., Zabjek, K. & Rivard, C. (2003). SpineCor- A non rigid brace for the treatment of idiopathic scoliosis: post-treatment results. *Eur Spine J.* 12, (2003), pp. 141–148

Coillard, C., Vachon, V., Circo, A., Beauséjour, M. & Rivard, C. (2007). Effectiveness of the SpineCor Brace Based on the New Standardized Criteria Proposed by the Scoliosis Research Society for Adolescent Idiopathic Scoliosis. *J Pediatr Orthop.* Volume 27, N° 4, (June 2007), pp. 375-379

Coillard, C., Circo, A. & Rivard, C. (2008). A new concept for the non-invasive treatment of Adolescent idiopathic Scoliosis: The Corrective Movement principle integrated in the SpineCor System. *Disability and Rehabilitation: Assistive Technology.* 3:3, (May 2008), pp. 112 - 119

Colot, T.; Verheyen, M. (2002). *Manuel Practique de Manipulations Osteopathiques,* Groupe Liaisons, ISBN 2-7160-0155-3, Rueil-Malmaison

Danielsson AJ, Nachemson AL. Radiologic findings and curve progression 22 years after treatment for adolescent idiopathic scoliosis: comparison of brace and surgical treatment with matching control group of straight individuals. Spine 2001;26(5):516-25.

del Campo, A. (2010). Physical Therapy in the Treatment of Adult and Paediatric Spinal Deformities: The SpineCor Method. *Scoliosis 2010.* Volume 5, Supplement 1:O31

Dolan LA, Weinstein SL. Surgical rates after observation and bracing for adolescent idiopathic scoliosis: an evidence-based review. Spine 2007;32(19 Suppl):S91-S100.

Dobosiewicz, K.; Durmała, J.; Kotwicki, T. (2005). Biodynamic method for 3-D correction of idiopathic scoliosis; a description of the method, *Ortopedia Traumatologia Rehabilitacja*, Vol. 7 (1), 2005; pp 49-54.

Dobosiewicz K, Durmala J, Czernicki K, Jendrzejek H 2002 Pathomechanic basics of conservative treatment of progressive idiopathic scoliosis according to Dobosiewicz method based upon radiologic evaluation. In: Grivas TB (eds) Studies in Health Technology Informatics 91, pp 336-341. Amsterdam, IOS Press.

Dobosiewicz K, Durmala J, Kotwicki T 2008 Dobosiewicz Method. Physiotherapy for idiopathic scoliosis. In: Grivas TB (eds) The conservative scoliosis treatment, pp 228-236. Amsterdam, IOS Press.

Durmala J, Dobosiewicz K, Jendrzejek H, Pilis W 2002 Exercise efficiency of girls with idiopathic scoliosis based on the ventilatory anaerobic threshold. In: Grivas TB (eds) Studies in Health Technology Informatics 91, pp 357-360.

Durmala J, Dobosiewicz K, Kotwicki T, Jendrzejek H 2003 Influence of asymmetric mobilisation of the trunk on the Cobb angle and rotation in idiopathic scoliosisin children and adolescents. Ortopedia Traumatologia Rehabilitacja 5: 80-85.

Durmala J, Dobosiewicz K, Piotrowski J 2007 Stabilisation of progressive IS in the period of maturation in girls treated using Dobosiewicz's method (period of the observation >36 months). 4th International Conference on Conservative Menagement of Spinal Deformities, SOSORT, Boston, USA, May 15-17, 2007.

Durmala J, Dobosiewicz K, Czernicki K 2007 A retrospective analysis (October 1999 - September 2004) of the effects of conservative treatment of in-patients with progressive scoliosis in the Department of Rehabilitation in Katowice, Poland. Annales Academiae Medicae Silesiensis 61: 1-3.

Dyner-Jama I, Dobosiewicz K 2000 Effect of asymmetric respiratory exercise therapy on respiratory system function; evaluation using spirometric examination in children with idiopathic scoliosis. Wiadomości Lekarskie 53: 603-610.

Dubousset J: Importance of the three-dimensional concept in the treatment of scoliotic deformities. In Dansereau J ed. International Symposium on 3D Scoliotic deformities joined with the VIIth International Symposium on Spinal Deformity and Surface Topography. Edited in Germany, Gustav Fisher Verlag; 1992:302-311

Ellenbecker, T.; Davies, G. (2001). *Closed Kinetic Chain Exercise. A Comprehensive Guide to Multiple Joint Exercises* (first edition), Human Kinetics, ISBN 0736001700,

Fallstrom K, Cochran T, Nachemson A. Long-term effects on personality development in patients with adolescent idiopathic scoliosis. Influence of type of treatment. Spine 1986;11(7):756-8.

Fusco C, Zaina F, Atanasio S, Romano M, Negrini A, Negrini S: Physical exercises in thetreatment of adolescent idiopathic scoliosis: An updated systematic review. *Physiotherapy Theory and Practice*, 2011, 27(1); 80-114.

Greenman, P. (2003). *Principles of manual medicine* (third edition), Williams & Wilkins, ISBN 0 7817 4187 4, New York

Hawes M. Impact of spine surgery on signs and symptoms of spinal deformity. Pediatr Rehabil 2006;9(4):318-

Jelacic M, Villagrasa M, Pou E, et al: Barcelona Scoliosis Physical Therapy School -BSPTS- Based on classical Schroth principles: short term effects on back asymmetry in idiopathic scoliosis. 2011 SOSORT Annual meeting, 8th International Conference on Conservative Management of Spinal Deformities. Barcelona, May 19-21

Lee, D. (2001). *Obręcz Biodrowa: Badanie I Leczenie Okolicy Lędźwiowo-miedniczno-biodrowej,* DB Publishing, ISBN 83-920324-0-3, Warszawa

Legaye J, Orban C: Evolution of scoliosis by optical scanner I.S.I.S. Studies in Helat Technology and Informatics 1995, 15: 415-421

Lehnert-Schroth Ch. Three-dimensional treatment for scoliosis. A physiotherapeutic method for deformities of the spine. 2007, The Martindale Press, Palo Alto, California

Lonstein JE. Scoliosis: surgical versus nonsurgical treatment. Clin Orthop Relat Res 2006;443:248-59.

MacLean WE, Jr., Green NE, Pierre CB, Ray DC. Stress and coping with scoliosis: psychological effects on adolescents and their families. J Pediatr Orthop 1989;9(3):257-61.

Myers, T. (2008). *Anatomy trains – myofascial meridians for manual and movement Therapists* (second edition), Churchill Livingstone, ISBN 978 0 443 10283 7, London

Negrini S, Grivas TB, Kotwicki T, Maruyama T, Rigo M, Weiss HR. Why do we treat adolescent idiopathic scoliosis? What we want to obtain and to avoid for our patients. SOSORT 2005 Consensus paper. Scoliosis 2006;1:4.

Negrini S, Antonini G, Carabalona R, Minozzi S. Physical exercises as a treatment for adolescent idiopathic scoliosis. A systematic review. Pediatr Rehabil 2003;6(3-4):227-35.

Negrini S, Fusco C, Minozzi S, Atanasio S, Zaina F, Romano M. Exercises reduce the progression rate of adolescent idiopathic scoliosis: results of a comprehensive systematic review of the literature. Disabil Rehabil 2008;30(10):772-85.

Negrini S, Zaina F, Romano M, Negrini A, Parzini S. Specific exercises reduce brace prescription in adolescent idiopathic scoliosis: a prospective controlled cohort study with worst-case analysis. J Rehabil Med. 2008 Jun;40(6):451-5.

Negrini S, Atanasio S, Zaina F, Romano M, Parzini S, Negrini A. End-growth results of bracing and exercises for adolescent idiopathic scoliosis. Prospective worst-case analysis. Stud Health Technol Inform. 2008;135:395-408.

Negrini A, Parzini S, Negrini MG, Romano M, Atanasio S, Zaina F, Negrini S. Adult scoliosis can be reduced through specific SEAS exercises: a case report. Scoliosis. 2008 Dec 16;3:20.

Parent S, Newton PO, Wenger DR. Adolescent idiopathic scoliosis: etiology, anatomy, natural history, and bracing. Instr Course Lect 2005;54:529-36.

Pehrsson K, Larsson S, Oden A, Nachemson A. Long-term follow-up of patients with untreated scoliosis. A study of mortality, causes of death, and symptoms. Spine 1992;17(9):1091-6.

Phillips B, Ball C, Sackett D, Badenoch D, Straus S, Haynes B, et al. Oxford Centre for Evidence-based Medicine Levels of Evidence. In. Oxford: Oxford Centre for Evidence-based Medicine; 2001.

Potaczek, T., Zarzycka, M., Lipik, E., Jasiewicz, B., Zarzycki, M & Kokot, A. (2008). The early results of the treatment of idiopathic scoliosis using the dynamic SpineCor brace. *Medical Rehabilitation.* 12 (2), (2008), pp. 1-6

Richardson, C.; Hodges, P.; Hides, J. (2004). *Therapeutic exercise for lumbopelvic stabilization* (second edition), Churchill Livingstone, ISBN 9780443072932, Edinbourgh

Szota M 2006 Analysis of effectivness biodynamic 3-D correction method of treatment of idiopathic scoliosis. PhD thesis, Medical University of Silesia.

Szwed, A., Kolban, M. & Jaloszewski, M. (2009). Results of SpineCor Dynamic Bracing for Idiopathic Scoliosis. *Ortopedia Traumatologia Rehabllitacja.* Volume 11, 5(6), (2009), pp. 427-432

Tsakiri, I., Vakaloglou, V., Karvounis, K., Kattou, P. & Soucacos, P. (2009). The use of the SpineCor Dynamic Corrective Brace in Greece: a preliminary report. *Scoliosis 2009.* Volume 4, Supplement 1:O35

Rigo M, Quera-Salvá G, Villagrasa M, et al: Scoliosis intensive out-patient rehabilitation based on Schroth method. Studies in Health Technology and informatics 2008, 135: 208-227

Rigo M, Villagrasa M, Gallo D: A specific classification correlating with brace treatment: description and reliability. Scoliosis 2010 5:1 (27 January 2010)

Romano M, Minozzi S, Bettany-Saltikov J, Zaina F, Chockalingam N, Weiss HR, Maier-Hennes A, Negrini S.. Exercises for adolescent idiopathic scoliosis (Protocol). Cochrane Database of Systematic Reviews 2009, Issue 2. Art. No.: CD007837. DOI: 10.1002/14651858.CD007837.

Romano M, Negrini A, Parzini S, Negrini S. Scientific Exercises Approach to Scoliosis (SEAS): efficacy, efficiency and innovation. Stud Health Technol Inform. 2008;135:191-207.

Rowe DE. The Scoliosis Research Society Brace Manual. Introduction. In: Manual of Brace Treatment for Idiopathic Scoliosis: Scoliosis Research Society; 2003.

Stokes IA, Burwell RG, Dangerfield PH: Biomechanical growth modulation and progressive adolescent scoliosis – a test of the 'vicious cycle' pathogenetic hypothesis: Summary of an electronic focus group debate of the IBSE. Scoliosis 2006, 1:16

Travell, J.; Simons, I. (1992). *Myofascial Pain and Dysfunction: The Trigger Point Manual* (first edition). Lippincott Williams & Wilkins, ISBN 978-0683083637, New York

Vachon, V., Coillard, C., Zabjek, K., Rhalmi, S. & Rivard, C. (2006) Analyse de survie d'une cohorte consécutive de 365 patients scoliotiques traités par le corset dinamique SpineCor a l'hôpital Sainte Justine. *Résonances Européenes du Rachis.* Volume 14, N° 43, (2006), pp. 1782-1786

Vera, A. (2009). Initial Results of SpineCor Treatment of Adolescent Idiopathic Scoliosis in Seville, Spain. *Scoliosis* 2009. Volume 4, Supplement 2: O51

Wan L, Wang G-x, Bian R. Exercise therapy in treatment of essential S-shaped scoliosis: evaluation of Cobb angle in breast and lumbar segment through a follow-up of half a year. Zhongguo Linchuang Kangfu (Chinese Journal of Clinical Rehabilitation) 2005;9(34):82-4.

Weinstein SL, Dolan LA, Cheng JC, Danielsson A, Morcuende JA. Adolescent idiopathic scoliosis. Lancet 2008;371(9623):1527-37.

Zaina F, Negrini S, Atanasio S, Fusco C, Romano M, Negrini A. Specific exercises performed in the period of brace weaning can avoid loss of correction in Adolescent Idiopathic Scoliosis (AIS) patients: Winner of SOSORT's 2008 Award for Best Clinical Paper. Scoliosis. 2009 Apr 7;4:8

# 2

# Kyphosis Physiotherapy from Childhood to Old Age

Jean Claude de Mauroy
*Clinique du Parc – Lyon, France*

## 1. Introduction

### 1.1 Definition: Morphotypology of the spine in the sagittal plane

Antero-posterior deviations of the spine are usually named kyphosis and lordosis.

The term kyphosis comes from the Greek: arched back applying to vertebral deviations with posterior convexity, usually thoracic.

The term lordosis means curved and is applied to vertebral deviations with anterior convexity, usually lumbar.

There in the sagittal plane a physiological thoracic kyphosis and a physiological lumbar lordosis which limits can be specified with the morphotypology study.

### 1.2 Defining a reference position

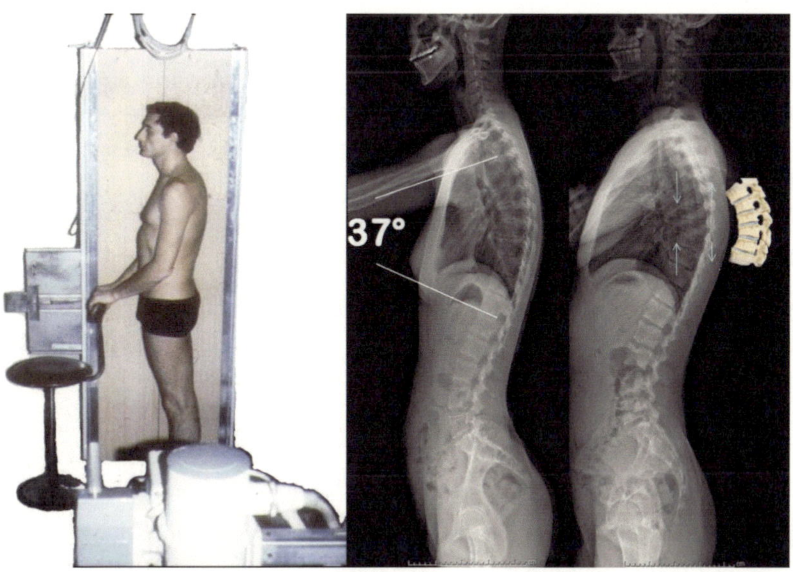


Fig. 1. Reference position, physiologic kyphosis and vicious circle of kyphosis

Unlike the frontal plane, in the sagittal plane, the upper limbs are projected onto the spine. The displacement of the upper limbs modifies the sagittal static.We chose to perform X-rays in a position closest to the clinical reference: lower limbs vertical, feet joint together, shoulder girdle and trunk released, looking horizontally. (Stagnara et all, 1982) With the agreement of some volunteers, we made two X-rays one in the clinical position (with upper limb overlaid on the spine), the other, in one of the positions normally used by radiologists ie: hands crossed in front of the chest, hands joint behind the neck, hands crossed on the head, horizontal upper limbs. None of these positions gave satisfaction so we described a specific position.The upper limbs are forward with hands resting on a support. The forearms are horizontal and the arms slightly tilted to avoid an overlapping with the spine. (Figure 1)

This position is now used by many American authors. (Voutsinas & MacEwen, 1986)

### 1.3 Statistical distribution of the angle of kyphosis: the gold standard

Kyphosis is the angle formed by the top plate of T4 and the lower plate of the vertebra, which is most inclined to the horizontal at the thoracolumbar junction.

The average angulation is of 37 ° ± 9°. Many authors have confirmed these parameters. (Korovessis & all, 1988; Fon & all, 1980; Vedentam & all, 1998)

The distribution of these parameters is Gaussian: the mode, the mean and the median are identical and there is a relative symmetry of the curve, which allows the use of statistical laws of normality.

There is no significant difference between men and women and finally there is no correlation between lordosis and kyphosis.

### 1.4 Correlation between clinical and radiological assessment

There is a correlation between clinical measures of C7 and L2 plumb line values (Figure 3) and radiological angulation. It is therefore possible to validate the evolution of a kypho-lordosis by clinical examination or better through a CAD CAM system of three-dimensional representation of the external shape of the trunk.

## 2. Biomechanics of the spine in the sagittal plane

The study of the sagittal aspect of the spine needs to include a consideration of the biomechanical aspects of the spine. One needs to consider the existing physiological curves in the sagittal plane as the spine is not straight in this plane as it is in the frontal plane.

### 2.1 A powerful brake: The anterior longitudinal ligament of the spine

The anterior longitudinal ligament inserts on the anterior vertebral body. Anatomically on a spine cadaver stripped of its muscles, it takes more than 27 kg of traction to cause a creep extension and an eventual rupture. (Figure 2)

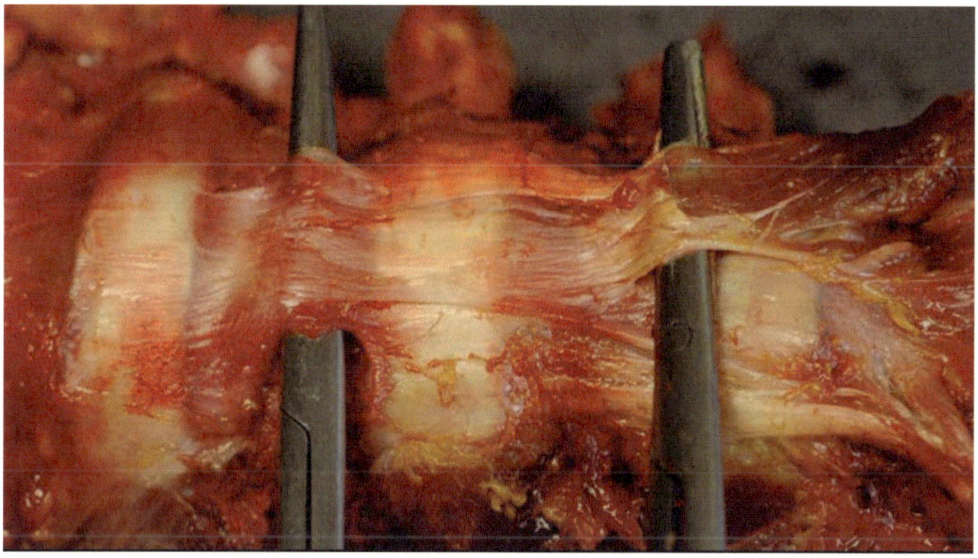

Fig. 2. The length modification of the common anterior vertebral longitudinal ligament requires a continuous traction of more than 27 Kg.

A retraction can be suspected when the patient reported pain when he is trying to correct the kyphosis spontaneously.

Clinically, this resistance can be assessed byconducting a test in hyper-extension. If the child is unable to correct their kyphosis one can measure the distance from the manubrium to a table to assess the effectiveness of physiotherapy. (Figure 4)

Radiologically, we could have a supine radiograph, the apex of the kyphosis resting on a hard pad. The lordosis is stabilized by bending the lower limbs; the upper limbs are in extension lying backward on the table. In our practice, we used x-ray in plaster cast for reducibility test and we do not take another X-ray to avoid excessive irradiation of the patient.

The concept of creeping is fundamental to understand treatment. Creeping can be defined as a definitive increase in the length of the ligament; it is opposed to the elasticity when the ligament regains its original length as the tension is released. For a creep, a continuous traction on the ligament must be performed for more than three weeks; this is the aim of the plaster cast in conservative orthopaedic treatment. The action of physiotherapy is limited in time; it remains at the level of elasticity. The removable brace does not cause a creep until it is worn continuously, which explains the importance of compliance in the results of treatment.

### 2.2 The vicious circle of adolescent and senescent kyphosis

#### 2.2.1 Fragility of the vertebral body

The fragility of the vertebral body can occur either during adolescence in Scheuermann's disease or in old age, with osteoporosis. (Scheuermann, 1921)

Scheuermann's disease is a defect in the growth plate at the cortex of the main core of the vertebral body. The cortical plate is altered and deformed, secondarily disk will enter the interior of the vertebral body (Schmorl's hernia), which causes a narrowing of the disc and stiffness. (Aufdermaur & Spycher 1986; Ippolito & all 1985)

Osteoporosis affects mainly the cortex of the vertebral body as a whole and the vertebral body collapses like a pancake. The quality of trabecular bone is important because it changes the modulus of elasticity of the entire vertebral body. Physiotherapy can improve the quality of trabecular bone.

#### 2.2.2 Poor sitting position and antalgic posture in Kyphosis

Three important points are discussed.

- In an open kinetic chain, the more the apical vertebral body moves away from the line of gravity while standing or sitting, the more the pressure will be important. Indeed the posterior musculature is forced to contract to maintain balance. In closed kinetic chain, such as when the forearms rely on the work plan, the pressure will be lower.
- A second element to be considered is the reduction of pressure on the vertebral body, when the back is applied to a rigid support. While sitting in the listening position, the spine properly applied on the backrest of the chair reduces pressure on the anterior part of the vertebra. Also sitting in the writing position when the thorax is applied on the anterior border of the table, the pressures are lower.
- The third element is the triangulation. While sitting in the listening position with both forearms and spine resting on a support, the brain will automatically correct the kyphosis. Sitting in the writing position with chondrocostal and forearms support allows the brain to correct the position of the spine in space. (Figure in scoliosis chapter)

#### 2.2.3 Apical vertebra wedging

On a weak vertebral bone, poor sitting position will place an increased pressure on the anterior wall of the vertebral body and kyphotic deformity with a reduction in the height of the anterior wall compared with the posterior wall. It is this asymmetry that causes kyphosis. It is less clear in adult osteoporosis, because the entire vertebral body is altered and the overall reduction in height is predominant. A correct posture maintained by physiotherapy may prevent the progression of kyphosis (Murray & all, 1993).

#### 2.2.4 Asymmetrical growth (Hueter-Volkmann laws)

During the teenage years, growth is inhibited when the pressures on the growth plate are too strong. On the contrary when the pressures are lower growth is stimulated. We, therefore, find a greater increase of the posterior arch compared to the vertebral body. In adulthood remodelling is much slower as bone mass is renewed every seven years, but the deformation is similar to that of puberty. (Figure 1)

## 3. Clinical examination and assessment of the spine in the sagittal plane

The order of examination of the spine in the sagittal plane will now be discussed

The standing position of the clinical examination needs to be reproducible. Ankles and toes are placed in a neutral anatomical posture. The lower limbs are stretched straight limiting an excessive recurvatum. The trunk and upper limbs are relaxed, palms of the hands on the lateral thighs, the eye is looking horizontally (de Mauroy, 1983).

### 3.1 Static: Kyphotic hump, sagittal plumbline

A plumb line is placed on the axis of the column, the extremity located at the apex of the sacral cleft (S2). The hand holding the plumb line is close to the occiput. The line is usually at a tangent to the apex of the kyphosis in T7 (corresponding to the tip of the scapula).

The arrow is measured at the high thoracic C7 level where the spinous process is more prominent. The arrow of lumbar lordosis is measured at the apex of clinical lordosis usually in L2 (L3 vertebra corresponding to a line passing horizontally through the two iliac crests).

The plumbline measure of kyphosis is half the sum of the C7 plumbline and the L2 plumbline.

In the sagittal plane, the overall harmonious balance of the spine is appreciated: tragus, acromion, trochanter, and ankles must be superimposed with the plumb line on the same vertical.

The examination in the sitting position is unclear because it can be very variable. We must examine the child when seated in front of us at the beginning of medical consultation. (Figure 3)

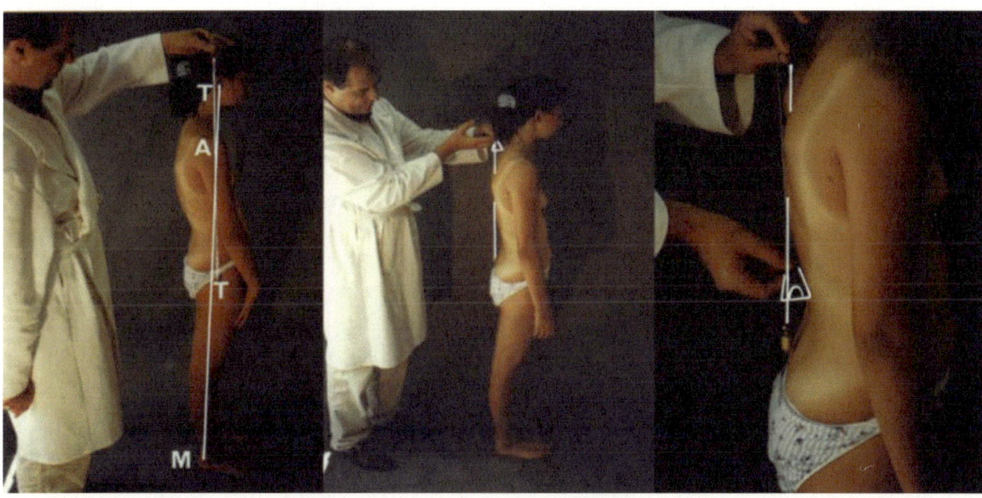


Fig. 3. Harmonious Kyphosis with TATM alignment | Cervical plumbline | Lumbar plumbline

### 3.2 Clinical diagnosis:

### 3.2.1 Angular & round kyphosis

**Angular kyphosis**

- Congenital kyphosis with bone defect is mostly seen at birth. When the discovery of the defect occurs after the age of 3 years, a night Milwaukee brace with crossbar centred on the defect can prevent the aggravation.

- Major forms found before the age of 3 years need surgery: an epiphysiodesis can be performed since the age of 6 months. (Winter, 1977; Lonstein, 1999; McMaster & all, 1999)
- The compaction of the vertebral body fractures, usually at the thoracolumbar junction is often angular. (Shen & Shen, 1999)
- Achondroplasia: after lumbar spinal stenosis it is the second problem of these patients. Kyphosis is located at the thoracolumbar junction and should be systematically prevented by the brace. In most severe cases anterior surgery performed with a distractor needs to be performed
- Morquio disease of autosomal recessive causes a significant thoracolumbar kyphosis with platispondily. In these patients there is also hypoplasia of the odontoid with unstable atlas / axis justifying caution in rehabilitation with neck mobilisation.

**Round or regular kyphosis**

Corporeal deformation is distributed over more than three vertebrae.

- Idiopathic: Stagnara and de Mauroy have defined this new entity of progressive kyphosis without Scheuermann's disease during puberty by analogy with idiopathic scoliosis. The vicious circle still described can explain idiopathic kyphosis (de Mauroy & Stagnara, 1979).
- Dystrophy or spinal growth or Scheuermann's disease is a defect of the growth plate of the vertebral body at the cortical level weakening the vertebral body and can cause wedging. The vertebral rim or secondary nucleus of ossification is inhibited only by hyperpressure related to kyphosis and in case of conservative orthopaedic treatment the brace can compensate the anterior wedging and restore a rectangular vertebral body (Sorensen, 1964; Ali & all, 1999; Lowe, 1999).
  This disease stops at the end of growth, when the growth cartilage is no longer active.
- Neurological: the paralysis of abdominal muscles is compensated by a lumbar kyphosis. Paralysis of the spinal muscles causes lumbar lordosis; the patient uses the remaining muscles as struts to keep the spine in a position where it can be effective. Alterations in psychomotor developments are accompanied by postural abnormalities: either global kyphosis in children who are on a wheel chair, or lordosis, which are sometimes very unstable with athetosis.
- Post-laminectomy: the extension of the laminectomy at facet joints and posterior ligaments results in 80% of kyphosis being very severe (Lonstein, 1987; Albert & Vacarro, 1998). When the Milwaukee brace is insufficient an anterior fusion should be considered.
- Radiotherapy in neuroblastoma and Wilms' tumour sterilizes the growth cartilage and retraction of soft tissues and can cause kyphosis; which usually reacts favourably to the brace.
- Rheumatology: Ankylosing Spondylitis has symptoms that can begin before the age of 20. As a result we can observe a disharmonious kyphosis with loss of lordosis and forward projection of the trunk. In addition to the anti-inflammatory therapy, rehabilitation includes daily prone postures and the practice of lordotic activity such as swimming. In extreme cases when the eyes can no longer look horizontally, one can realize posterior osteotomies to balance the head on the line of gravity.

- Osteoporosis in adults: the high thoracic kyphosis is often painful. This type of kyphosis is worse when associated with a loss of physiological lordosis.

### 3.2.2 Clinical forms: High thoracic, thoracic, thoraco-lumbar

The apex of the kyphosis will be located on x-ray and by palpation and used to determine the appropriate physical therapy and conservative orthopaedic treatment. We can distinguish three different types of kyphosis:

- The high thoracic kyphosis or cervico-thoracic. It is accompanied by a forward projection of the neck and the shoulder in internal rotation.
- The medium thoracic kyphosis which is the more common type of kyphosis. It can be harmonious when accompanied by a lumbar hyperlordosis; however, it is disharmonious when the lordosis decreases.
- The thoracolumbar kyphosis is the most common form of non congenital angular kyphosis.

## 3.3 Dynamic clinical assessment

### 3.3.1 Dynamic balance

At our centre we use the rating of Russe and Gerhard quantifying the amplitudes of the spine in three planes (Russe & Gerhard, 1976).

In the sagittal plane, the axis of motion is a horizontal line across the two femoral heads. A virtual line joining the acromion to the trochanter is drawn; the position of this line from the vertical to maximal extension and flexion will be recorded. The finger-floor distance corresponds to the stiffness in the spine and pelvis.

For example: S 30-0-90 (which means 30° extension to 90° of flexion).

In the frontal plane, a virtual line joins the top of the medial gluteal fold (S2) to the C7 spinous process. The extreme positions of the virtual line from left to right are noted: F 35-0-30 F (from 35° lateral left tilt to right lateral 30°).

In the horizontal plane, the left to right rotation of the shoulder girdle is measured in relation to the fixed pelvic girdle. A first horizontal virtual line is joining the two femoral heads, at shoulder level, one second horizontal line is joining the two humeral heads: R 45-0-45 (45 ° rotation of the shoulder girdle from left and right). (Figure 4)

### 3.3.2 Testing vertebral reducibility: Direct rigidity

In the prone position, we make a systematic examination of the spine:

- Palpation of the skin in search of fibromyalgia at the cervico-scapular.
- The percussion of the spinous process can cause pain usually corresponding to Scheuermann's disease at the apex of the kyphosis and at the thoracolumbar junction.
- The muscle contractions will be found at the lumbar painful areas and at the apex of the kyphosis.
- The painful transverse pressure at the lumbosacral junction evokes a spondylolysis, common in cases of hyper-lordosis (7% of the population).

- A dehiscence at the spinous process evokes a spina bifida.
- The rigid spine will be assessed in this position, the upper limbs are positioned laterally along the body and the patient is asked to move with hyperextension of the spine. The distance manubrium-table is measured appreciating the reducibility of kyphosis in this position and we note the location of the rigid zone (apex of the kyphosis or thoracolumbar junction).

### 3.3.3 Testing scapular and pelvic girdle reducibility: The indirect rigidity

In the supine position, we evaluate the stiffness of the pelvic and scapular girdles.

Under the pelvis, the hamstrings are often retracted and the popliteal angle is measured: if the thigh is vertical, the popliteal angle is the angle between the vertical and the leg when trying to extend the entire leg.

In order to measure the retraction of anterior plane including the ilio psoas, the patient is asked to hold his knee; starting from the vertical the leg is lowered in extension asking the patient to relax the trunk and to bend the opposite limb. At the end of the movement a pelvic tilt is seen and the distance heel-table is measured corresponding to the start of the pelvic tilt.

At the shoulder level, the upper limbs are lowered in extension and anterior retraction is assessed by the distance table-elbow ensuring that the spine remains straight without hyper-lordosis. (Figure 4)

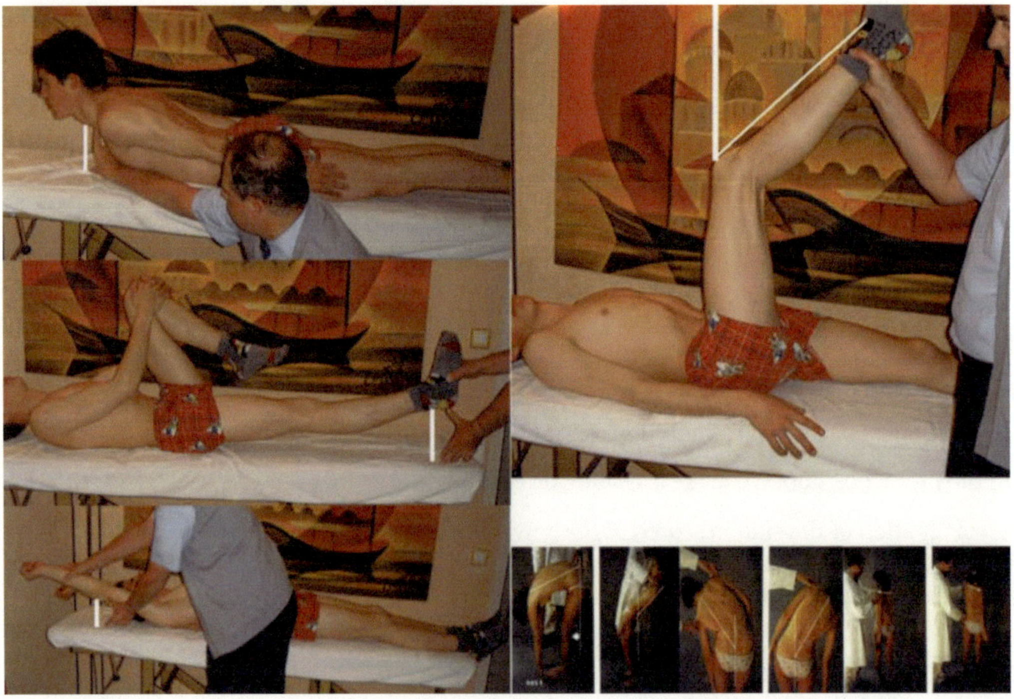

Fig. 4. Direct and indirect rigidity testing | Russe and Gerhard dynamic assessment

### 3.4 Muscle impairment

- The flexibility of anterior thoracic muscles is decreased in patients with kypholordosis. The role of the ilio psoas is difficult to assess, usually its retraction causes a lordosis and sometimes a compensatory kyphosis.
- The thoracic extensor muscles must be stretched. The posterior muscles are weak and mostly hypotonic, which means that they cannot maintain their contraction over time and they get tired very quickly.

### 3.5 Evaluation of thoracic pain in adolescents

Pain has special characteristics in children. The child does not memorize the pain he is experiencing, he lives in the present. In the absence of cortical representation of the spine, it does not somatise. Pain is classified into six stages:

- Stage 0: no pain,
- Stage+: pain found only with percussion of the spinous process during clinical examination,
- Stage ++: mechanical pain during or after sport,
- Stage +++: static positional pain: in prolonged sitting or standing position,
- Stage ++++: pain at rest,
- Stage +++++: using analgesics, which is exceptional in children.

After puberty some back pain are related to neuromuscular hyper excitability. The pain is the result of contractures which occurs when the neuro-muscular junction is poorly vascularized. Pain is associated with asthenia, tremor of eyelids and microcirculatory disorders: ends of the feet and hands cold.

### 3.6 Psychological assessment: body image: Quasimodo, Pulcinella...

Deviations of the spine were in the past of tuberculosis origin (Pott's disease) or secondary to poliomyelitis with kyphosis and scoliosis. Currently a flat back is mostly seen with idiopathic scoliosis and humps are the consequence of the asymmetric rib rotation of the vertebral body when the trunk is bending forward in the Adams test.

The round back is inscribed in a cultural context that assigns moral values in relation to defined standards: "Stand up straight, or you will become hunchbacked." The right axis is an axis of will, the verticality of the body is an element of expression. The relationship between body and mind is established on the basis of identity: a twisted body means twisted ideas.

The "bump" is often the seat of ridicule, malice, and is a cause of mockery. The evil spirit or devil seats in the hump. Saturn, the patron of sorcerers, is often depicted with a kyphosis. If it is the sign of the curse it can also bring good luck "touch the bump brings good luck."

This evocation of the cultural round back deformity can explain the attitude of the family of kyphotic children. The child who misbehaves is described as different and his body bearing a physical disgrace is devalued. The child may have kyphotic feelings of aggression, frustration and withdrawal. Hunched posture of old age, those who have "rolled their hump" destabilizes the adolescent. The kyphotic child escapes this negative attitude by taking refuge in an imaginary world. The child avoids having feelings and as a consequence the child will grow more and more frustrated when facing reality: “It's always me who is accused, but now it does nothing to me, I am accustomed to. “

Psychological tests indicate insecurity, shyness and introversion.

The child's perception of the body is difficult and emotional control is worse. Support and encouragement in their actions is needed (Lindeman & Behm, 1999).

### 3.7 Adult camptocormia

Camptocormia is a medical condition that is characterized by forward flexion of the spine. The camptocormia kyphosis has become more common due to the aging population. It is a weakness of the paraspinal posterior muscles. This weakness is often of pyramidal origin and clinical signs of Parkinson's are often noted as well. Because of this weakness, the patient stretches the last muscle fibres with a forward bending tension. The characteristic of camptocormia is that kyphosis desappears in supine position. Some antigravity braces can delay the progression of the kyphosis (Ryan & Fried, 1997; Kado & all, 2007).

In conclusion after this examination we can already distinguish between different types of kyphosis:

- Kyphotic or lordotic attitudes, related to ligamentous laxity or muscular hypotonia. These deviations can be reduced during the hyperextension test and improved with specific therapy.
- The paramorphism can be the result of a sport activity encouraging the kyphosis, such as: swimming (dolphin, butterfly).
- The constitutional kyphosis or lordosis, corresponding to a family morphotype is usually stable.
- The kyphosis adaptation: for instance the myopia determines a significant kyphotic attitude that can stiffen during growth. Similarly the long persistence of kyphosis in children with infantile encephalopathy promotes progressive structural deformation of the vertebral bodies.
- The structural kyphosis that usually justifies specific management.

## 4. Radiological assessment and aetiology

Radiography is unfortunately necessary to clarify the diagnosis and measure the morphometric parameters which will guide physiotherapy (Van Rosen & all, 1998). The classical radiological assessment includes a profile X-ray done on cassette filter 30cm by 90cm in teleradiography tube located 2 meters from the patient. The focus is on T6 to avoid excessive deformation of the vertebral body. Currently we use the EOS technique which has the advantage of less irradiation for the patient and getting better visibility of the end plates, because the dose is adjusted for each horizontal plane and the ray is tangent to the plate, except of course in case of excessive kyphosis. Whatever the technique, patient positioning as described in the morphotypology section, is crucial for kyphosis.

### 4.1 Scheuermann disease

The diagnosis of Scheuermann's disease is radiological.

The radiographic signs are classified into four stages:

- Stage +: irregular cartilage plates,
- Stage + +: intraspongious hernia,
- Stage + + +: 7° to 10° of vertebra wedging at the apex of the kyphosis,
- Stage + + + +: wedging greater than 10 ° on one vertebra or dystrophic lesions spread over 5 vertebrae. (Tribus, 1998, Lowe & Line, 2007))

### 4.2 Congenital malformation

Congenital kyphotic malformations are less frequent but more severe than those causing scoliosis, the risk of progressive paraplegia is important. There are two different types:

- Type I: hypoplasia of the vertebral body, the posterior arch is preserved and leads to kyphosis with growth.
- Type II: aplasia of the vertebral body. It is a congenital defect of the anterior segment of the vertebral body called "bars" or "block" (Lonstein, 1999).

### 4.3 Vertebral osteoporosis

Bone densitometry is used to quantify osteoporosis, but for the same bone mineral density (BMD), vertebral consequences may differ depending on the patients. The x-ray shows the characteristic deformations of spinal osteoporosis: decreased height, more or less symmetrical compaction and deformation like a pancake.

Vertebral indices are used to quantify the progression of osteoporosis.

### 4.4 Ankylosing spondylitis

Symptoms of ankylosing spondylitis can begin before the age of 20. It results in a disharmonious kyphosis with a loss of lordosis and a forward projection of the trunk. In addition to the anti-inflammatory therapy, rehabilitation includes:

- Daily prone postures,
- Practicing lordotic activity such as swimming.

In extreme cases we can make posterior vertebral osteotomies to improve lumbar lordosis and balance the head on the gravity line.

### 4.5 Other aetiologies

Nutritional kyphosis can result from nutritional deficiencies, especially during childhood, such as vitamin D deficiency (producing rickets), which softens bones and results in curving of the spine and limbs under the child's body weight (Hensinger, 1977).

Myelomeningocele like laminectomies can evolve in kyphosis. (Banta & Hamada, 1976).

## 5. Typical exercises for adolescent hyperkyphosis

### 5.1 Awareness of the deformation with a mirror or camcorder

The child must become aware of the deformed image of his back and gain a better representation of his shape, his position and his dynamics in space. This static and dynamic awareness is achieved through a video camera placed laterally to get a profile view. It is necessary to explain to the child the ideal position in the sagittal plane. The child also needs to be told the means of achieving this ideal position in order to learn and correct disharmonious balance first in a segmental way from feet to head and then globally in static, and after in dynamic during the movement of walking. The difficulty lies in correcting sequentially:

- Anteversion or retroversion of the pelvis
- Lumbar lordosis,
- Thoracic kyphosis,
- Forward projection of the neck,
- Shoulders antepulsion.

In some structural cases, the correction is difficult because of stiffness or poor muscle quality. Correcting a kyphotic posture takes time and compliance with physiotherapy is vital for the best possible outcome.

### 5.2 Mobilization of the thoracic spine in hyperextension especially in case of direct stiffness: Segmental exercises

Direct rigid kyphosis is very often the result of Scheuermann's disease which has as a consequence height loss of the intervertebral disc with segmental mobility limitation.

The Stretching and the relaxing of the spine in extension are done with:

- Passive postures in prone or quadruped position
- Stretching of the anterior intervertebral ligament in a supine position with apex of kyphosis on a block,
- Passive postures with at the end active extension,
- Facet joint mobilization in the three planes by combining active and hyperextension lateral bending and rotation with active hyperextension. The relaxation must be global and three-dimensional.

The segmental stiffness at the kyphosis apex favours limiting the amplitudes of the chest if breathing is deep and wide. Breathing exercises especially diaphragmatic and lateral-costal are done in maximal deep inspiration and expiration. (Figure 5)

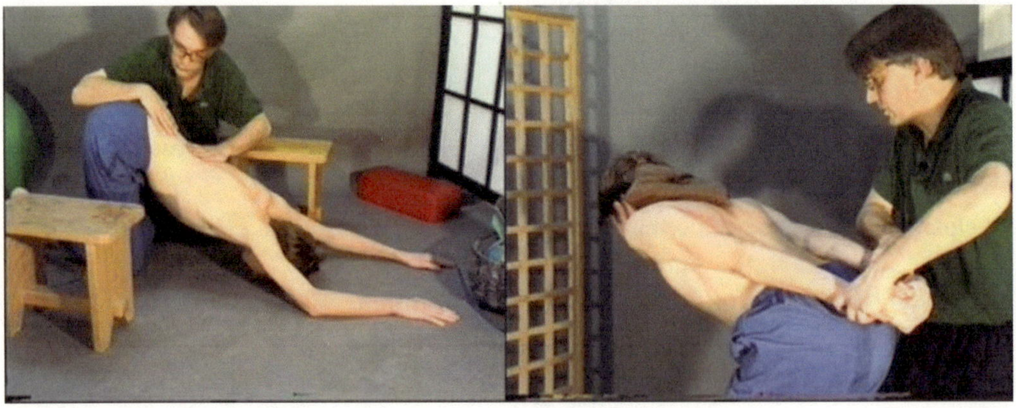

Fig. 5. Segmental mobilisation

The Schroth method is useful for lumbar kyphosis. (Lehnert-Schroth & Weiss, 1992)

### 5.3 Stretching of anterior thorax muscles and hamstrings in case of indirect stiffness

Passive stretching can increase range and reduce tension. For instance you can use the shoulder blade squeeze: with your back straight squeeze your shoulder blades together as hard and as far as possible pain-free. Hold for 7 seconds relax 7 seconds and repeat 10 times. To stretch the tight and overly developed chest or pectoral muscles it is necessary to lie on a mat with knees bent and hands placed at the base of the skull, shoulder blades and ribs connected to the mat, elbows down towards the floor are pressed feeling a pull across the front of the chest and under the armpits.

For hamstrings, place your foot on a step or chair, keep your knee and back straight and lean forward at your hips until you feel a stretch in the back of your thigh. Lying on your back with a rope or band wrapped around one foot and the other leg long and anchored to the mat; strap is tensioned with exhalation as the leg is brought up until feeling a pull in the back of the upper thigh.

### 5.4 Global stretching with RPG or Mézières exercises

Françoise Mézières (1909-1991), a French physiotherapist was teaching and practising classical segmented physiotherapy. She gradually discovered the value of a global stretch of muscle chains in a position with lower limbs and trunk flexed at 90 °. For the spine deformation is most often used in closed kinetic chain, feet on the ground and hands resting in front of a support in horizontal extension of the trunk (Mezieres, 1978). (Figure 6)

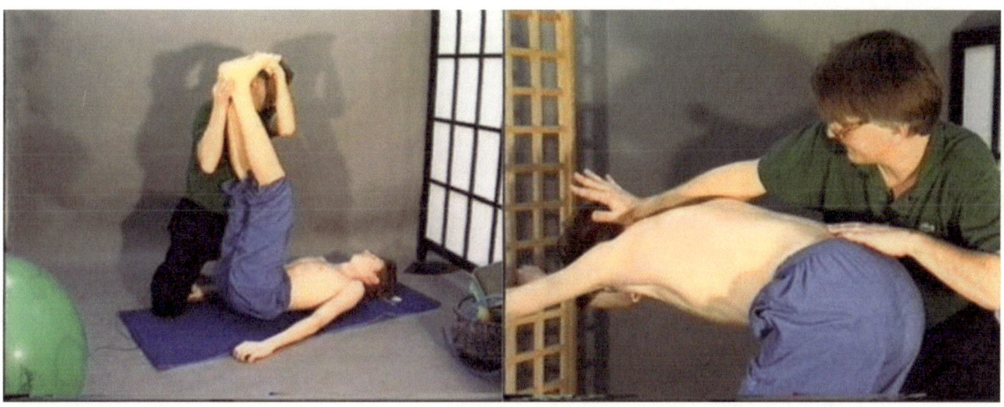

Fig. 6. Meziere's postures

Pilates' exercises are also useful for length/tension imbalance in the body with weak? Abdominal muscles, tight chest and hamstrings muscles and a weak overstretched upper back.

### 5.5 Integration of postural correction

We described the main stages of the postural correction with a mirror. Once the stiffness has disappeared and the muscles have been strengthened, this correction must be perfect and maintained in all the activities of daily life, what we call the 24 hours of the spine.

### 5.6 Extensor spinal and abdominal muscles strengthening in corrected position

Strengthening exercises are given for abdominal muscles and back extensors to stabilise the spine and maximise its function. These reinforcements are made with small weights of a maximum of 5 kg with a spine correctly positioned. The duration is 7 seconds, which is equal to the resting time. The direction of motion is extension. The positions need to be gradual, the first being the prone position used for testing reducibility in hyperextension. The progression is: sitting, standing, moving, for instance by walking. The pace should be relatively slow with breath control for aerobic metabolism and a preferential development of slow fibres muscle type I. (Lam & all, 1999)

### 5.7 Integration of the corrected posture during imbalance exercises (neuromuscular control and equilibrium)

At first we can use the supine position on a Swiss ball. The physiotherapist maintains the legs of the patient seeking to balance only with the upper limbs in extension and external rotation, the trunk being released. In a second step, the child is lying in prone position and he must straighten at maximum the spine, making a paravertebral muscles active strengthening in instable posture. (Figure 7)

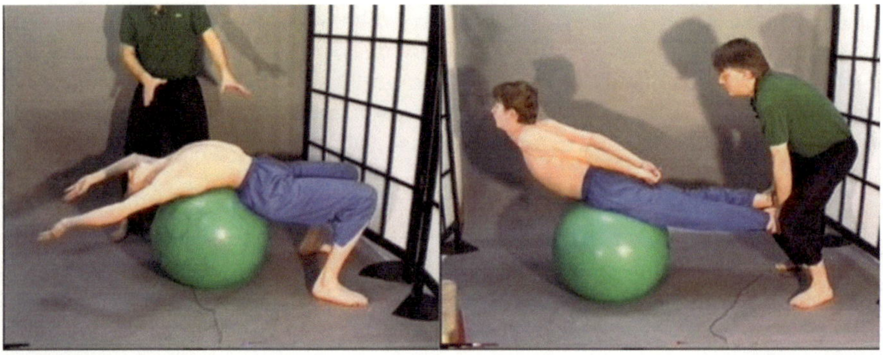

Fig. 7. Exercises on a Swiss ball

### 5.8 Practice of `extension` sport activities

No sports activities are contraindicated in kyphosis, because almost all activities involve spinal extension. Sport activities performed in a sitting position like cycling and rowing are indicated in cases of harmonious kypho-lordosis ie hyperkyphosis balanced by hyperlordosis. The position of the spine needs to be controlled. With swimming some styles such as the dolphin and butterfly styles which increase curves in the sagittal plane must be avoided. Sport activities complete physical therapy and in case of conservative orthopaedic treatment, best results are obtained in patients regularly practicing a sport.

Hypotonia is physiological in teenagers and sport is one of the best ways to fight against this physiological hypotonia.

### 5.9 Back education

The back school is mostly used in cases of adult back pain and does not significantly modify any morphological parameters of the spine. An educational program for adolescents is very useful for trainees (Troussier & all, 1999). Scheuermann first described apprentice watchmakers kyphosis… (Scheuermann, 1921)

## 6. Typical exercises for adult hyperkyphosis

In adult cases, kyphosis keeps on evolving. When the kyphosis is the main reason to consult a doctor and when a specific aetiology is eliminated; the following can be distinguished:

- Kyphosis in young adults, is mostly painful with a slightly stiffness of the paraspinal structures, but the muscles remain strong. The aesthetic aspect is important, the patient no longer accepts a hunchback image and often requires surgical morphologic correction.
- Old people kyphosis is characterized by bone fragility, often with osteoporosis. The charge on the osteoporosis pain which becomes increasingly resistant to conventional treatments, the musculature is less powerful and the spine is stiffer.
- Very old people kyphosis is characterized by progressive atrophy of the muscles that can lead to the extreme situation of camptocormia.

When the kyphosis occurs in a specific etiologic context such as a fracture, orthopaedic conservative treatments are generally more accepted.

### 6.1 Static prone corrective posture, relaxation

This is the main treatment of ankylosing spondylitis. It is very important that the stiffening of the spine occurs in a position closest to the line of gravity. The ideal position occurs when the direction of the gaze is stabilized in a slightly oblique line down from the horizontal. The patient should be able to see where he is going.

### 6.2 Analgesic and deep tissue massage

Painkillers for the patient are the easiest and fastest way to solve the pain problem. It is justified if the assessment reveals a precise aetiology associated with kyphosis. In other cases, it is a palliative solution for the short term that does not treat the actual cause of chronic mechanical pain. In addition to the side effects of painkillers, pain is often a protective mechanism for the spine, a kind of seatbelt. The excessive use of painkillers can be compared to driving without a seatbelt.

Massages are performed in the supine position and an anti-kyphosis posture achieved by adding pads under the pelvis and at the sterno-clavicular area.

Analgesic massages are also always preferable to palliative painkillers.

### 6.3 Global postural physiotherapy

Far from suffering from weaknesses, excessive strengths can be found in groups of paravertebral muscles called 'muscular chains' by Mézières. Being multi-joint, these muscles are banded together in a chain-like manner in which they overlap like tiles on a roof. There are four muscular chains in our bodies, the main one being situated at the back, running from head to feet. Although made of several muscles, muscular chains always behave like a single muscle. Sometimes they end up too tight and too short. The shortening of the anterior muscular chains causes kyphosis. Any departure from the normal, ideal shape will inevitably cause pain and malfunction, restoring shape to its primitive condition has a powerful therapeutic value as it addresses the primary cause of our musculo-skeletal ills. The normalisation of morphology is the most efficient way, the only lasting way, to 'cure' back and related pains.

We have described for the teenager` the square` in closed kinetic chain with feet on the ground and fixed point on the Swedish ladder. For adolescents, the trunk is not applied on a support, in order to be accessible. This position is often painful for adults and we will use the position "at the extremity of the table" that allows a true relaxation of the paraspinal musculature. The table height is adjusted to maintain the right angle between the legs and the trunk. The upper limbs are extended in line with the trunk. (Figure 6)

### 6.4 Muscular reharmonisation

We have already considered the disharmony within the anterior and posterior muscle chains. There is also a disharmony between anterior and posterior chain, i.e between extensors and flexors of the trunk. In concentric isokinetic, the typical ratio is 0.7 with more strength for the extensors. This ratio can be changed in case of kyphosis with decreased extensor strength compared to the flexors.

### 6.5 Breathing exercises for costovertebral mobility

The exercises are executed in maximum magnitude using the upper limbs to facilitate the mobilization of the thorax. Movements will be slow if possible with an abdominal strap to avoid lordosis and focus on mobilizing the thoracic area (Priftis & all, 2003).

## 7. Physiotherapy if bracing is necessary

### 7.1 Exercise with a soft brace

A soft back brace like spinecor can be used as an aid to bending and lifting properly when the kyphosis is flexible. They are worn during the day; they can never be worn at night. They have the disadvantage of increasing lordosis. In case of disharmonious kyphosis their use completes physiotherapy treatment

### 7.2 Exercises in plaster cast: Flexibility, range of motion

The plaster cast is made in a Cotrel's frame with a transversal band at the apex of the kyphosis (Cotrel & all, 1964). The superior limbs are in extension. The knees are bent if we want to correct the lordosis at the same time. In some cases of disharmonious kyphosis, the inferior limbs are in extension to maintain lumbar lordosis.

Immediately after the plaster cast the physiotherapist needs to achieve significant therapeutic patient education with:

- Advice for skin care under the plaster cast .Skin should be cleaned daily by a parent with a strip of gauze and rubbing alcohol.
- Advice to avoid digestive abdominal dilatation. The risk of gastric dilatation is not negligible. Meals need to be taken in small portions and, and soft drinks avoided
- Advice for sitting positions with increased trunk-thigh angle need to be provided

The exercises include:

- Breathing needs to be controled. The body cast limits inspiration and it is necessary to focus on expiration. The swelling of a balloon or practice of the flute is recommended. Hyperventilation which decreases pulmonary oxygenation is avoided.
- Mobilization in plaster cast. It should be remembered that the movement is fundamental during the plaster cast time. It is this movement which will gradually creep and lengthen the anterior longitudinal ligament.
- Mobilizations of pelvic and scapular girdles are facilitated by the fixed point of the plaster on the trunk.
- A global strengthening of the paravertebral muscles in active auto-axial elongation is performed, but the weight of the plaster alone of about 7 kg is in itself a strengthening exercise.

Physiotherapy should be monitored daily and at least twice a week by a physiotherapist.

### 7.3 Exercises with Milwaukee brace

The Milwaukee brace worn at night with posterior transversal bar cross centred on kyphosis is used routinely before the pubertal growth mainly in cases of congenital malformation. The brace guides the growth that occurs mainly at night. Physiotherapy is carried out in the evening keeping the brace on. The child stands with his hands on the anterior bar and performs two basic movements. The first is a self-active axial elongation to lift the chin from the cervical collar. The extension is held during 7 seconds and the release time is also 7 seconds. The second exercise is performed in sagittal plane; it is an anterior shift with distance from the transverse posterior pad. The corrective movement is also maintained 7 seconds; the same relaxation time is allowed (Blount & Moe, 1973; Sachs & all, 1987; Winter & all, 1987).

### 7.4 Exercises with adolescent Lyon Kyphosis brace

The plexidur bivalve brace has a posterior shell usually T7-S3. The upper part is just under the apex of kyphosis. A screw box is a turning point enabling the adjustment in the standing and sitting position.

The anterior shell has a manubrial thrust reinforced by a metallic bar. (Figure 8)

Physiotherapy is a fundamental part of the Lyon orthopaedic treatment. There is no rigid brace without associated physiotherapy (Stagnara & all, 1966; Stagnara & de Mauroy, 1975).

In the initial phase, the exercises performed with the brace and identical to those described with the plaster cast.

When the shoulder and pelvic girdles have been relaxed and strengthened, physiotherapy can be carried out without brace. The aim is to maintain an identical position to that of the spine under brace in various static positions of daily life: sitting, standing, leaning forward...

Subsequently, the same position will be maintained in dynamic situations of imbalance, such as walking on a beam.

The third time is the integration of the corrected position in sport.

### 7.5 Exercises with adult polyethylene bivalve overlapped brace

Two major factors determine the bracing of kyphosis in adults:

1. The difficulty of setting up the brace which often requires a third party.
2. The significant imbalance in the sagittal plane.

The first task of the physiotherapist will be to teach the patient to set up his brace alone. It will sometimes be necessary to relax the shoulders or to make some ergonomic changes to the brace. Then, the physiotherapist will insist on walking with the brace. The correction of kyphosis can change the eyes direction and the patient can be worse when he walks, he will teach him how to use adherent and flexible sports shoes closed by Velcro, as the traditional lacing will be very difficult. Proprioceptive stimulation is global and the technique of blindfolded can be used to increase inferences and enhance the patient's confidence.

The brace can also reduce vital capacity by 20%. For this reason a very important thoraco-abdominal window is made in the brace to change as little as possible the abdomino-diaphragmatic range of mobility. Respiratory physiotherapy aims to promote the expiratory time and control the breathing rate (de Mauroy, 1995).

## 8. Ergonomics: The sitting position depending on the lumbopelvic morphotype

### 8.1 Radiological parameters

To determine the lumbar-pelvic morphotype, we must measure and compare, in our radiological reference position, the inclination of the sacral slope to the horizontal and the pelvic incidence. The sacral slope is highly variable and there are often changes associated with kyphosis, the pelvic incidence does not depend on the posture and does not change with anteversion or retroversion of the pelvis.

In 80% of cases, the sacral slope is close to 37° and the pelvic incidence close to 53°. Standard sitting position resumes the Staeffel's 90° rule: Feet rest horizontally on the floor, vertical legs, horizontal thighs on a seating base itself horizontal and a vertical trunk.

In case of thoracolumbar kyphosis, there is often a horizontalisation of the sacral slope with stable pelvic incidence at 53°. In these cases, we may use an ergonomic kneeling chair.

In case of accentuated kypho-lordosis, there is often an increase of the inclination of the sacral slope. If the pelvic incidence is normal at 53°, we can use a triangular cushion slightly tilted backward.

In some cases, the pelvic incidence will be modified reflecting a constitutional defect in the lumbopelvic static. It is this change in lumbar-pelvic incidence which is responsible of overlying deviations. A conservative treatment with orthopaedic brace will be realized in these cases.

### 8.2 Using ergonomic seats

A kneeling chair is an ergonomic chair with tights dropped to an angle of 60° from vertical. The intended purpose of a kneeling chair is to reduce lower back strain by promoting proper spinal alignment. This spinal alignment can be useful for thoraco-lumbar kyphosis (Dury & Francher, 1985; Lander & all, 1987; Bettany-Saltikov & all, 2008).

In other cases, a classic ergonomic chair can be used, avoiding wheels in cases when the chair does not have to be moved often during the exercice. The seat height will be determined by the size of the patient so that we find the Staeffel's 90° rule. The seat is usually horizontal and the back close to the vertical. The height of the work plan will be determined by the distance of the horizontal forearms to the ground.

The patient will differentiate the listening position and the writing position which are exactly opposite.

In the listening position, the coccyx is backward as possible, the back spine is pressed on the back of the chair, thighs are horizontal and the feet are in front of the seat. They may also rest on a support. Ideally forearms are on the armrest.

In the writing position, the feet are behind the chair, the coccyx is in front of the seat, thighs in forward tilt. The chest is pressed against the front edge of the work plan and forearms rest on the work planWhen there is a high cervico-thoracic kyphosis of more than 25° between T1 and T4, a lectern will be used with a desk tilted at 15° on the horizontal work plane.

Poor lighting of the work plan can promote kyphosis.

The computer screen should be vertical, the centre at eye level.

For technical professional reasons, it is not always possible to use these reference positions, and in these cases an ergonomic adaptation of the workplace will be conducted.

## 9. Results, consensus and best evidence

(Figure 8)

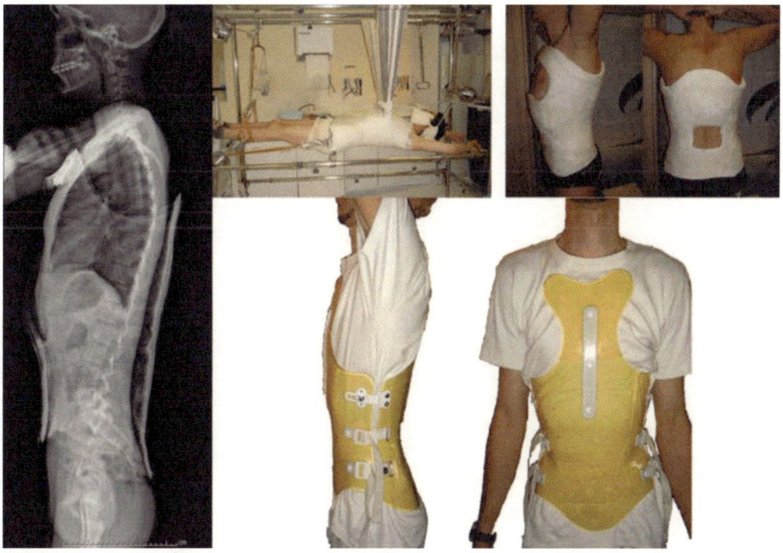

Fig. 8. Lyon kyphosis orthopaedic management

### 9.1 Results

272 patients (142 boys and 130 girls) with an average age of 13.6 months were treated from 1987 to 2005 and reviewed 2 years after removal of the brace, even if compliance with bracing was not satisfactory. Treatments were conducted according to the same historic

protocol in the same structure and followed by the same author. Braces were madeby the same orthotic Company.

The Lyon orthopaedic treatment is usually carried out at the end of pubertal growth which is the more common indication. The plaster cast was systematically made in the day hospital before bracing. According to the importance of the angulation and the rigidity of the curvature, we made one or two plaster casts, each cast for one month time. The specific physiotherapy programme began when the plaster cast was made and carried onduring the bracing phase. The brace was worn in sitting position at school or only the night for the most favourable cases. The brace was removed at the end of growth. A test of temporary removal of 6 months during summer allows the doctor to make sure the angle is stabile (de Mauroy, 2010).

The diagnosis was Scheuermann in 119 cases and idiopathic in 153 cases. There were 10 cases with thoracolumbar patterns.

Patients were evaluated both by clinical or radiological examination. The first clinical and radiological assessment was made before the plaster cast. With a plaster cast, the reducibility of the curve needs to be measured radiologically. The successive controls take place at the time of the delivery of the brace then every 6 months until the removal, then 2 years and 5 years after the removal of the brace. All the clinical and radiological data are written according to a precise form on the computer programme. Since 1998 an automatic update of a specific spreadsheet is performed by the secretary during the medical control. All the data was analyzed by SPSS 18 IBM software.

After a description of the main parameters: means, SD, normal distribution. We shall use the parametric paired tests to compare the data, usually the Pearson correlation coefficient. The main criterion variable is the Stagnara angular correction between the initial radiography and that realized 2 years after the end of the treatment.

### 9.1.1 Angular evolution during treatment

Results are presented in table 1. X-ray at 1 year and removal are done without brace.

By comparing Cobb angles at the beginning of treatment and two years after the brace removal:

- 0 patient had an angular worsening of more than 5° (bad results)
- 16 patients had a stabilization between more or less 5° (stabilization)
- 256 Patients had an improvement of more than 5° (good results)

| | Initial | In Plaster | in brace | 1 year | Removal | Removal + 2 years |
|---|---|---|---|---|---|---|
| Kyphosis | 59,43<br>+- *8,00* | 32,09<br>+- *8,21* | 37,12<br>+- *11,01* | 39,58<br>+- *8,34* | 39,78<br>+- *7,81* | 40,84 (31%)<br>+- *8,15* |
| Lordosis | 48,90<br>+- *11,83* | 38,21<br>+- *9,17* | 41,76<br>+- *8,91* | 44,71<br>+- *9,09* | 43,74<br>+- *9,63* | 44,20 (10%)<br>+- *9,23* |
| Sacral slope | 39,13<br>+- *7,88* | 34,89<br>+- *7,34* | 36,36<br>+- *6,89* | 37,26<br>+- *7,14* | 36,89<br>+- *6,87* | 37,27 (5%)<br>+- *7,37* |

Table 1. Angular evolution of Kyphosis, lordosis and sacral slope

### 9.1.2 Correlation between the measure of the clinical plumbline measures and the radiological angulation

The correlation is very significant between the clinical measurement and radiological evaluation. (Table 2)

| | | | | |
|---|---|---|---|---|
| Arrow Kyp / Cobb Kyph init | **0,322** | | | significant at the 0,01 level (2-tailed) |

Table 2. Correlation between the measure of the plumline measures and the radiological angulation

### 9.1.3 Correlations between Scheuermann, sex, pain and thoraco-lumbar localization

Our statistics analysis encompasses 142 males and 130 females. Males have more Scheuermann's disease. 116 patients present a Scheuermann's disease with more pain (126/140 cases) and thoraco-lumbar localization (8/10 cases). (Table 3)

| | | |
|---|---|---|
| Sex / Etiology | **0,194** | significant at the 0,01 level (2-tailed) |
| Etiology / TL | **-0,177** | significant at the 0,05 level (2-tailed 0,034) |
| Etiology / Pain | **0,142** | significant at the 0,05 level (2-tailed) |

Table 3. Correlations between Scheuermann, sex, pain and thoraco-lumbar localization

### 9.1.4 Study of the correlations influencing the final angular correction

We shall present the significant results according to the value of the Pearson's correlation coefficient. The immediate reduction in plaster cast is an excellent predictive criterion of the final correction (table 4).

| | | |
|---|---|---|
| Cobb Kyp plaster / Cobb Rem + 2 | **0,536** | significant at the 0,01 level (2-tailed) |
| Plaster cast correction / Final Cobb | **0,497** | significant at the 0,01 level (2-tailed) |
| Cobb Kyph init/Cobb correction | **0,455** | significant at the 0,01 level (2-tailed) |

Table 4. Most significant correlations with the final result.

### 9.1.5 Many parameters are not significant:

The final result does not depend on: the initial height and weight, the sex, the pain, the aetiology and the tight hamstring (table 5).

| | | |
|---|---|---|
| Height/Cobb K Correction | -0,115 | non significant : sig, (2-tailed 0,057) |
| Weight / Cobb K Correction | -0,039 | non significant : sig, (2-tailed 0,518) |
| Sex/Cobb K Correction | 0,043 | non significant : sig, (2-tailed 0,480) |
| Pain / Cobb K Correction | 0,026 | non significant : sig, (2-tailed 0,676) |
| Aetiology / Cobb K Correction | -0,039 | non significant : sig, (2-tailed 0,521) |
| Tight Hamstrings/ Cobb K Correction | 0,034 | non significant : sig, (2-tailed 0,585) |
| Lord init / Cobb K Correction | 0,007 | non significant : sig, (2-tailed 0,914) |

Table 5. Non significant parameters

### 9.1.6 Correlation between the reduction in plaster cast and bivalve plexidur brace

The correction in plaster cast is an excellent predictive criterion of the 'in brace' correction (table 6).

| | | | |
|---|---|---|---|
| Cobb Kyp plaster / in brace | **0,476** | | significant at the 0,01 level (2-tailed) |

Table 6. Correlation between the reduction in plaster cast and bivalve plexidur brace

### 9.1.7 Stability 5 years after the removal of the brace

In 90 cases revised 5 years after the brace removal, we confirm the excellent stability (table 7).

| | | |
|---|---|---|
| Kyp R / Kyp R+ 5 (N=90) | **0,545** | significant at the 0,01 level (2-tailed) |
| Kyp R+2 / Kyp R+5 (N=90) | **0,798** | significant at the 0,01 level (2-tailed) |

Table 7. Stability 5 years after the removal of the brace

It is difficult to express the obtained correction, because contrary to scoliosis, the angulation of reference is not 0°, but 37°[7] .The result of 40° on average corresponds practically to a total correction. Some studies use a percentage of angular correction. It is to be compared with the results obtained by Platero (17,55 % with exercises only ¦ 25,21 % with brace only ¦ 30,88 % plaster cast + brace) (Platero & all, 1997). We obtain an identical correction percentage of 31 % for the protocol associating plaster cast and brace. The corrections of sacral slope and lordosis correspond to a balance of the spine in the sagittal plane. All the cases of this study concern patient treated in pre-pubertal phase. As for the scoliosis, we prefer the Milwaukee modified brace before puberty.

The Kyphosis plumline measure is the half-sum of the distance measured in mm with a plumb line in C7 and L2. The good correlation with the radiological angulation makes of this measure a reliable element to monitor.

The Scheuermann's disease concerns boys more and the thoraco-lumbar localization. The results of the treatment are similar. In every case, the pain is improved by the conservative treatment, usually during the plaster cast.

As for the scoliosis, the angular correction obtained initially in plaster cast is predictive of the final result. That is why we do not hesitate to make the second plaster cast if we think we can obtain an additional correction. The elements, which were not significant initially (aetiology, pain, tight hamstring…), prove the mechanical character and the reliability of the treatment.

The correction in plaster cast must be reproduced in plexidur bivalve brace it constitutes a reference element.

The stability between 2 and 5 years is excellent. It is connected to the improvement of the vertebral wedging reported by numerous authors (Bradford & all, 1974; Bradford, 1977; Gutowski & Renshaw, 1988; Weiss & all, 2009).

### 9.2 Consensus

The experts from SOSORT are convinced of the usefulness and appropriateness of conservative treatment for the management of Kyphosis and they use this treatment on a daily basis in their clinical practice (de Mauroy, 2010).

The main rehabilitation techniques used are: self postural control and self-elongation. Back school does not seem useful. Physiotherapy exercises should be repeated at home daily for 20 minutes; they are indeed useful before bracing.

The main indications are Scheuermann and pain especially if the kyphosis is rigid.

The biomechanical base for conservative treatment is to decrease mechanical stress on the anterior wall of the vertebral body.

The main indications for early treatment are: rigidity, size of the curve and the Cobb angle.

The best time is at the onset of puberty. The brace should be worn for about 2 years and removed at the end of growth without skeletal maturity at Risser 5.

For a Thoracic Kyphosis: the brace must be worn all night and for part of the day. The most appropriate brace is a 4 point system or a 5 point system in case of muscular imbalance.

For a Thoraco-lumbar kyphosis: the brace must be worn during the day in the sitting position and the ideal brace is a 4 point system.

For a juvenile kyphosis: the brace must be worn part time, and the ideal brace is the Milwaukee.

## 10. Conclusion

Physiotherapy and more generally orthopaedic management of kyphosis is not as well known as scoliosis management because the breathing consequences of kyposis are less disabling. With the aging population kyphosis often becomes a major problem due to the pain and imbalance in standing and walking. The loss of autonomy becomes crucial and often it is too late to treat effectively. The management needs to be effective during adolescence; there is no "light" treatment of kyphosis. Regarding adult cases, treatment is more of a lifestyle issue and an adaptation of the workplace, which makes it very difficult to prove the efficacy of physiotherapy. The results are better with conservative orthopaedic treatment and the retrospective study is encouraging. Long term prospective studies are expected to confirm the interest of these treatments, as for regular kyphosis, neurological surgical risk is greater than in scoliosis due to the risk of stretching of the Adamkievitz' artery.

## 11. References


Albert, TJ.; Vacarro, A.(1998). Postlaminectomy kyphosis. *Spine* 1998 ; 23 : 2738-2745, ISSN 0362-2436

Ali, RM.; Green, DW.; Patel, TC.(1999). Scheuermann's kyphosis. *Curr Opin Pediatr* 1999 ; 11 : 70-75, ISSN 1040-8703

Aufdermaur, M.; Spycher, M.(1986). Pathogenesis of osteochondrosis juvenilis Scheuermann. *J Orthop Res* 1986 ; 4 : ISSN: 0736-0266

Banta, JV.; Hamada, JS.(1976). Natural history of the kyphotic deformity in meningomyelocele. *J Bone Joint Surg* 1976 ; 58-A : 279, ISSN 0021-9355

Bettany-Saltikov, J.; Warren, J.; Jobson, M.(2008). Ergonomically designed kneeling chairs are they worth it? : Comparison of sagittal lumbar curvature in two different seating postures. *Studies in health technology and informatics* 140: 103–6. ISSN: 0926-9630

Blount, WP.; Moe, JH.(1973) The Milwaukee brace. *The William and Wilkins Co* Baltimore, ISBN 0683008691

Bradford, DS.; Moe, JH.; Montavaldo, FJ.; WINTER, RD.(1974). Scheuermann's Kyphosis and Roundback Deformity. Results of Milwaukee Brace treatment. *J Bone Joint Surg,* 1974;56:740-758, ISSN 0021-9355

Bradford, SD.(1977). Juvenile Kyphosis. *Clin Orthop* 1977 ; 128 : 44-55, ISSN: 0009-921X

Cotrel, Y.; Morel, G.; Rey, JC.(1964). Le traitement orthopédique des cyphoses structurales en cours de croissance. *Rev Rhum* 1964 ; 31 : 445-451, ISSN:0301-8474

Drury, CG; Francher, M (1985). Evaluation of a forward-sloping chair. *Applied ergonomics* 16 (1):41–7, ISSN: 0003-6870

Fender, P.; de Mauroy, JC.; Sengler, J.; Bourderont, D.(1995). Jusqu'à quel âge peut-on envisager le traitement orthopédique d'une hypercyphose dorsale de l'adolescent. *Les cyphoses : de l'enfant à l'adulte.* Masson 1995 ; Col Pathol Locom 30 : 141-147, ISBN 2-225-84775-4

Fon, GT.; Pitt, MJ.; Thies, ACJ.(1980). Thoracic kyphosis: Range in normal subjects. *AJR Am J Roentgenol* 1980, 134:979-83, ISSN: 0361-803X

Gutowski, WT.; Renshaw, TS.(1988). Orthotic results in adolescent kyphosis. *Spine* (Phila Pa 1976). 1988; 13(5):485-9, ISSN: 0362-2436

Hensinger, R.(1977). Kyphosis secondary skeletal dysplasia and metabolic disease. *Clin Orthop* 1977, 128, 113-127, ISSN: 0009-921X

Ippolito, E.; Bellocci, M.; Montanaro, A.; Ascani, E.; Ponseti, IV.(1985). Juvenile Kyphosis: An Ultrastructural Study. *J Pediatr Orthop* 1985:5:315. ISSN: 1060-152X

Kado, DM.; Prenovost, K.; Crandall, C.(2007). Narrative review: hyperkyphosis in older persons. *Ann. Intern. Med.* 147 (5): 330–8, ISSN: 0003-4819

Korovessis, PG.; Stamatakis, MV.; Baikousis, AG.(1998). Reciprocal angulation of vertebral bodies in the sagittal plane in an asymptomatic Greek population. *Spine* 1998 ; 23 : 700-705, ISSN 0362-2436

Lam, KS.; Mehdian, H.(19999). The importance of an intact abdominal musculature mechanism in maintaining spinal sagittal balance. Case illustration in prune-belly syndrome. *Spine* 1999 ; 27 : 719-722, ISSN 0362-2436

Lander, C.; Korbon, GA.; Degood, DE.; Rowlingson, JC.(1987). The Balans chair and its semi-kneeling position: an ergonomic comparison with the conventional sitting position. *Spine* 12 (3): 269–72, ISSN: 0362-2436

Lehnert-Schroth, C: Dreidimensionale Skoliosebehandlung - *Atmungs-Orthopädie System Schroth.* 7. Auflage. Urban & Fischer in Elsevier, München 2007, ISBN 978-3-437-44025-0.

Lindeman, M.; Behm, K.(1999) Cognitive strategies and self-esteem as predictors of brace-wear noncompliance in patients with idiopathic scoliosis and kyphosis. *J Pediatr Orthop* 1999 ; 19 : 493-499, ISSN: 1060-152X

Lonstein, JE.(1999). Congenital spine deformities : scoliosis, kyphosis, and lordosis. *Orthop Clin North Am* 1999 ; 30 : 387-405, ISSN: 0030-5898

Lonstein, JE.(1987). Post laminestomy kyphosis *Clin Orthop* 1987 ; 128, 93-100, ISSN: 0009-921X

Lowe, TG.(1999). Scheuermann's disease. *Orthop Clin North Am* 1999 ; 30 : 475-487, ISSN: 0030-5898

Lowe, TG.; Line, BG.(2007) Evidence based medicine: analysis of Scheuermann kyphosis. *Spine.* 2007; 32(19 Suppl):S115-119, ISSN: 0362-2436

de Mauroy, JC.; Stagnara, P.(1979). Cyphose idiopathique : entité pathologique. Masson 1979 ; *Actual Rééd Fonct Réadapt* 4 : 216-219, ISSN 0223-9183

de Mauroy, JC.(1983). La cyphose de l'adolescent. *J Readapt Med* 1983 ; 3 : 144-148, ISSN 0242-648X

de Mauroy, JC.(1995) Traitement orthopédique des cyphoses. *Les cyphoses de l'enfant à l'adulte.* Masson edit. 1995 ; Collection de pathologie locomotrice 30 : 133-141, ISBN 2-225-84775-4

de Mauroy, JC. ; Weiss, HR. ; Aulisa, A. ; Aulisa, L. ; Brox, J. ; Durmala, J. ; Fusco, C. ; Grivas, T. ; Hermus, J. ; Kotwicki, T. & all. (2010) 7th SOSORT consensus paper: conservative treatment of idiopathic & Scheuermann's kyphosis. *Scoliosis.* 2010 May 30;5:9. DOI 10.1186/1748-7161-5-9

de Mauroy, JC.; Vallèse, P.; Fender, P. ; Lecante, C. (2010) Historical Lyonaise brace treatment for adolescent hyperkyphosis. Results of 272 cases reviewed 2 years minimum after removal of the brace. *Scoliosis.* 2010; 5(Suppl 1): O69. DOI 10.1186/1748-7161-5-S1-O69

Mc Master, MJ.; Singh, H.(1999). Natural history of congenital kyphosis and kyphoscoliosis. A study of one hundred and twelve patients. *J Bone Joint Surg Am* 1999 ; 81 : 1367-1383, ISSN 0021-9355

Mezières, F.(1978) Retour à l'harmonie morphologique par une rééducation spécialisée. Sur des notions nouvelles reconstruisons. la cinésiologie. *Kinésit Scient* 1978 ; 157 : 45-54, ISSN 0023-1576

Mollon, G.; Violay, M.(1978). Kinésithérapie et appareillage des cyphoses traitées par chirurgie. *Kinesit Scient* 1978 ; 157 : 13-21, ISSN 0023-1576

Murray, PM.; Weinstein, SL.; Spratt, KF.(1993). The natural history and long term follow-up of Scheuermann's kyphosis. *J Bone Joint Surg* 1993 ;75-A : 236-248, ISSN 0021-9355

Platero, D.; Luna, JD.; Pedraza, V.(1997). Juvenile kyphosis : effects of different variables on conservative treatment outcome. *Acta Orthop Belg* 1997 ; 63(3) : 1194-201, ISSN: 0001-6462

Priftis, KN.; Hager, J,; Vlachou, M.; Anthracopoulos, MB.(2003). Effects of bracing on lung function in idiopathic juvenile kyphosis. *Pediatr Pulmonol.* 2003; 35(2):83-6, ISSN: 8755-6863

Russe, OA.; Gerhardt, J.(1976). International SFTR method of measurement end recording joint motion. pp81, Hans Huber, Berne 1976

Ryan, SD.; Fried, LP.(1997). The impact of kyphosis on daily functioning. *J Am Geriatr Soc* 1997 ; 45(12) : 1479-1486, ISSN: 0002-8614

Sachs, B.; Bradford, SD.; Winter, RB.; Lonstein, J.; Moe,J.(1987). Scheuermann Kyphosis, follow-up of Milwaukee brace treatment. *J Bone Joint Surg* 1987 ; 69 : 50-57, ISSN 0301-620X

Scheuermann, H.(1921). Kyphosis Dorsalis Juvenilis. *Zeitschr Orthop Chir* 1921 ; 41 : 305-317, ISSN 0044-3220

Shen, WJ.; Shen, YS.(1999). Nonsurgical treatment of three-column thoracolumbar junction burst fractures without neurologic deficit. *Spine* 1999 ; 24 : 412-415, ISSN: 0362-2436

Sorensen, KH.(1964). Scheuermann juvenile kyphosis. Clinical appearence, radiography, aetiology and prognosis. *Munksgaard,* Copenhagen 1964

Stagnara, P.; de Mauroy, JC.; Dran, G.; Gonon, GP.; Costanzo, G.; Dimnet, J.; Pasquet, A.(1982). Reciprocal angulation of vertebral bodies in a sagittal plane : approach to references for the evaluation of Kyphosis and Lordosis. *Spine* 1982 ; 7 : 335-342, ISSN: 0362-2436

Stagnara, P.; de Mauroy, JC.; Villard, B.(1975). Traitement des cyphoses régulières. *Ann Med Phys* 1975 ; 18 : 481-496, ISSN 0168-6054

Stagnara, P.; du Peloux, J.; Fauchet, R.(1966). Traitement orthopédique ambulatoire de la maladie de Scheurmann en période d'évolution. *Rev Chir Orthop* 1966 ; 52 (7) : 585-600, ISSN 1877-051

Tribus, CB.(1998). Scheuermann's kyphosis in adolescents and adults : diagnosis and management. *J Am Acad Orthop Surg* 1998 ; 6 : 36-43, ISSN: 1067-151X

Troussier, B.; Marchou-Lopez, S.; Pironneau, S.; Alais, E.; Grison, J.; Prel, G.; Pequegnot, C.; Degaudemaris, RRR.; Phelip, X.(1999). Back pain and spinal alignment abnormalities in schoolchildren. *Rev Rhum Engl* Ed 1999 ; 66 : 370-380, ISSN 1169-8446

Van Rosen, BJ.; Toussaint, HM.; Kingma, I.; Bot, SD.; Caspers, M.; Harlaar, J.; Wuisman, PI.(1998). Accuracy of the sagittal vertebral axis in a standing lateral radiograph as a measurement of balance in spinal deformities. *Eur Spine* 1998 ; 7 : 408-412, ISSN 0940-6719

Vedantam, R.; Lenke, LG.; Keeney, JA.; Bridwell, KH.(1998). Comparison of standing sagittal spinal alignment in asymptomatic adolescents and adults. *Spine* 1998 ; 23 : 211-215, ISSN: 0362-2436

Voutsinas, SA.; Mac Ewen, GD.(1986). Sagittal Profiles of the Spine. *Clin Orthop Res* 1986 ; 210 : 235-242, ISSN 0009-921X

Weiss, HR.; Turnbull, D.; Bohr, S.(2009). Brace treatment for patients with Scheuermann's disease - a review of the literature and first experiences with a new brace design. *Scoliosis*. 2009, Sep 29;4:22. PMCID: PMC2761858, DOI 10.1186/1748-7161-4-22

Winter, RB.; Hall, JE.(1978). Kyphosis in Childhood and adolescence. *Spine* 1978 ; 3 : 285-308, ISSN: 0362-2436

Winter, RB.; Lonstein, J.; Moe, J.; Willson, S.(1987). Scheuermann kyphosis. Follow-up of Milwaukee brace treatment. *J Bone Joint Surg* 1987 ;69-A : 50-57, ISSN 0021-9355

Winter, RB.(1977). Congenital kyphosis. *Clin Orthop* 1977 ; 128 : 26-32, ISSN 0009-921X

Zaina, F.; Atanasio, S.; Ferraro, C.; Fusco, C.; Negrini, A.; Romano, M.; Negrini, S.(2009). Review of rehabilitation and orthopedic conservative approach to sagittal plane diseases during growth: hyperkyphosis, junctional kyphosis, and Scheuermann disease. *Eur J Phys Rehabil Med*. 2009 Dec;45(4):595-603, ISSN: 1017-6721

3

# A Review of Non-Invasive Treatment Interventions for Spinal Deformities

Clare Lewis
*California State University, Sacramento, USA*

## 1. Introduction

Spinal deformities of the scoliotic type, affect approximately 2% of the population.[1] When an individual has an s-shaped curve of greater than 20 degrees, treatment is usually indicated. For curves that are not (severe is defined as greater than 45 degrees; requiring surgery) various physical therapy treatment interventions are often tried. These include Electrical muscle stimulation, bracing, exercise programs, and manipulation. All of these interventions are considered conservative treatments that have been commonly used to treat scoliosis.[2-4] The impact of various interventions has been scrutinized in the literature, and their outcomes have been found to have variable levels of effectiveness.[3] In addition to the physical impact of scoliosis, the emotional impact of scoliosis must also be addressed since having scoliosis will often dramatically affect a patient's self-image, which in turn can impact their perceived pain levels. Ultimately, scoliosis can result in a number of secondary impairments, which if left untreated, can be severe.[5,6]

A new conservative treatment device for scoliosis that has recently been written about in the literature is the ATM2™ (active therapeutic movements version 2).[7] The ATM2™ is a vertical treatment table; a concept that was developed by a physical therapist trained in "Mulligan Techniques" (Mobilization with Movement).[8] Clinical evidence has been documented for the efficacy of the ATM2™ when used with patients who have types and causes low back pain, hip problems, shoulder problems and cervical problems. The ATM2™ was found to be effective in helping to reverse the curves in a patient with scoliosis who was actually being treated for a frozen shoulder in a recently published case study.[7] Since this publication, continued study has been undertaken to determine if the ATM2™ is effective in treating mild to moderate scoliosis, particularly its effectiveness in reversing the curves of subjects with scoliosis.

This chapter will review traditional non-invasive physical therapy interventions for scoliosis. It will discuss the effectiveness or lack thereof of exercises, manipulation, bracing, and electrical stimulation. It will introduce the ATM2™ as a new intervention for mild to moderate scoliosis and discuss published research thus far along with ongoing research currently being conducted.

## 2. Background: Types, causes and diagnosis of scoliosis

### 2.1 Definition of scoliosis

Scoliosis is used to describe an abnormal curvature of the spine, but it is not in itself a disease or a diagnosis. The curvature of the spine from scoliosis is easily seen as a deviation on the frontal plane and may develop as a single curve (shaped like the letter C) or as two curves (shaped like the letter S).

### 2.2 Types and causes of scoliosis

Scoliosis usually develops in the upper back (thoracic spine) or area between the upper back and lower back (the thoracolumbar area of the spine). It may also occur only in the lower back (lumbar spine). Scoliosis can be caused by congenital, developmental or degenerative problems, but most cases of scoliosis have no known cause, and are considered to be idiopathic scoliosis. While there are many forms of scoliosis, four of the most common ones include:

- Congenital scoliosis. This is a relatively rare form of congenital malformation of the spine. Patients with congenital scoliosis will often develop scoliotic deformities in their infancy.
- Neuromuscular scoliosis. This may occur when the spine curves laterally due to weakness of the spinal muscles or from other neurologic problems. This form of scoliosis is especially common for individuals who cannot walk due to their underlying neuromuscular condition and are confined to a wheelchair for locomotion (such as muscular dystrophy or cerebral palsy). This may also be called *myopathic scoliosis.*
- Degenerative scoliosis. Scoliosis can also develop later in life, as joints in the spine degenerate and the intervertebral discs lose height, leading to a bending in the back. This condition is sometimes called *adult scoliosis.*
- Idiopathic scoliosis. By far the most common form of scoliosis is idiopathic scoliosis, which most often develops in adolescents and typically progresses during the adolescent growth spurt. Because it most often occurs during adolescence, this condition is sometimes called *adolescent scoliosis.*

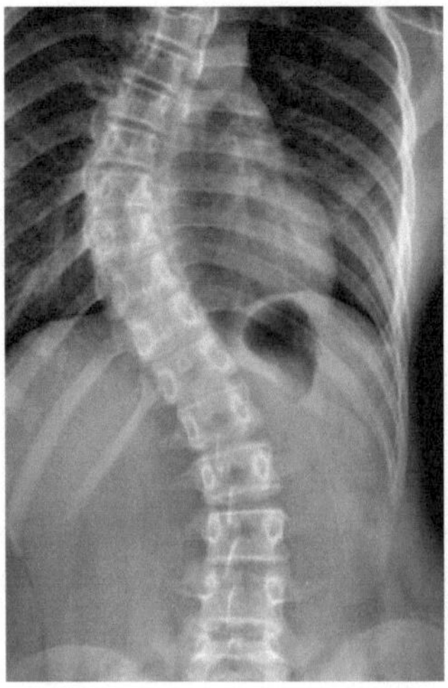

Fig. 1. Idiopathic Scoliosis

Idiopathic scoliosis occurs to some degree in approximately one half million adolescents in the US. There is no known cause of idiopathic scoliosis although it does tend to occur in families.

Idiopathic scoliosis is usually categorized into three age groups:

- From birth to 3 years old - called *infantile scoliosis*
- From 3 to 9 years old - called *juvenile scoliosis.*
- From 10 to 18 years old – called *adolescent scoliosis.*

This last category of scoliosis, which occurs from 10 to 18 years old, comprises approximately 80 percent of all cases of idiopathic scoliosis.

The risk of curvature progression is increased during puberty, when the growth rate of the body is the fastest. Scoliosis with significant curvature of the spine is much more prevalent in girls than in boys, and girls are eight times more likely to need treatment for scoliosis because they tend to have curves that are much more likely to progress. Still, the majority of all cases of scoliosis are mild and do not require treatment.

It is important to note that idiopathic scoliosis results in spinal deformity, and is not typically a cause of back pain. Of course, people with scoliosis can develop back pain, just as most of the adult population can. However, it has never been found that people with idiopathic scoliosis are any more likely to develop back pain than the rest of the population.[9-11]

In children and teenagers, scoliosis often does not have any noticeable symptoms and scoliosis is often not noticeable until the curve has progressed significantly.

Scoliosis does not come from any types of sports involvement, the use of backpacks, sleeping positions, posture, or minor leg length differences.

When viewed from the side (in the sagittal plane), a healthy spine curves inward in the lower back (lordosis) and outward in the upper back (kyphosis). When viewed from the back (posteriorly), a spine with normal curves appears as a straight line down the back. A person with scoliosis however, will appear to have a lateral curve in their spine.

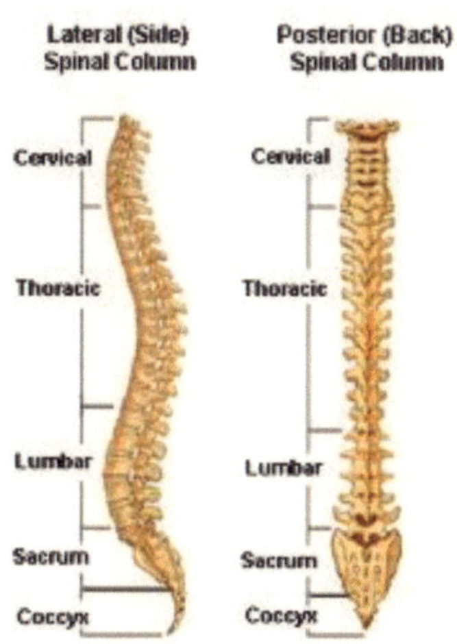


Fig. 2. Normal Spine

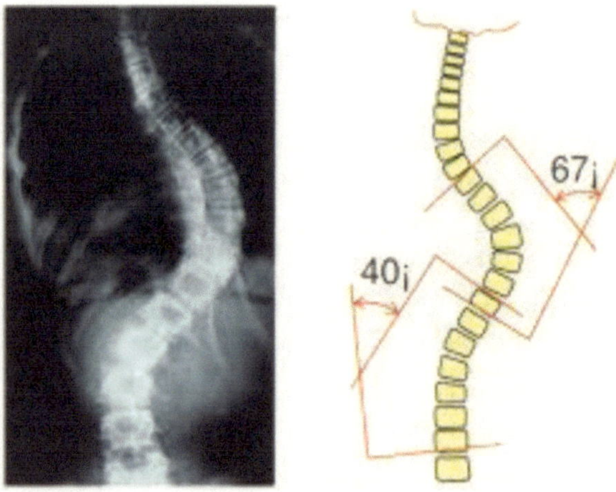


Fig. 3. Spine with Scoliosis

There are several common physical symptoms that may indicate scoliosis. Typically, symptoms of scoliosis may include one or several of the following:

- One shoulder is higher than the other
- One shoulder blade protrudes further than the other
- One side of the rib cage appears higher than the other
- One hip appears higher or more prominent than the other
- The waist appears uneven
- The body tilts to one side
- One leg may appear shorter than the other

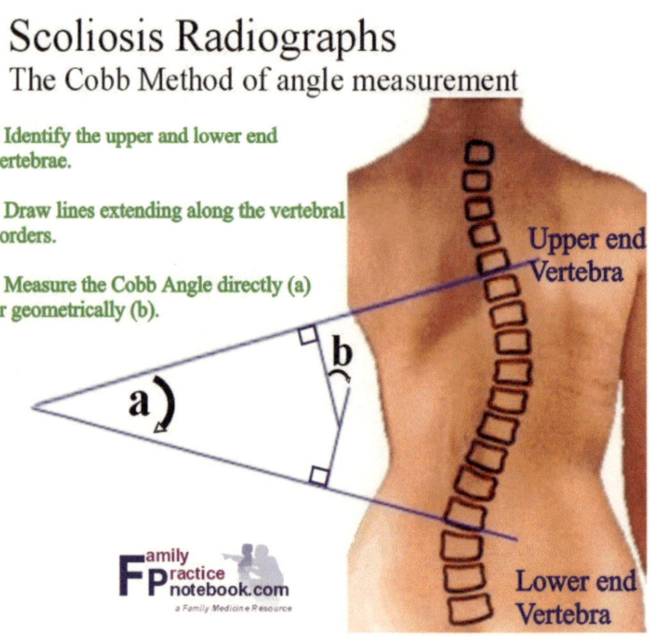


Fig. 4. Cobb Angle

As one can see by reviewing the list above, pain is not a typical symptom of scoliosis. Back pain in a child or teen who has scoliosis may indicate another problem, and the child should be evaluated by a paediatrician and/or spine specialist. If a child or teen has back pain and also has scoliosis, it is very important that he or she see a doctor to determine the cause of the pain, since it is probably something other than the scoliosis and therefore may require additional treatment. Note that neurologic injury is not included in the list above since it is a very rare presenting symptom in most cases of scoliosis.

Early detection is essential for scoliosis treatment to be most effective. In general, people with a family history of spinal deformity are at greater risk for developing scoliosis.

### 2.3 Diagnoses

Frequently, a scoliosis curve in the spine is first diagnosed in school exams or in a regular check up with a paediatrician. Most students are routinely given the *Adam's forward bend test* in school when they are in fifth and/or sixth grade to determine whether or not they may have scoliosis. The test involves the student bending forward with arms stretched downward toward the floor and knees straight, while being observed by a healthcare professional. This angle most clearly shows any asymmetry in the spine and/or trunk of the adolescent's body.

Because a scoliosis curvature is usually in the thoracic or thoracolumbar spine (upper back or mid back), if a rib hump or asymmetry of the lumbar spine is found, or if the shoulders are at different heights, it is possible that the individual has scoliosis. If this is the case, a follow-up with a physician for a clinical evaluation and an x-ray is the next step.

- Physician's exam. The clinical evaluation with the physician will usually include a physical exam, during which the physician will also test to make sure that there are no neurological deficits. Neurological deficits due to scoliosis are uncommon but it is still necessary to check, because there are rare causes of scoliosis that may have spinal cord involvement.
- X-ray. The x-ray is ordered to both confirm the scoliosis diagnosis and check on the magnitude of the spinal curvature. The x-ray will also give some indication of the skeletal maturity of the patient, which in turn may influence treatment decisions.

In rare instances a physician may also request an MRI scan of the thoracic and/or cervical spine. If there are any neurological deficits that would indicate impingement of the spinal cord (e.g. brisk reflex), if there is a left-sided thoracic curvature (they are almost always right sided), or if the child is very young (8 to 11 years old), an MRI scan is advisable to examine the possibility of an intracanal spinal lesion, which can cause scoliosis.

Depending on the results of the physician's clinical evaluation and the diagnostic tests, a scoliosis treatment plan may be recommended which may include observation, bracing, or possibly surgery to reduce or correct the spinal curve.[9-13]

### 2.4 Alternative to x-rays for diagnoses

The Formetric was invented in the 1980's to periodically evaluate the progress in scoliosis treatment without subjecting the patient to potentially harmful x-ray radiation. It has been widely used since 1996 in Germany, and other European countries by orthopedic doctors and surgeons. The Formetric is FDA approved and it works by using a harmless white light to deliver a fast (40 milliseconds) high-definition optical measurement of the surface of the back to produce graphical, clinical, and analytical information on the spine. It does this and can be used to evaluate scoliosis both in and out of braces.[14]

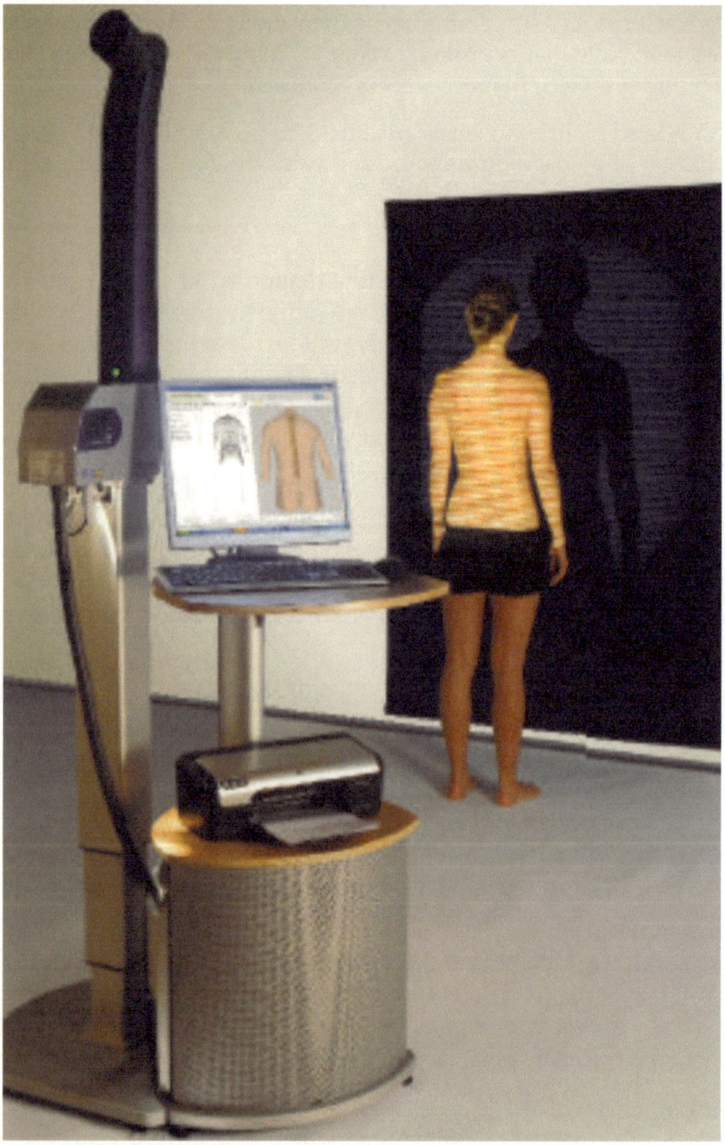

Fig. 5. Formetric System for Scolisis Measurement

Another non-X-ray method used to analyze the scoliotic curve is the ISIS (Integrated Shape Imaging System). This technique uses structured light to illuminate a patient's back from an angle while a digital photograph is taken. The height of the surface is calculated using Fourier transform profilometry with an accuracy of ±1 mm. The surface is related to body axes using bony landmarks on the back that have been palpated and marked with small colored stickers prior to photographing.[14]

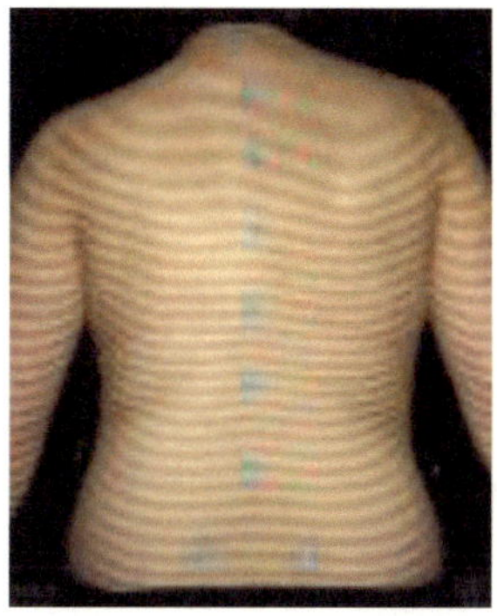

Fig. 6. ISIS

## 3. Scoliosis treatment

Decisions for scoliosis treatment are primarily based on two factors:

- The *skeletal maturity of the patient* (or rather, how much more growth can be expected)
- The degree of spinal curvature.

Although the cause of idiopathic scoliosis is unknown, the way scoliosis curves behave is fairly well understood. Basically, the younger the patient and the bigger the curve, the more likely the curve is to progress.

Three common scoliosis treatment options used in the United States and the United Kingdom for adolescents include:

- Observation
- Back braces (with and without exercises)
- Scoliosis-specific exercises[9,10]

### 3.1 Observation

The scoliosis curvature is measured on x-rays by what is known as the Cobb method. This form of measurement is accurate to within 3 to 5 degrees.

Curves that are less than 10 degrees are not considered to represent scoliosis but rather are referred to as spinal asymmetry. These types of curves are extremely unlikely to progress and generally do not require treatment. If the child is very young and physically immature, then the progress of the curve can be followed during the child's regular check up with his or her paediatrician. If the curve is noted to progress beyond 20 degrees, then the child should be referred to an orthopaedic surgeon with expertise in scoliosis for continued treatment. [9-11]

Curves that are between 20 to 30 degrees in a growing child can be observed at 4 to 6 month intervals. Any progression of less than 5 degrees is not considered significant. If the curve progresses more than 5 degrees, then the curve will need treatment. Additionally, any curve over 30 degrees in a skeletally immature individual will need treatment. Scoliosis treatment for patients with progressing curves, or curves over 30 degrees are often treated with the use of a back brace.[1]

It is important to note that the treatment for curves of 15-30 degrees in continental Europe differs from that used in the United States. Many physical therapy approaches like SEAS, Schroth, DOBOMED, SpinCore PT, Barcelona method are used for mild curves in Poland Spain, Italy and France. [15-17]

### 3.2 Back braces

Bracing is designed to stop the progression of the spinal curve, but it does not usually reduce the amount of angulation already present. The majority of curve progression takes place during a child's growth phase, and once the growth has ended, there is little likelihood of progression of a curve. Therefore, bracing is continued until the child is skeletally mature and has finished growing. Curves that are greater than 50 degrees in angulation, tend to continue to progress after skeletal maturity. As a result, the treatment objective is to try to get the child into adulthood with less than a 50-degree curvature.[9]

There are two types of commonly used scoliosis braces in the USA: a thoracolumbar sacral orthosis (TLSO) and a Charleston bending brace.[16] Braces such as the Cheneau, Sforzesco and others are used in continental Europe.[17]

The TLSO is a custom molded back brace that applies three-point pressure to the curvature to prevent its progression. It can be worn under loose fitting clothing, and is usually worn 23 hours a day. This type of scoliosis brace can be taken off to swim or to play sports.

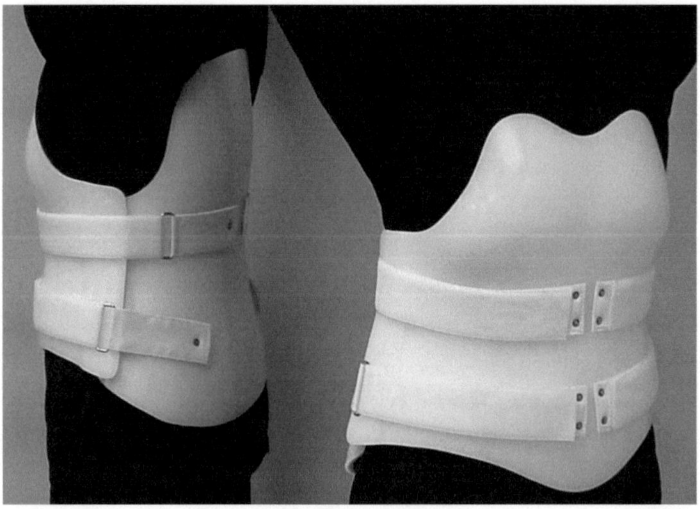

Fig. 7. TLSO Brace

A Charleston bending back brace applies more pressure and bends the child against the curve. This type of scoliosis brace is worn only at night while the child is asleep.

Since bracing normally only works to stop the progression of the curvature in a growing child, it is not used for those children who are already skeletally mature or almost mature. It is only used for younger children (e.g. girls who are about 11 to 13 years old, and boys who are about 12 to 14 years). If an older child has a curve greater than 30 degrees and is almost mature, his or her curvature will be treated with observation only, since there is little growth left and bracing will be unlikely to have much impact on the curve structure.[16]

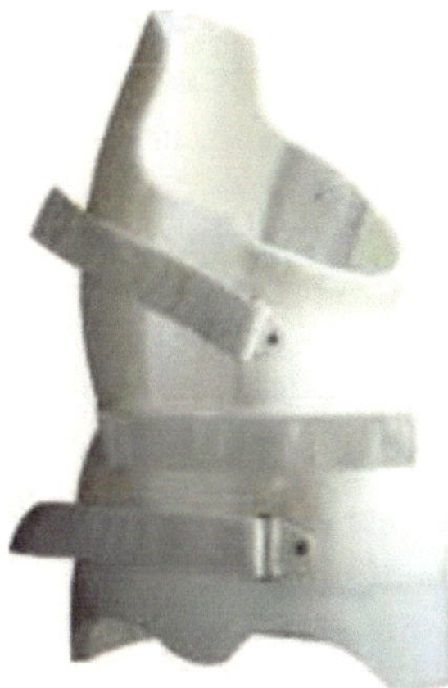

Fig. 8. Charleston Brace

Unfortunately, even with appropriate bracing, some scoliosis spinal curves will continue to progress. For these cases, especially if the child is very young, bracing may still be continued to allow the child to grow before fusing the spine.[9,10]

### 3.3 Adult Scoliosis Brace by spinecor

Due to the nature of the adult spine, treatment should be based on a comprehensive evaluation which should include an Orthopedic/Neurological examination, recent upright X-rays, and functional assessment exams. The Spinecor Adult Brace can be used to create spinal translation designed to de-weight the adult spine. However, chronic inflammatory conditions, arthritis, stenosis, and spondylosis must also be carefully considered when applying the orthosis. Principles of neuro-muscular-re-education are employed to reduce chronic pain cycles, and improve muscle memory in postural restoration. Neuromuscular

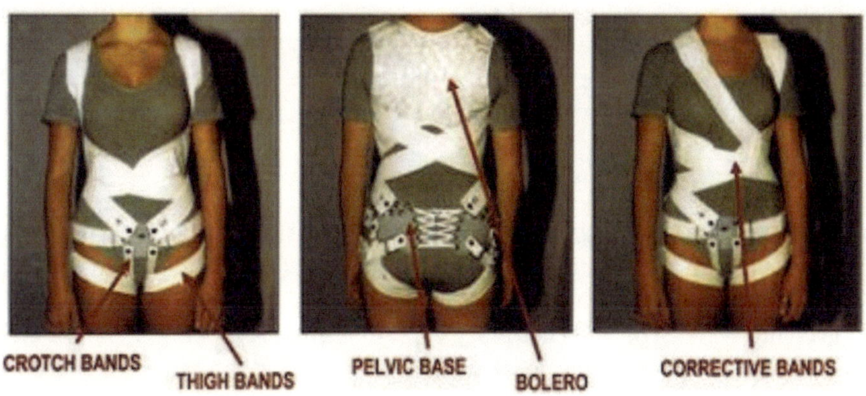


Fig. 9. Spincor Brace for Adults

re-education involves teaching the patient how to correctly use agonist and antagonist muscle groups and how to overcome faulty habitual patterns of use. Strengthening of weak muscle groups and stretching of tight muscle are incorporated to help assist with proper muscle usage which otherwise might not be attainable due to muscular dysfunction.

*Treatment Objectives-* Adult Scoliosis treatment is designed to improve activities of daily living (ADLs) and to provide pain relief. Curvature reduction through improved postural stability may also be considered as an objective. Changes in bone shape are not realistic in the adult spine.

*Treatment Outcomes-* Initial studies of Adult Spinecor Bracing (Deutchman, Lamantia 2004) reported reduction in Cobb angle measurements in all types of curvatures regardless of curvature location or age. However longitudinal studies are still needed.

*Treatment Protocols-*The Adult Spinecor brace wearing schedule varies depending upon the patient's co-morbid ities and treatment objectives. Typical wearing schedules are daytime, only to tolerance. Fatigue may limit brace usage in the first few months of treatment, however the majority of adults find they can build up tolerance to wearing the brace fulltime within the first 3 months. Patients with chronic active pain syndromes are typically fitted in an Adult configuration designed to de-weigh the spine and reduce pain. Changes in brace configuration are based on clinical measure ments specific to each patient. Brace fittings designed to restrict and re-educate movement patterns are typically used initially in patients without pain. Patients are typically under care for 24 months, however, some patients continue to be cared for and monitored for many years when degeneration and spinal instability are an issue.

*Indications-* The Spinecor Adult Scoliosis is indicated for Adults with De-Novo (adult onset) Scoliosis, and Adolescent Idiopathic Scoliosis in Adulthood. Patients who are suffering from pain, fatigue or weakness as a result of scoliosis, are typically considered candidates for bracing. Indications also include those patients interested in improving the cosmetic appearance of their waist angles and ribcage. Patients with respiratory conditions secondary to Scoliosis may also be considered for treatment if there is no significant reduction in vital capacity or blood oxygen levels with the brace properly fitted.

*Contraindications-* Patients with severe stenosis resulting in neurological deficit may not respond well to Adult Scoliosis Bracing. There are contraindications for patients with recent pathological fracture or unstable spondylolysthesis with neurological deficit. Patients may also be refused treatment if expected outcomes are not reasonably possible.[18]

### 3.3.1 The Spinecor Brace for adolescent scoliosis

The objective of the Spinecor Brace in Adolescence is to stabilize the spine by re-training movement patterns while holding the spine in a corrected (de-rotated) position. Treatment is applied with the compressive force of gravity and the brace in order to influence the shape of growing bones. The Spinecor brace is custom fitted and configured specifically for each patient depending upon the location and direction of the scoliosis. Brace wearing schedules are similar for most adolescence; 20 hours of brace wearing with two breaks separated by at least 4 hours between breaks. Full time wearing is expected within two weeks of the initial fitting. The length of treatment depends upon when the diagnosis is made. Females typically wear the brace beginning at diagnosis and ending two years beyond their menses, or when the growth plates are completely fused. Males typically continue to grow up until their 17th year, and may benefit from brace wearing into their late teens.

*Treatment Outcomes-* The Spinecor brace is the only orthosis which has been reported to permanently reverse the curvature associated with Scoliosis. Due to the ease of use, compliance is much higher with Spinecor, as is the psychological well being of patients compared to those who did not receive treatment, or were prescribed a rigid orthosis.[19,20]

*Treatment Indications-* Patients diagnosed with Scoliosis from ages 1 thru adolescence with curvatures measuring from 12-50 degrees at the time of diagnosis are considered candidates for bracing. Candidates with smaller curvatures who are less mature are considered ideal; however, individual cases may be accepted on an elective basis with informed consent.

*Contraindications-* Neuromuscular scoliosis is considered a contraindication, however individual cases are sometimes considered for intervention as long as the patient understands the limitations. Patients who experience significant progression (5 degrees or more) on three consecutive follow-up evaluations may also be released from care, however individual cases may again be exempt from release as long as the patient clearly understands the limitations of treatment effectiveness.[21]

### 3.3.2 Effectiveness of bracing

In one extensive review of the literature, authors concluded that the effectiveness of bracing is not yet established, but may be promising.[3,4]

## 3.4 Electrical stimulation

The usual protocol for using electrical stimulation to treat scoliosis is to place the electrodes so that they are on the convex side(s) of the spinal curve. The convex side of the curve is considered to be the long and weak side, while the concave side of the curve is considered to be strong and tight. The intention of the treatment is to stimulate those long and weak muscles and thereby result in the ability of the newly strong muscles to pull the spine into a more vertical position.

### 3.4.1 Effectiveness of electrical stimulation

Clinical studies have not shown efficacy for electrical stimulation as a treatment intervention for scoliosis.[3,4]

## 3.5 Exercise methods

Scoliosis specific physical therapy exercises that were developed in the 1950's consist of moves ments that are designed to help stretch out the concave side of the curve and to strengthen the convex side of the curve. These are often done on the floor on a mat starting from a quadruped or side lying position. Hanging from a bar is also included in these exercises.[22]

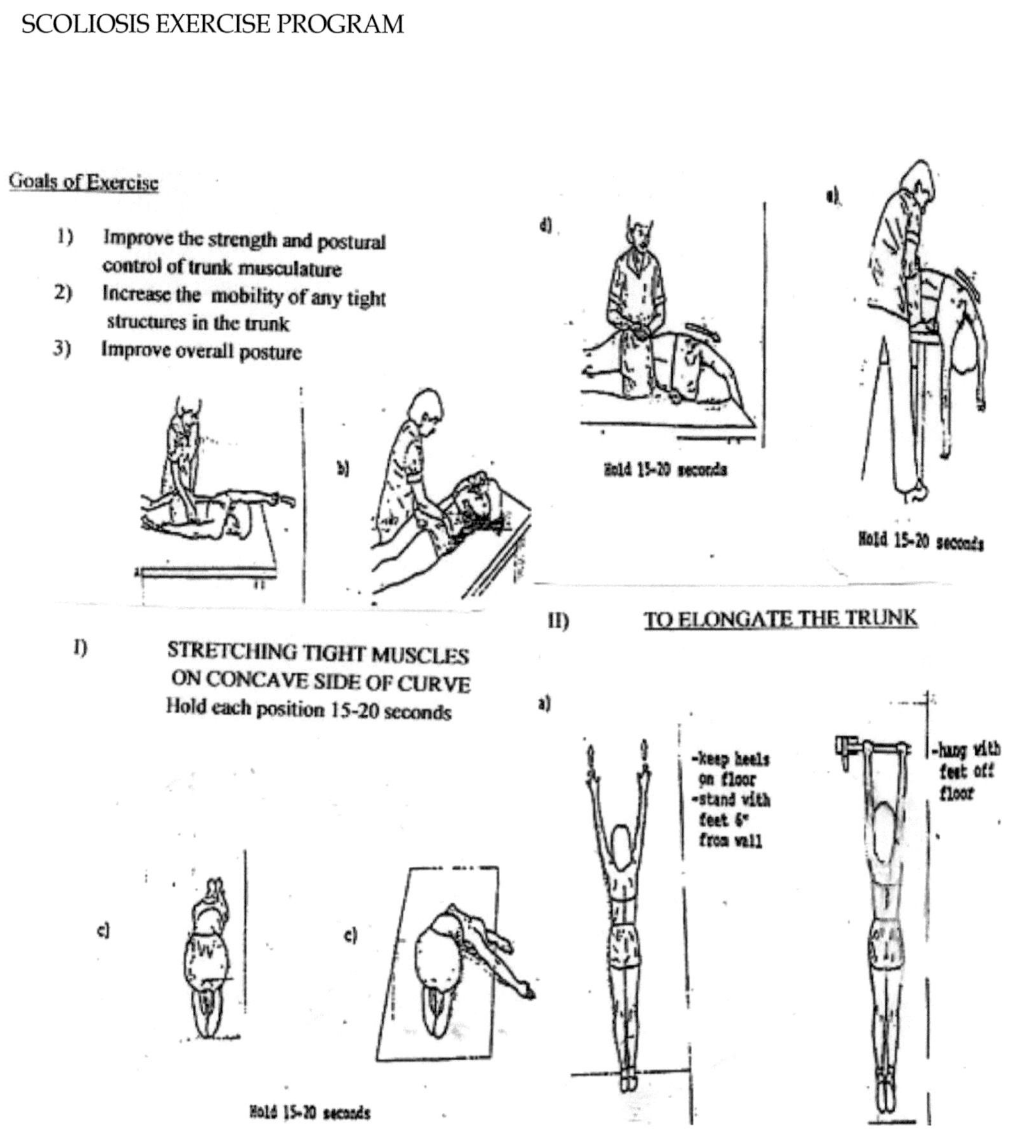


Fig. 10. Scoliosis Exercise Program

Scoliosis exercise

Abdominal Strengthening

a) pelvic tilt

-hold low back flat to floor for 10 seconds, relax and repeat

b) Repeat exercise III a (above) with knees extended

c) Partial sit up - hold 10-15 seconds

Progress difficulty by.
a) crossing arms across ches
b) fold hands behind head

d) Double leg lowering

-extend knees
-slowly lower legs - only as long as you are able to maintain posterior pelvic tilt

BACK EXTENSION STRENGTHENING

Hold each position 10-15 seconds

Lie on stomach, arms at side, tuck chin into chest. Pull shoulder blades together, lift head and shoulders off floor.

Do the above exercise with arms out to sides at 90°.

Do the above exercise with arms out to sides at 90° and the shoulder and elbows bent 90°.

continued

d) Do the above exercise with arms outstretched and slightly out to the side, lift one arm at a time with your hand positioned so the thumb is up. Alternate arms for each repetition.

e) Lie on stomach, clasp hands behind head, tuck chin into chest. Raise elbows and chest off floor and hold. Slowly return to starting position.

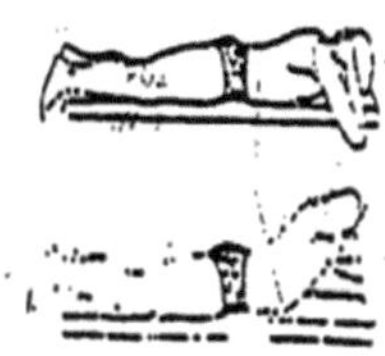

f)

*Patient Exercise Hand-out, Adapted from Therapeutic Exercise for body alignment and function[14]*
*Note: most of these exercises address only one plane of deformity*

Fig. 11. Scoliosis Exercise

### 3.5.1 Effectiveness of exercise

In one extensive review of the literature, authors concluded that the effectiveness of exercise is not yet established, but may be promising.[3] In another extensive review of the literature, only one study demonstrated efficacy for exercises in reducing both the rate of progression or the magnitude of the Cobb angle at the end of treatment. However, this study was categorized as poor quality, and failed to provide solid evidence for or against the efficacy of physical exercises in the treatment of adolescent idiopathic scoliosis. Nevertheless, considering that exercises could also be proposed on the basis that benefits rather than to avoid progression have been shown in the literature, and since results contained in published studies suggest an effect on the primary goal of preventing progression, there is a basis for including exercises. [23]

With the advent of newer *exercise methods*, especially those using a more three-dimensional approach, more research will be needed to determine the effectiveness of these interventions.

## 3.6 Schroth Method

The Schroth Method was developed in Germany in 1927 by Katarina Schroth to treat her own scoliosis. This exercise method incorporates a three dimensional therapy to reshape the ribcage and reduce the deformity associated with Scoliosis. The aim of these scoliosis specific exercises for adults and children should reduce the incidence of scoliosis progression, reduce postural rotation, improve mobility, improve postural stability, reduce pain, and improve cardiopulmonary function.

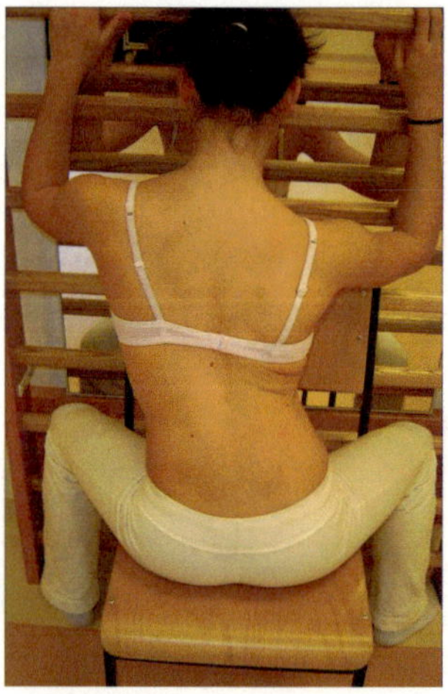

Fig. 12. Schroth Method Exercises to Elongate the Spine

### 3.6.1 Schroth breathing (focused inspiration)

Through focused expansion of the flattened portion of the ribs (within the thoracic concavity) during inspiration, and forced contraction of the rib hump(s) on exhalation, the Schroth Method has been shown to successfully reshape the rib cage, decrease rotation of the vertebral bodies, and improve vital lung capacity (Ottman et al 2005) in patients with Scoliosis. Patients are taught to breathe while in seated, lying and specific standing postures.

### 3.6.2 Schroth exercises (isometric postures)

Scoliosis is not a primary problem with the muscles, however, the stabilizer muscles of the trunk, pelvis and spine do become imbalanced, and, can be reconditioned to help slow, stop and reverse the scoliosis curvature. This is especially true of the lumbar spine. The Schroth Exercises are designed to be isometric contractions of the core while in mechanically advantageous positions. In other words, patients are shown postures which would help reduce the postural deformity associated with their scoliosis, and are asked to hold those postures during Schroth breathing.

Fig. 13. Scroth Exercise with Mirror Feedback

### 3.6.3 Schroth walking (dynamic postural control)

The pelvis can rotate in Scoliosis causing lumbar rotation, hip height discrepancy, pain and progression of the Scoliosis. Often Schroth therapy and manually applied treatment are necessary to improve flexibility in this area. When a patient can create pelvic corrections, which improve the alignment, he or she can learn to move with the pelvis in this more desirable position. This is done in the very beginning by having the patient consciously hold the pelvis during controlled walking ("walking through tall grass"). Schroth described 5 pelvic corrections, basically all degrees of freedom should be considered, 3 rotations and 3 translations, and with every combination in between. For the most part the corrections can be visualized without diagnostic measures.

Vestibular Rehabilitation: Neurological connections between the eyes and the spine have long been neglected in non-surgical treatment of scoliosis, despite the many studies which identify vestibular and oculomotor dysfunction in the majority of patients with Scoliosis. Habituation training and vestibular enhancement exercises may be appropriate in patients with vestibular involvement in their Scoliosis. Therapies and exercises to improve the central neurological controls of posture may include home, computer usage, or in office procedures. Following a functional neurological evaluation, including the use of state-of-the-art diagnostic equipment to evaluate vestibular function, a neurological retraining program is designed for patients on an individual basis.

Scoliosis Exercises for Pain Relief: Scoliosis pain in adults can be relieved with proper exercises designed to help activate neglected muscles; however, not all exercises are appropriate for patients with chronic pain. Often the use of an Elastic Tension Orthosis is more appropriate for these patients, and when pain relief is evident, exercises may be more effective for stabilization.

Scoliosis Exercises for Lumbar Scoliosis: Lumbar Scoliosis can result in pain in adults and can be progressive if measuring over 30 degrees Cobb angle. Scoliosis exercises for lumbar scoliosis are designed to activate muscles which straighten and de-rotate the scoliosis.

Scoliosis Exercises for Thoracic Scoliosis: Scoliosis exercises for Thoracic scoliosis are designed to expand the flat area of the thoracic concavity and de-rotate the rib hump using respiration and posture correction.[24-28]

### 3.6.3 Effectiveness of the Schroth Method

As indicated above, the Schroth Method has been found to be effective using the forced inspiration method. Another study of 813 patients who completed the Schroth in-patient program, showed an increase in chest expansion of more than 20%. The conclusion of the study author was that a course of inpatient treatment of the Schroth method resulted in an increase in vital capacity and chest expansion.[25] In another study by Weis, 118 inpatients with an average Cobb measurement of 31 degrees were treated with the Schroth method and tracked for four years. Results showed that 16.1% had progression of their curve, 68.7% had stabilization of their curve and 15.2% had an improvement of more than 5 degrees. The author concluded that in patients with idiopathic scoliosis, the Schroth method appeared to retard the natural progression of scoliosis compared to the natural history of scoliosis without intervention.[26]

### 3.7 Yoga and scoliosis

Yoga practice can be instrumental in potentially limiting the progression of a functional scoliosis and possibly a structural scoliosis. A yoga practice may also help reduce pain and bring more balance and alignment to a body with scoliosis. With a functional scoliosis, an appropriate yoga practice can release tightened muscle groups that distort the spine, thus alleviating an underlying cause of scoliosis. In yoga for both functional and structural scoliosis, this may be accomplished in asana (poses/postures), through breath work, and through the process of being mindfully aware of the body's position and movements. In doing so, one gains insight and information from the body and proceeds in the practice with an increasing awareness and understanding of what one's body truly needs. Yoga can also develop (build-up) weakened muscle groups. This brings greater balance to the body's muscles. With asana and breath-work, the yoga practitioner can create more separation between the vertebrae and therefore in the body in general. By elongating the spine, the disks between the vertebrae can be nourished and relieved of pressure from the collapsing which takes place when the curved spine is fighting gravity and is being drawn down. Strengthening the muscles along the spine can help prevent the lateral curvature from increasing. In various asana, one can manipulate the spine in the direction opposite its lateral curves, and also de-rotate the spine in the direction opposite its twist. Developing the deeper abdominal muscles provides support to the spine that is learning symmetry, and strengthening the legs provides the same benefit. The yoga practitioner can utilize the breath to release areas of tension in the body and also to bring a "filling" into spaces where the body is more collapsed, such as a lung or one side of the ribs. In yoga for scoliosis, or a yoga practice that emphasizes alignment, one learns to align the body with the structural system; in other words, the bones. In practicing poses and learning to stand in a manner that utilizes the bones for support (finding the body's "plumb line"), rather than overtaxing muscles, joints, and ligaments by moving or standing with the bones misaligned, one can re-organize the body and function more harmoniously with gravity, rather than working against it. Through yoga, one can discover a more refined symmetry in the body, rather than the constant experience of the body's compensating act for imbalance and asymmetry. When the yoga practitioner with scoliosis begins to discover this more refined manner of aligning the body, one's curvatures may then more easily co-exist with gravity, and in this the yoga practitioner may find they experience less pain and more balance.[29]

### 3.7.1 Effectiveness of yoga for scoliosis

Few studies have been conducted about the effectiveness of yoga in the treatment of scoliosis but several testimonials from people who have scoliosis have found it to be effective in improving their posture and other factors such as improved vital capacity. One case study was published which involved a 46-year-old woman who was diagnosed with idiopathic scoliosis. She had been recommended as a candidate for spinal fusion during her early adolescent years. She instead chose to work with conservative therapy sessions for the period of one year. She then opted to practice somatic movement reeducation techniques and yoga, mainly Body-MindCentering™, The Feldenkrais Method™, swimming, and Iyengar Yoga. The yoga regimen was shown to help the patient to stabilize weak muscles; enhance her range of respiration; increase postural tone in standing, sitting, and lying; and strengthen and lengthen the spine through specific postures used as traction. Movement reeducation based on the integration of concepts of bio-mechanics and motor development improved the subject's proprioception, sensibility, and awareness of functional movement. At the time of publication, the subject was pain-free, living a functional and active life.[30]

Another study done in Korea[31] examined the effect of yoga on twenty women with scoliosis. These women ranged in age from twenty to fifty. The results of this study showed a statistically significant decrease of 3.7-degrees correction in the Cobb angles of this sample. It is important to note that these results must be interpreted with caution as the results are within the margins of measurement error (5 degrees) as mentioned above.

## 3.8 The ATM2™ in the treatment of scoliosis

The ATM2™ was never intended as a treatment specifically for treating scoliosis, so it was with much surprise when it was discovered by accident, that a patient who was being treated for a frozen shoulder was also getting as a secondary benefit, an improvement of her longstanding scoliosis and kyphosis. After her first treatment, the patient explained, that she immediately noticed when she sat in her car, that her head was touching the headrest of her car seat. She had never been able to do this before because her thoracic kyphosis prevented her head from going back far enough to touch the headrest. At her second treatment session, documentation of the patients posture began, along with other measurements related to her scoliosis. After several treatment sessions aimed at treating her postural faults, significant improvements were evident in the patient's posture from the ATM2™ interventions alone. This was the first time such documentation had taken place for postural improvements noted while using the ATM2™, although many clinicians have noted similar improvements in their patients' posture after treatment with the ATM2™.[7]

Since publication of the above case study, further study was undertaken to determine if the ATM2™ would be effective in treating scoliosis by reversing the curves of other subjects with mild to moderate scoliosis.

The following information is from the current study:

## 4. Methods and materials

Thirty-eight subjects (34 female, 4 male) with a diagnosis of mild to moderate scoliosis were recruited from a scoliosis support group and by advertising to local physicians. Inclusion criteria were as follows: subjects were between the ages of 12-65 with a diagnosis of scoliosis curve between 20° and 50°. Exclusion criteria were as follows: a previous surgery to correct the curve, current pregnancy, any serious medical co-morbidities, any history of claustrophobia or currently suffering from severe back pain that required medication. Subjects were put on the ATM2™ and performed *pain free* movements in several directions. All subjects performed resisted extension and side bending into their convexity, in order to move into the direction that reversed their curve. The subjects were progressed from 1 set of 10 repetitions to 3 sets of 10 repetitions in each direction. All subjects completed at least 1 round of treatments, which consisted of between 8-10 sessions. Some subjects who were available completed a second round of sessions for a total of 20 treatments. Objective measures utilized in this study included:

- Numeric Pain Rating Scale
- SF-36
- Oswestry Disability Index
- Cobb angles from pre and post intervention x-rays
- Height
- Trunk ROM measurements including:
  - Flexion and Extension (measured using double inclinometer method)
  - Side Bending (measured fingertips to floor, with measuring tape)
  - Rotation (measured with goniometer)

### 4.1 Results

All outcome measures were analyzed using a paired t-test. Multiple outcome measures demonstrated statistically significant change at the 0.05 Level of Significance including:

- Right Side Bending (p=0.007)
- Right Rotation (p=0.003)
- Left Rotation ($p=3.07x10^{-5}$)
- Trunk Flexion (p=0.1 *approaching significance, note: majority of subjects able to touch fingers to floor at the beginning of the study so no change expected in this variable)
- Height (p=.005)
- Numeric Pain Rating (p=0.0002)
- Oswestry Disability Index (p=0.01)

Five subjects were able to obtain pre and post treatment x-rays. Decreases in angle of curvature were seen in 3 subjects when determining Cobb angles. Statistical analysis was not run because of the small sample size. Cobb angles improved or were unchanged for the 5 subjects for whom x-rays were obtained. The Minimal Clinically Significant Difference (MCSD) for Cobb angles is ~7 degrees in adolescents (MCSD not available for adults). SF-36 data showed statistical significance for the majority of variables.

### 4.2 Discussion and conclusion

This study was done to determine if the ATM2™ can treat scoliosis by reversing spinal curvature. The hypothesis for efficacy is that by stabilizing a scoliotic curve, and then moving into resistance, the apex of the curve would receive mobilization while the active movement from the subject would achieve neuromuscular re-education of the supporting musculature. X-rays taken before and after intervention showed decreases in Cobb angles for three subjects who were able to provide post treatment x-rays. Before and after photos also showed changes in those who began with greater spinal curvatures. Some subjects reported they received comments about improved posture from family and friends. Subjects also reported improved self-image as a result of the study. In addition, several objective measures of ROM, as well as height, pain rating and the Oswestry Disability Index have been found to be statistically significantly improved after treatment with the ATM2™. The ATM2™ was shown to be effective in reducing the Cobb angles of three subjects, as well as observationally with pre and post intervention posture pictures. Based on the results of this study, treatment with the ATM2™ positively affects perceived level of health, self-image, visible posture correction, ROM, height, and self-reported disability measures.

Further study is ongoing with the goal of improving the power of current the study by increasing the sample size, especially with regard to Cobb angle data. The ATM2™ intervention continues to provide significant positive results for the subjects who have been involved in our study.[32]

Fig. 14. The ATM2™

## 5. Emotional and psychological considerations

Although this chapter deals with non-surgical physical methods of treatment for scoliosis, the emotional and psychological impact of such physical interventions cannot be ignored. Physical impairments and disabilities often result in low self-evaluation, reduced self-worth, poor self-image and lessened self-respect. Emotional fatigue, hopelessness, helplessness and depression may ensue. In turn, a weakened mental state will down-regulate general functional abilities. Such is often the case with individuals suffering from scoliosis. This spinal affliction not only causes discomfort, and sometimes pain, but often self-consciousness from the associated physical deformities that are very evident to the patient. These include difference in shoulder height, difference in hip height position, difference in shoulder blade height or position, with the head not centered with the rest of the body. These deviations from normal cumulate in obvious (to the person with scoliosis) or not so obvious deformity. In addition the required physical examinations for many years may also result in further self-consciousness for the adolescent. The application of bracing which is most often the treatment of choice can further add to poor body image. As can be expected, having scoliosis has significant psychological implications.[5] Many of our subjects have talked about the emotional aspect of their scoliosis and the huge benefit that they gained in improved self-esteem as a result of the intervention.

As an example of an intervention to specifically address the emotional component of scoliosis, an education program called "success coaching" was developed. It has been touted as an important component in the treatment of scoliosis. The famous psychologist and personality theorist, Alfred Adler, stated that the primary goal of all human behavior is *self-perfection*. He further stated that when we face our weaknesses with optimism and courage, we "will then make the necessary effort to *compensate* for such weaknesses." Such efforts often result in superior achievement. Feelings of inferiority and inadequacy must be combated at all costs. The concept of *Success Coaching* focuses on self-acceptance and places emphasis on the identification of individual strengths. Success coaching incorporates realistic goal setting for high achievement in all aspects of life. Specific aims include the following:

1. Overcoming feelings of inadequacy
2. Overcoming social isolation
3. Overcoming of "feeling different and disabled"
4. Learning psychosocial coping skills
5. Learning realistic goal setting
6. Establishing internal locus of control (self-reliance and motivation for high achievement)
7. Peak performance training (using one's potentials to the upper limit)
8. Realistic self-appraisal in both cognitive and social domains
9. Elimination of negative and self-defeating behaviors

Success coaching also entails informative lectures on stress management and coping strategies associated with all aspects of daily life (work, school, social and family). Supportive individual and group discussion sessions are tailored to specific areas of difficulty and adjustment problems.[33]

In summary, many different physical interventions have been tried for the treatment of scoliosis. Some have shown better efficacy than others. As new effective interventions are being discovered, the future of non-surgical success for scoliosis looks bright and promising.

## 6. References

[1] Scoliosis Research Society Web site. Available at: http://www.SRS.org/htm/glossary/medterms.htm
[2] Skaggs DL, Bassett GS. Adolescent idiopathic scoliosis: an update. Am Fam Physician.1996 ;53:2327–2335.
[3] Lenssinck M., et al. Effect of Bracing and Other Conservative Interventions in the Treatment of Idiopathic Scoliosis in Adolescents: A Systematic Review of Clinical Trials. Physical Therapy. 2005 ;85:1329-1339
[4] Surface Electrical Stimulation Versus Brace in Treatment of Idiopathic Scoliosis. Durham, J., Moskowitz A., Whitney J. Spine. 1990 :15;888-892
[5] Payne B, Ogilvie J, Resnick M, et al. Does scoliosis have a psychological impact and does gender make a difference? Spine.1997 ;22:1380–1384.
[6] Sapountzi-Krepia DS, Valavanis J, Panteleakis GP, et al. Perceptions of body image, happiness, and satisfaction in adolescents wearing a Boston brace for scoliosis treatment. J Adv Nurs. 2001 ;35:683–690.
[7] Lewis, C., Erhard, R., Drysdale, G. Kyphoscoliosis Improvement WhileTreating a Patient for Adhesive Capsulitis Using the Active Therapeutic Movement Version 2. JMPT. 2008 :31;715-722
[8] Vicenzion B., Hing W., Rivett D., Hall T. Mobilisation with Movement: The art and the science. Churchill Livingstone, Australia, 2011 ISBN: 9780729538954
[9] http://www.webmd.com/osteoarthritis/guide/arthritis-scoliosis
[10] www.spine-health.com/conditions/scoliosis/scoliosis-what-you-need- know
[11] www,srs,org
[12] www.spine-health.com/conditions/scoliosis/scoliosis-surgery
[13] http://www.nlm.nih.gov/medlineplus/ency/article/001241.htm
[14] www.biometrixmedica.com/en/products/formetric.3D.html
[15] http://www.eurospine.org/p31000269.html
[16] http://www.spine-health.com/conditions/scoliosis/types-scoliosis-braces
[17] Negrini S. et al. Effectiveness of complete conservative treatment for adolescent idiopathic scoliosis (bracing and exercises) based on SOSORT management criteria: results according to the SRS criteria for bracing studies - SOSORT Award 2009 Winner. Scoliosis. 2009 :4;19
[18]www.spinecorportation.com/English/ScientificInformation/%20Vachon%20-%20Effectiveness%20of%20the%20SpineCor%20brace.pdf
[19] http://www.spinecorbrace.us/scientificinformation.htm ref scoliosissystems.com/Spinecor_Brace.html
[20] http://www.spinecorresearch.com/
[21] www.gillettechildrens.org/fileupload/vol11no3.pdf
[22] Daniels L and Worthingham C. Therapeutic Exercise for body alignment and function. WB Saunders, Philadelphia, 1977.
[23] Negrini S. et al. Physical exercises as a treatment for adolescent idiopathic scoliosis. A systematic review. 2003 :6;227-235
[24] http://www.schrothmethod.com/clinical-scoliosis-studies
[25] Weiss, R. The Effect of an Exercise Progrm on Vital Capacity and Rib Mobility in Patients with Idiopathic Scoliosis. Spine. 1991:88-93.

[26] Weiss, R. The Progression of Idiopathic Scoliosis Under the Influence of a Physiotherapy Rehabilitation Exercise Program. Physiotherapy. :78
[27] www.scoliosissystems.com/Schroth_3_Dimensional_Scoliosis_ Rehabilitation.html
[28] Lehnert-Schroth. The Schroth Method: Three-Dimensional Treatment for Scoliosis. ISBN: 978-0-914959-02-1
[29] http://yogaquest.wordpress.com/yoga-for-scoliosis/
[30] Monroe, M. Yoga and movement re-education for the treatment of idiopathic scoliosis. From 7th International Conference on Conservative Management of Spinal Deformities Montreal, Canada, 20-22 May 2010.
[31] Hoe-Won, K. The Effect of a Yoga Correction Exercise Program for Scoliosis. KACEP Annual Meeting, Korean Association of Certified Exercise Professionals; 2007.
[32] Lewis study in progress
[33] http://www.ehow.com/about_5568171_psychological-effects-scoliosis.html

# Part 2

# Physical Therapy for Biomechanical and Musculoskeletal Conditions

# 4

# An Overview on the Efficacy of Manual Therapy (Manipulations and Mobilisations) on Nonspecific Cervical Pain: A Systematic Review in Adults

José Antonio Miralias Martínez
*Head of Physical Medicine and Rehabilitation, Castellón General Hospital Castellón, Spain*

## 1. Introduction

Cervical pain is a common problem that is associated with significant morbidity and costs. We know relatively little about the effect of manual therapy on neck pain, and its efficacy remains unclear[1]. Manual therapy includes the following different forms of treatment:

1. "spinal manipulation", which is defined as low-amplitude, high-speed manual operations that are short, precise and selective for one vertebral segment; these manipulations are applied until the normal physiological range of motion is exceeded but without reaching the anatomical limit[1];
2. "spinal mobilisation", which is defined as passive, low-speed movements of the vertebral segments within anatomical limits[2].

Manual therapy is applied primarily in cases of pain with a spinal origin, particularly nonspecific and mechanical pain, and it can be used either as a sole therapeutic option or in association with other types of drug treatment or physiotherapy[3].

In this study, we evaluated meta-analyses and systematic reviews of randomised controlled trials (RCTs) because they represent the best scientific evidence in the hierarchy of evidence-based medicine (EBM).

## 2. Objective

The aim of this study was to assess the efficacy and safety of manual therapy (manipulation and mobilisation) for nonspecific cervical pain.

## 3. Search strategy

A search was conducted of the following electronic bibliographic databases from their respective starting dates to December 2010: Central Medline (March 2000), Embase (1947), Mantis (October 2008), Cinahl (May 2006), Icl (September 2008), Amed (December 2007), Pedro (November 1989), SciSearch (March 2005), the Cochrane Controlled Trials Register

(April 1993), the Cochrane Musculoskeletal Review Group Specialised Trials Database (April 2004), and the Web of Science (1991).

A manual search was also conducted in the *Journal of Manipulative and Physiological Therapeutics, Manual Therapy, Physiotherapy, Spine* and *Rehabilitation* (Madrid). The following keywords were used: neck pain (cervicalgia), cervical spine (columna cervical), manual therapy (terapia manual), manipulation (manipulación), mobilisation (movilización), manipulation/mobilisation, and cervical manipulation versus mobilisation in adults. The first author of each study was used in subsequent searches to avoid missing relevant studies.

## 4. Selection of studies

We selected only meta-analyses and systematic reviews of RCTs that investigated the use of mobilisations and manipulations as treatments for nonspecific mechanical neck disorders. At least one of the following parameters was measured: pain, range of movement, pain on palpation, and overall or functional improvement. We excluded RCTs that analysed cervical pain with other aetiologies. The levels of evidence were classified in various ways by the authors[4], as shown in Table 1. The definition of RCT quality ranks the level of evidence as low, medium or high for scores below 25%, between 25% and 50%, and higher than 50%, respectively, of the total maximum[5].

| |
|---|
| 1. Strong Evidence: Multiple high-quality RCTs. |
| 2. Medium Evidence: One high-quality RCT or multiple low-quality RCTs. |
| 3. Limited Evidence: One low-quality RCT. |
| 4. Inconclusive Evidence: Inconsistent or contradictory results in multiple trials. |
| 5. Absence of evidence: No studies. |

RCT = randomised controlled trial.

Table 1. Levels of evidence[4].

## 5. Results

The role of manual therapy in nonspecific mechanical cervical pain was determined by searching the literature and examining the results by year of publication.

Five low-quality RCTs were obtained from 1991 and earlier; therefore, it was not possible to draw conclusions, and further work to produce higher- quality studies is needed[6].

From 1992 to 1996, 24 RCTs met the selection criteria; they were categorised by the type of intervention used (12, physical medicine; 9, manual therapy; 4, more than one form of intervention; 4, drug treatment; and 3, educational). We concluded that the various treatment techniques have not been studied in sufficient detail to properly allow for an assessment of their efficacy, and that the results were contradictory[7].

From 1997, we identified 14 RCTs totalling 892 patients. In these studies, we found the most explicit systematic reviews on the distinctions between mobilisations and manipulations and among acute, subacute and chronic pain. There were no RCTs on manipulation and only 3 low-quality RCTs on mobilisation (two of which dealt with cervical whiplash)[8].

From 1998 to 2002, 20 medium-quality RCTs were found; these RCTs showed better results for manual therapy and exercise (manipulation or mobilisation, manipulation and mobilisation or massage) than for the control groups (waiting list or placebo). There was no evidence that treatment by manipulation was better than the control [9].

Among the 33 selected RCTs from 2003, 42% ranked as high quality[10]. There was no evidence that treatment by manipulation was better than the control treatment[11].

From 2004 to 2010, 12 RCTs met the selection criteria. Using the criteria developed by Koes et al.[6] (and later adapted by Sarigiovannis and Hollins[5]), the RCTs had quality scores between 25 and 67 (out of a maximum of 100). Eight were medium-low quality, of which 6 reported positive results [12, 13-17] and 2 reported negative results[18, 19], and 4 were high quality, of which 2 reported positive results[20] [21] and 2 reported negative results[22 *,23*,24] (* are from the same study; see Table 2). Thus, the evidence for the efficacy of manual cervical spine therapy remains inconclusive[5]. To evaluate the evidence for manual therapy, an analysis of the various processes yielded the following results.

| *Study* | *Conclusion* | *Score (Scale: 0–100)* |
|---|---|---|
| Bronfort[22*] | Negative | 67 |
| Evans[23*] | Negative | |
| Hoving[20] | Positive | 64 |
| Hurwitz[21] | Positive | 54 |
| Yurkiw[24] | Negative | 51 |
| Wood[13] | Positive | 49 |
| Pikula[14] | Positive | 47 |
| Jordan[18] | Negative | 44 |
| Parkin-Smith[15] | Positive | 41 |
| Nordemar[19] | Negative | 36 |
| Modley[16] | Positive | 32 |
| Vernon[12] | Positive | 29 |
| Brodin[17] | Positive | 25 |

*22 and 23 Both refer to the same study.

Positive: A statistically significant difference was observed in the efficacy of manual therapy (manipulation/mobilisation) and other treatments.

Negative: No statistically significant differences were observed in the efficacy of manual therapy or other treatments.

Table 2. The methodological scores of the RCTs[4] according to the adopted criteria[10].

### 5.1 Acute neck pain

There were no RCTs for the treatment of acute neck pain by vertebral manipulation[25] (an absence of evidence)[4]. There were 3 low-quality RCTs for mobilisation[25] (moderate evidence)[4] and two for cervical whiplash[26,27]. In a randomised group of patients with acute neck pain (all of whom were previously treated with collars and analgesics), there were no differences among the mobilisation, TENS or control groups evaluated at 1, 6 and 12 weeks[19]. There was no evidence supporting the use of spinal manipulation[25], and there was limited evidence against passive spinal mobilisation for acute neck pain[25].

### 5.2 Cervical whiplash

For cervical whiplash, there were better pain reduction and mobility-recovery results at 8 weeks in the group treated with early active mobilisation than in those treated with conventional therapy (analgesics, advice and home exercise)[26]. There was less pain after 2 years with early active mobilisation compared with physiotherapy (cold or shortwave) or collars[27]. There was moderate evidence in favour of early active mobilisation[26,27] and no evidence supporting the use of spinal manipulation in whiplash[25].

### 5.3 Chronic subacute neck pain

There were 2 RCTs[12, 28] (moderate evidence)[4] comparing manual therapy with mobilisation or spinal manipulation[7,25]. No differences were observed in short-term pain and range of movement in patients with chronic neck pain who were randomised to treatment groups for manipulation or mobilisation[28]. This result differs from other studies that showed better short-term results with manipulation[12]. Thus, there is unclear evidence on the difference in the efficacy between manipulation and spinal mobilisation[5, 25].

Four RCTs[17,31,32,33] (moderate evidence)[4] compared manipulation and/or mobilisation with other treatments[25]. There was less pain in the patients treated with mobilisation and salicylates (compared with massage and salicylates or traction/electrical stimulation and salicylates) at 3 months, but not over the long term[17].

There were better initial results with manual therapy (manipulation and mobilisation) than with physiotherapy (short wave, electrotherapy and ultrasound), conventional therapy (analgesics, home exercises and advice) or placebo, but there were no differences at 3 or 12 months[31]. Increased muscle relaxation was achieved by manipulation, but this effect was not significant[31,32,33] (TE Global 0.42 (95% confidence interval, -0.005 to 0.85). Therefore, there is inconclusive evidence regarding the role of manual therapy (manipulation and mobilisation) in treating chronic neck pain[5, 25].

There was a better outcome in the manual therapy group (manipulation and/or mobilisation) than in the physical therapy and general medical treatment groups (analgesics, education and advice) at 7 and 26 weeks, but not after 1 year[20, 29]. There was less pain, disability and drug consumption at 4 and 12 months after treatment, but with no differences between the treatment groups (physiotherapy and mobilisation, manipulation or intensive training)[18,29]. Mobilisation and strengthening exercises (isometrics with elastic bands) and mobilisation and resistance exercises (cephalic elevations in the prone and supine positions) produced better results than did the control treatment (recreational activity) in terms of reducing pain and increasing functional recovery after a year[30].

When randomised to three treatment groups (i.e., manipulation (I), manipulation plus conventional exercise (II), or high-technology exercises, developed by MedX corporation, in addition to cervical extension isokinetic exercises (III)), no difference was observed among the groups at 3 months[22,29]. However, there was higher satisfaction at the end of 3 months in Group II [22,29], better results and higher satisfaction at 12 months in Groups II and III [22,29], better results at 24 months in Groups II and III[10, 23] and higher satisfaction at 24 months in Group II[10, 23].

There was moderate short- and long-term evidence in favour of stretching programmes plus strengthening for chronic mechanical neck pain with or without headache[29], and for stretching programmes for patients with chronic mechanical neck pain[10]. There was inconclusive evidence supporting the relative benefits of a programme of stretching plus strengthening exercises compared to manual therapy (mobilisation and manipulation) or to other therapeutic approaches [21, 25, 26, 29].

Studies or subcategories
(Note: Of the 960 eligible patients, only 336 agreed to participate.)

| |
|---|
| 1. *Manipulation (1 session) versus Control/Placebo*[12, 33] |
| Subtotal IC 95 % |
| Heterogeneity Test: Chi2=0.74 gl 1 (P=0.72)2 = 0% |
| Test for Overall Effect: Z=1.79 (P=0.07) |
| 2. *Manipulation plus Mobilisation versus Placebo*[31] |
| Subtotal IC 95 % |
| Heterogeneity Test: Not applicable. |
| Total Effect Test: Z=0.26 (P=0.80) |
| 3. *Manipulation plus Mobilisation versus Controls (Waiting List)*[34, 35] |
| Subtotal IC 95 % |
| Heterogeneity Test: Chi2=0.13 gl 1 (P=0.39)2 = 0% |
| Total Effect Test: Z=1.72 (P=0.08) |
| *4. *Manipulation/Mobilisation plus Exercises versus Short-Term Controls* [35-37] |
| Subtotal IC 95 % |
| Heterogeneity Test: Chi2=1.38 gl 2 (P=0.50)2 = 0% |
| Total Effect Test: Z=4.72 (P<0.00001) |
| **5 *Manipulation/Mobilisation plus Exercises versus Long-Term Controls* [35] |
| Subtotal IC 95 % |
| Heterogeneity Test: Not applicable. |
| Total Effect Test: Z=2.77 (P<0.006) |

There was strong evidence against manipulation[12,33] and manipulation plus mobilisation[31,34, 35] in isolation compared with controls (placebo/waiting list) for function and the general perceived effect in subacute or chronic mechanical neck disorders with or without headaches.

There was strong short-[35-37]* and long-term[35] ** evidence in favour of multimodal treatment (manipulation/mobilisation + exercise) compared with controls in subacute or chronic mechanical neck disorders with or without headaches for the following measures:

1) pain reduction [pooled SMD -0.85 (95% CI: -1.20 to -0.50)];

2) improved function [pooled SMD -0.57 (95% CI: -0.94 to -0.21)]; and

3) general perceived effect [pooled SMD -2.73 (95% CI: -3.30 to -2.16)].

Table 3. External Validation[10]

There was inconclusive short- and long-term evidence on the effect of strengthening exercises for the relief of chronic mechanical neck pain[10] and on the role of manual therapy in chronic neck pain [5, 10,25,29]. There were no clear differences between exercises and manual techniques or other physical therapies, or between strengthening and resistance exercises [11]. It was thus not possible to determine which technique or dosage is most effective or whether certain groups benefit more from a given form of therapy [10, 25].

Regarding external validity, there have been comparative metaanalysis reviews of treatment by manipulation and mobilisation in mechanical neck disorders. These studies have used the resulting pain after treatment as a measure[10, 21], and the most significant data are shown in Table 3. These studies demonstrate that there is inconclusive evidence for the efficacy of manual therapy (manipulation and mobilisation) in chronic neck pain[5, 10, 25,28].

### 5.4 Radiating neck pain

There was greater improvement in function and pain with manual therapy (manipulation/mobilisation) directly on the cervical spine and indirectly on the shoulder and dorsal spine than without treatment[37]. We found limited evidence in favour of exercise and manual therapy (mobilisation/manipulation) in radiating chronic neck pain cases [9, 10, 25, 37]. There was no evidence for the role of manual therapy (manipulation and mobilisation) in radicular cervical cases[10]. There was moderate evidence against stretching programmes plus strengthening for myofascial pain in the neck and shoulder[5, 29,31]. Consequently, the evidence for the efficacy of manual therapy (manipulation and mobilisation) was inconclusive [5,29,31].

### 5.5 Cervicogenic headaches

At both 7 weeks and after 1 year, the intensity and frequency of pain decreased more with manipulation, exercise, and manipulation with exercise than it did with no treatment[35], and combining the treatments (manipulation and exercise) did not change the results[35]. There was strong short- and long-term evidence in favour of multimodal treatments that included exercise and mobilisations in subacute or chronic mechanical neck pain with headache, as assessed by pain reduction, improved function and general perceived effect[29]. There was moderate short- and long-term evidence in favour of strengthening exercises alone or with other treatments for pain, function and general perceived effect on chronic neck pain with headache[35]. There was also evidence against the efficacy of manipulation and/or mobilisation alone in the treatment of cervicogenic headaches[10,11]. Thus, there is inconclusive evidence regarding the efficacy of manual therapy (isolated manipulation and mobilisation) in the treatment of cervicogenic headaches[10,11].

## 6. Safety

Most sources indicated that the incidence of serious accidents during treatment by cervical spine manipulation is low (approximately 1 per million per year)[1,3]. The most commonly described injuries were Wallenberg's syndrome, dissection or thrombosis of the vertebral or carotid arteries and brainstem injury[1]. Adverse reactions were more likely to occur after manipulation than after cervical spinal mobilisation[3]. Therefore, iatrogenic sequelae may be reduced, outcomes may be improved, and satisfaction and security may be increased by using mobilisation[1]. Only one study reported adverse effects from manual therapy[21].

Spinal manipulation and mobilisation are commonly used in the treatment of cervical spine disorders[35]. Their use has been associated with serious complications, including an increased incidence of cerebrovascular accidents (CVAs)[39-42] and minor side effects, such as headache, stiffness, and symptom worsening[43-45]. In a systematic review of the adverse effects of spinal manipulation, Ernst[41] suggested that spinal manipulation is associated with frequent mild and transient adverse effects, as well as more serious complications that can lead to permanent disability or death. The incidence of reported adverse effects has varied between studies (ranging, for example, between 1 per 50,000 manipulations[44] and 1 per 228,050[43, 46]).

A review of the literature related to cervical artery dysfunction and manual therapy suggests that due to reporting bias, inferences about the magnitude of the risks of manipulative therapy should be conservative in relation to the surveys[46]. Other authors have also stated that due to concerns about the validity of the calculations applied to these data, it is not currently possible to estimate the risk of complications after treatment in a meaningful way without reporting the incidence of risk for cervical manipulation[47].

Ernst concluded that incidence figures cannot be reliably estimated at present, due to the lack of sufficiently broad and rigorous prospective studies[41]. Thiel and Bolton have suggested the need for a system to record adverse effects on a routine basis that is not based on the practitioner's subjective recall[48]. Several tests also exist to gauge the risk of adverse effects, with and without the use of mobilisation or high-velocity thrust (HVT) techniques, which have not been as widely reported in the literature[43].

Beca (2002) reported a higher incidence of minor adverse reactions with the use of non-HVT techniques (27.5%) compared with HVT techniques (16.1%). Magarey reported a higher rate of adverse effects associated with the use of non-HVT techniques (1 out of 180 therapists per week of treatment) compared with HVT techniques (1 out of 177.5 therapists per week of treatment)[44]. Magarey also reported that adverse effects were caused by the particular test procedures, which involved rotation. In contrast, Hurwitz reported that patients who received spinal manipulation were more likely to experience adverse effects than were patients treated with mobilisation; however, his risk estimates were imprecise. The reported side effects associated with mobilisation included increased pain, headache and fatigue[49].

The use of functional testing of the position of the cervical spine has been proposed as part of the evaluation of vertebrobasilar insufficiency (VBI) before the application of HVT and non-HVT techniques to the cervical spine[50]. However, functional position tests have been criticised for their "lack of sensitivity, specificity[51] and validity[42]". The poor validity of the functional position tests for the detection of alterations in blood flow has also been noted [52-54].

It appears that the risk of adverse reactions is associated with the testing procedures themselves[39] coupled with the time consumed by the testing[44], suggesting that the clinical utility of functional position testing is questionable. However, these tests are currently defended for VBI assessment as part of a comprehensive assessment protocol that also includes a detailed subjective evaluation and places special emphasis on the therapists' clinical reasoning in the process[50].

## 7. Discussion

The definition and concept of manual therapy varies according to different authors. For example, the study with the highest-scoring methodology includes mobilisation of the spine and soft tissue through coordination and stabilisation exercises[20]. The wide variety of manipulative techniques used and the qualifications of the professionals involved make comparing studies difficult [20]. Virtually all authors have agreed on the need for high-quality and long-term RCTs to establish precisely the efficacy and safety of manual therapy[1,2], thereby facilitating meta-analyses rather than only systematic reviews[5]. It is encouraging to note that the three papers with the highest scores for methodological quality were published after the year 2000[20,21,23]. However, none of these RCTs included pre-randomisation, and only one included post-randomisation for psychosocial assessment of the patients[21].

Additionally, the qualifications or professional experience of the manual therapist were not considered; these qualifications are important for the proper indication and application of cervical spinal manipulations[5].

It would also be desirable to implement placebo treatments that are as similar as possible to manual therapy techniques but without any specific activity[5]. However, the absence of evidence for the efficacy of physical treatment does not mean that such treatments are not effective (according to evidence based medicine), although the evidence does suggest that manual spinal therapy has a definite placebo effect[1,2].

There is a need for higher-quality and longer-term RCTs to demonstrate the efficacy and safety of manual therapy in general, and of its main techniques (manipulation and mobilisation) for mechanical cervical spine disorders in particular[55]. There should be a national notification system for adverse effects, applied on a routine basis, that utilises a protocol for collecting the adverse effects associated with the use of these techniques and the therapist's VBI assessment"56".

There is no evidence to suggest that physiotherapists are better qualified and are more effective in the application of cervical spinal manipulations than are other healthcare professionals"57".

The populations with neck pain, with or without headaches, in the RCTs were quite homogeneous.

Howe"32" reported a rapid and significant improvement of symptoms in patients with a painful or rigid neck, pain or paresthesia in the shoulder, or pain or paresthesia in the hand. The main weaknesses of this study include the following: im sub-optimal randomisation and a failure to mention drop-outs Bitterli"34" reported an improvement of 35% in the group receiving early active mobilisation, but found no improvement after spinal manipulation. This study has a high risk of bias due to the low quality of the methodological design (non-randomised trial, small sample).

Jull (2002) "35" reported a reduction of the frequency and intensity of headache and neck pain when using spinal manipulation, and the effect lasted until the 12-month follow-up. However, the inability to control the placebo effect could increase the risk of bias (see Tables 4, 5, and 6).

| Author (year) | Study design | Characteristics of participants (n) | Diagnostic criteria | Experimental intervention (therapeutic) | Control intervention | Primary outcome measure | Main results | Length of follow-up | Comments |
|---|---|---|---|---|---|---|---|---|---|
| Bitterli (1977)[34] | Quasi-randomised trial, 3 groups | 30 patients with chronic neck pain, headaches or degenerative changes | None | (I) SM (mean of 6.2 sessions with more than 3.2 manipulations) and mobilisation (3 sessions), (II ) SM (mean of 7.2 manipulations in more than 3.8 sessions) (doctor) | Waiting list | Pain (VAS 100 mm) | No differences between groups | 12 weeks | High risk of bias due to the low quality of the methodological design |
| Howe (1983)[32] | RCT, two groups | 52 patients with subacute to chronic neck pain, radicular symptoms and headaches | None | SM (up to 3 manipulations in a single session (doctor) | Azapropazone | Pain | Significantly less pain in the treatment group | 3 weeks | - |
| Jull (2002)[35] | RCT, four groups | 200 patients with chronic cervical neck pain and headaches | Sjaastad[58] | (I) SM, (II) EX, (III) SM + EX (minimum of 8 and maximum of 12 treatments over 6 weeks, each session not lasting more than 30 minutes) (physiotherapist) | No intervention | Change in headache frequency from baseline to immediately after treatment and at 12 months | Significant reduction in headache frequency during the 12-month follow-up for SM and EX compared with the control ( $P < .05$) | 12 months | No control of placebo effect |

EX = exercise; RCT = randomised clinical trial; SM = spinal manipulation, transcutaneous electrical nerve stimulation; VAS = visual analogue scale; - = inconclusive results.

Table 4. Spinal manipulation for the treatment of cervicogenic headaches in three of the included RCTs

| Author (year) | Random sequence generation | Suitable randomisation | Blinding of subjects or personnel | Blinding of the outcome reviewers | Withdrawals and dropouts | Total (Jadad score) |
|---|---|---|---|---|---|---|
| Bitterli (1977) [34] | 0 | 0 | 0 | 0 | 1 | 1 |
| Howe (1983) [32] | 1 | 0 | 0 | 0 | 0 | 1 |
| Jull (2002) [35] | 1 | 1 | 0 | 1 | 1 | 4 |

Table 5. Quality Evaluation in three of the included RCTs (Jadad score)[59]

| Author (year) | Sequence generation | Concealment of allocation | Blinding of subjects, personnel or outcome reviewers | Incomplete outcome data | Selective outcome reporting | Other sources of bias | Total |
|---|---|---|---|---|---|---|---|
| Bitterli (1977) [34] | -1 | -1 | -1 | 0 | 0 | -1 | -4 |
| Howe (1983) [32] | 0 | -1 | -1 | -1 | -1 | -1 | -5 |
| Jull (2002) [35] | 1 | -1 | 1 | 1 | 1 | 0 | 3 |

> 0 = low risk of bias; 0 = unclear risk of bias; < 0 = high risk of bias.

Table 6. Quality Evaluation in three of the included RCTs (Cochrane tool)[59]

This review has some limitations. Although we used broad search criteria, we cannot guarantee that we did not miss any relevant publications. Due to the number of RTCs reviewed, the total number of subjects, and the low design quality, it is difficult to draw clear conclusions. Although the study populations in the RCTs are quite homogenous, it is not possible to perform a meta-analysis.

This overeview had the advantages of spanning the available literature on nonspecific cervical pain, included only the highest-quality studies, and used recommended methods for systematic reviews.

In the future, studies evaluating the efficacy and safety of manual therapy should be designed according to the international CONSORT recommendations. Furthermore, investigators need be very careful when performing sample size calculations in order to avoid sources of bias.

The information in the studies should be sufficient to allow researchers to reproduce the results independently. The data could suggest a bias in favour of physiotherapists for the treatment of neck pain. However, this bias does not mean that physiotherapists are better qualified or that they are more effective in the application of cervical spinal manipulations compared with other healthcare professionals (see Tables 7, 8, and 9).

| **Author (year)** | **Details of SM treatment (direct quote where applicable)** |
|---|---|
| Bitterli (1977) [34] | Patients in group B were treated by a doctor who was also a qualified massage therapist. They received an average of 7.2 manipulations on the cervical spine using the technique described by Maigne. |
| Howe (1983) [32] | The techniques are similar, with only small differences from those described by Bourdillon. The essence of manipulation is to move the joint(s) as comfortably as possible and then apply moderate, high-velocity but very low-amplitude thrusts in the same direction. |
| Jull (2002) [35] | Manipulative therapy (MT) described by Maitland. This therapy includes the joint mobilisation technique (in which the segment is moved passively) and the high-velocity technique. |

Taking into account the variability and lack of standardisation of SM treatments, it is difficult to replicate these studies independently and/or draw firm conclusions.

Table 7. Details of spinal manipulation (SM) treatment in three of the included RCTs

| **Author (year)** | **Details of adverse events** |
|---|---|
| Bitterli (1977) [34] | Manipulation and mobilisation were well tolerated, typically with a minimal, mild reaction lasting less than 24 hours. |
| Howe (1983 [32] | NIP |
| Jull (2002) [35] | Headache as a minor, transient side effect caused by treatment was reported by 6.7% of subjects during the 6-week intervention period. |

NIP = no information provided. Two of the three RCTs reported adverse effects (AE)[34, 35], and one RCT did not provide this information [32].

Table 8. Adverse effects (AE) reported in three of the included RCTs

| **Author (year) Profession** | **Positive** | **Negative** | **Inconclusive** |
|---|---|---|---|
| Howe (1983) [17] MD | = | - | - |
| Bitterli (1977) [34] MD | - | = | - |
| Jull (2002) [18] PT | - | - | = |

MD = doctor of medicine; PT = physiotherapist; - = inconclusive results.

Table 9. Positive versus negative studies by type of health professional in three of the included RCTs

## 8. Conclusions

There is no evidence to support the use of spinal manipulations for acute neck pain, and there is limited evidence against passive spinal mobilisation. However, there is strong evidence against manipulation alone or manipulation in addition to mobilisation in isolation compared with control groups (placebo/waiting list) in terms of improving function and the general perceived effect for the treatment of subacute or chronic mechanical neck disorders with or without headache. There is strong short- and long-term evidence in favour of multimodal treatments (manipulation/mobilisation plus exercise) compared with control groups for reducing pain, thereby improving function and the general perceived effect in subacute or chronic mechanical neck disorders with or without headaches.

There is no evidence to support the role of manual therapy in cervical radicular conditions. There is evidence against the efficacy of isolated manipulation and/or mobilisation in the treatment of cervicogenic headaches. There were no serious adverse effects associated with the use of HVT techniques; a number of minor adverse effects were reported. The adverse effects associated with the use of non-HVT techniques were more serious and included a transient ischaemic attack, a fall due to this attack, and a fainting episode. The adverse effects associated with the use of non-HVT techniques justify a specific investigation, especially in view of their widespread use on the upper cervical spine.

There was a low utilisation of VBI assessment protocols, and the questionable utility of VBI assessment protocols in clinical practice was highlighted in one study. Positional VBI tests cannot detect all of the patients at risk of adverse effects associated with the use of manual therapy. Additional large-scale studies are needed to investigate the risk of serious adverse reactions associated with the use of both HVT and non-HVT techniques. Ideally, this research should not depend solely on subjective information obtained from providers, as was the case in this study. Finally, a notification system for adverse effects should be used on a routine basis. This system should incorporate protocols for collecting the adverse effects associated with the use of these techniques and the therapist's VBI assessment.

## 9. References

[1] Mirallas-Martínez JA. Cerebral vascular complications after cervical spinal manipulation. Rehabil (Madr) 2003; 37: 33–9.

[2] Gemmell H and Miller P. Comparative effectiveness of manipulation, mobilization and the activator instrument in treatment of nonspecific neck pain: A systematic review. Chirop & Osteop 2006; 14: 1–7.

[3] Vastravers P, Maigne JY. Cervical manipulations and precautionary principles. Rev Rhum. (Ed. Fr) 2000; 67: 349–54

[4] Sackett DL, Straus SE, Richardson WS, Rosenberg W, Haynes RB. Evidence-Based Medicine: How to Practice and Teach EBM. Second Edition. Churchill Livingstone: Edinburgh 2000.

[5] Sarigiovannis P, Hollins B. Effectiveness of manual therapy in the treatment of nonspecific neck pain: A review. Phys Ther Rev 2005; 10: 35–50

[6] Koes BW, Assendelft WJ, van der Heijden GJ, Bouter LM, Knipschild PG. Spinal manipulation and mobilization for back and neck pain: A blinded review. BMJ 1991; 303:1,298–303

[7] Aker PD, Gross AR, Goldsmith CH, Peloso P. Conservative management of mechanical neck pain: Systematic overview and meta-analysis. BMJ 1996; 313: 1,291–6.

[8] Hurwitz EL, Aker PD, Adams AH, Meeker WC, Shekelle PG. Manipulation and mobilization of the cervical spine: A systematic review of the literature. Spine 1997; 22: 1,676–7.

[9] Gross AR, Kay T, Hondras M, Goldsmith C, Haines T, Pelosos P, Kennedy C, Hoving J. Manual therapy for mechanical neck disorders: A systematic review. Man Ther 2002; 7:131–49

[10] Gross AR, Hoving JL, Haines TA, Goldsmith CH, Kay T, Aker P, Bronfort G, Cervical Overview Group. A Cochrane review of manipulation and mobilization for mechanical neck disorders. Spine 2004; 29:1,541–8.

[11] Ernst E. Chiropractic spinal manipulation for neck pain: A systematic review. J Pain 2003; 4: 417–21.

[12] Vernon H, Aker P, Burns S, Vijakaanen S, Short L. Pressure-pain threshold evaluation of the effect of spinal manipulation in the treatment of chronic neck pain: A pilot study. J Manip Physiol Ther 1990; 13:13–6.

[13] Wood TG, Colloca CJ, Matthews R. A pilot randomised clinical trial on the relative effect of instrumental (MFMA) versus manual (HVLA) manipulation in the treatment of cervical spine dysfunction. J Manipul Physiol Ther 2001;24:260–71.

[14] Pikula JR. The effect of spinal manipulative therapy (SMT) on pain reduction and range of motion in patients with acute unilateral neck pain: A pilot study. J Can Chiropractic Assoc 1999;43:111–9

[15] Parkin-Smith GF, Penter CS. A clinical trial investigating the effect of two manipulative approaches in the treatment of mechanical neck pain: A pilot study. J Neuromusculoskel System 1998;6: 6–16

[16] Moodley M, Brantingham JW. The relative effectiveness of spinal manipulation and ultrasound in mechanical pain: Pilot study. Chiropractic Technique 1999;11:164–8.

[17] Brodin H. Cervical pain and mobilization. Manual Med 1985;2:18-22.

[18] Jordan A, Bendix T, Nielsen H, Hansen FR, Host D, Winkel A. Intensive training, physiotherapy, or manipulation for patients with chronic neck pain: A prospective, single-blinded, randomised clinical trial. Spine 1998; 23: 311–9.

[19] Nordemar R, Thörner C. Treatment of acute cervical pain: A comparative group study. Pain 1981;10: 93–101.

[20] Hoving JL, Koes BW, de Vet HCW, van der Windt DAWM, Assendelft WJJ, van Mameren H et al: Manual therapy, physical therapy, or continued care by a general practitioner for patients with neck pain. A randomised, controlled trial. Ann Inter Med 2002;136:713–59.

[21] Hurwitz EL, Morgenstern H, Harber P, Kominski GF, Yu F, Adams AH. A randomised trial of chiropractic manipulation and mobilization for patients with neck pain: Clinical outcomes from the UCLA neck-pain study. Am J Public Health 2002; 92:1,634–41.

[22] Bronfort G, Evans R, Nelson B, Aker PD, Goldsmith CH, Vernon H. A randomised clinical trial of exercise and spinal manipulation for patients with chronic neck pain. Spine 2001; 26:788–99.

[23] Evans R, Bronfort G, Nelson B, Goldsmith CH. Two-year follow-up of a randomised clinical trial of spinal manipulation and two types of exercise for patients with chronic neck pain. Spine 2002; 27: 2,383–9.

[24] Yurkiw D, Mior S. Comparison of two chiropractic techniques on pain and lateral lesion in neck pain patients: a pilot study. Chirop Tech 1996; 8: 155–62.

[25] Hurwitz EL, Aker PD, Adams AH, Meeker WC, Shekelle PG. Manipulation and mobilization of the cervical spine: a systematic review of the literature. Spine 1996; 21:1,746–60.

[26] Mealy K, Breman H, Fenelon GC. Early mobilization of acute whiplash injuries. BMJ 1986; 292: 656–7.

[27] McKinney LA. Early mobilization and outcome in acute sprain of the neck. BMJ 1989; 299:1,006–8.

[28] Koes BW, Bouter LM, van Mameren H, Esser AH, Verstegen GM, Hofhuizen DM, Houben JP, Knipschild PG. Randomized clinical trial of manipulative therapy and physiotherapy for persistent back and neck complaints: results of one year follow up. BMJ 1992; 304:601-5.

[29] Kay TM, Gross A, Santaguida PL, Hoving J, Goldsmith C, Bronfort G. Cervical Overview Group. Exercises for mechanical disorders of the neck. (translation of the Cochrane Revision). In The Cochrane Library Supplement, 2005, Number 1. Oxford: Update Software, Ltd. Available at: http//www.update-software.com (translation of The Cochrane Library, 2006, Issue 1. Chichester, UK: John Wiley & Sons, Ltd.).

[30] Ylinen J, Takala E-P, Nykänen M, Häkkinen A, Mälkiä E, Pohjolainen T, Karppi SL, Kautiainen H, Airaksinen: Active neck muscle training in the treatment of chronic neck pain in women. A randomized controlled trial. JAMA 2003; 289: 2,509–16.

[31] Cassidy JD, Lopes AA, Yong-Hing K. The immediate effect of manipulation versus mobilization on pain and range of motion in the cervical spine: a randomised controlled trial. J Manipulative Physiol Ther 1992; 15:570–5

[32] Howe DH, Newcombe RG, Wade MT. Manipulation of the cervical spine-a pilot study. J R Coll Gen Pract 1983; 33: 574–9.

[33] Sloop PR, Smith DS, Goldenberg E, Dore C. Manipulation for chronic neck pain: a double-blind controlled study. Spine 1982; 7: 532–5.

[34] Bitterly J, Graf R, Robert F, Adler R, Mumenthaler M. Evaluation of physical therapy for the control of spondylogenic headaches. Nervenarzt 1977; 48: 259–62.

[35] Jull G, Trott P, Potter H, Zito G, Niere K, Shirley D, et al: A randomized controlled trial of exercise and manipulative therapy for cervicogenic headache. Spine 2002, 27: 1,835–43.

[36] Karlberg M, Magnusson M, Eva-Maj M, Melander A, Moritz U. Postural and symptomatic improvement after physiotherapy in patients with dizziness of suspected cervical origin. Arch Phys Med Rehabil 1996; 77: 874–82.

[37] Allison GT, Nagy BM, Hall T. A randomized clinical trial of manual therapy for cervicobrachial pain syndrome - a pilot study. Man Ther 2002;7: 95–102

[38] Cagnie B, Vinck E, Beernaert A, Cambier D, How common are side effects of spinal manipulation and can these side effects be predicted?, *Manual Therapy* 2004; 9: 151–6

[39] Di Fabio RP. Manipulation of the cervical spine: risks and benefits, *Physical Therapy* 1999; 79: 50–65.

[40] Rothwell DM, S.J. Bondy and J.L. Williams, Chiropractic manipulation and stroke: a population-based case-control study. Stroke 2001; 32: 1,054–60.

[41] Ernst E. Adverse effects of spinal manipulation: A systematic review, Journal of the Royal Society of Medicine 2007; 100: 330–8.

[42] Haldeman S, Kohlbeck FJ, McGregor M. Unpredictability of cerebrovascular ischemia associated with cervical spine manipulation therapy: A review of sixty-four cases after cervical spine manipulation, *Spine* 2002; 27: 49–55.

[43] Michaeli A. Reported occurrence and nature of complications following manipulative physiotherapy in South Africa, *Australian Journal of Physiotherapy* 1993; 39: 309–15.

[44] Magarey ME, Rebbeck T, Coughlan B, Grimmer K, Rivett DA, Refshauge K. Premanipulative testing of the cervical spine review revision and new clinical guidelines. *Manual Therapy* 2004; 9: 95–108.

[45] Thiel HW, Bolton JE, Docherty S, Portlock JC. Safety of chiropractic manipulation of the cervical spine. *Spine* 2007; 2375–8.

[46] Kerry R, Taylor A,. Mitchell J,. McCarthy C. Cervical arterial dysfunction and manual therapy: a critical literature review to inform professional practice, *Manual Therapy* (2008), pp. 278–88.

[47] Stevinson C, Honan W, Cooke B, Ernst E. Neurological complications of cervical spine manipulation, *Journal of the Royal Society of Medicine* 2002; 94: 107–10.

[48] Thiel HW, Bolton JE, Docherty S, Portlock JC. Safety of chiropractic manipulation of the cervical spine, *Spine* 2007; 2,375–8.

[49] Hurwitz EL, Morgenstern H, Vassilaki M, Chiang LM. Frequency and clinical predictors of adverse reactions to chiropractic care in the UCLA neck pain study. *Spine* 2005; 30: 13, 1,477–84.

[50] Australian Physiotherapy Association, Clinical guidelines for assessing vertebrobasilar insufficiency in the management of cervical spine disorders http://www.physiotherapy.asn.au (2006) [accessed 11.11.08].

[51] Richter R, Reinking M. How does evidence on the diagnostic accuracy of the vertebral artery test influence teaching of the test in a professional physical therapy education program? *Physical Therapy* 2004; 85: 589–99.

[52] Côté P, Kreitz BG, Cassidy JD and H. Thiel. The validity of the rotation-extension test as a clinical diagnostic procedure before manipulation of the neck: A secondary analysis. *Journal of manipulation and physiological therapeutics* 19 (1996), pp. 159–164.

[53] Rivett D, Reid D. Risk of stroke for cervical spine manipulation in New Zealand, *New Zealand Journal of Physiotherapy* 26 (1998), pp. 14–17.

[54] Haynes M, Milne N. Colour duplex sonographic findings in human vertebral arteries during cervical rotation. *Journal of Clinical Ultrasound* 2000; 29: 14–24.

[55] Mirallas-Martínez JA. Effectiveness of manual therapy (manipulations and mobilisations) on nonspecific cervical pain: Scientific evidence. Rehabilitation (Madr.) 2007; 41(2): 81-7.

[56] Sweeney A, Doody C. Manual therapy for the cervical spine and reported adverse effects: A survey of Irish Manipulative Physiotherapists. Manual *Therapy.* 2010); 15: (1), 32-6.

[57] Ernst E, Canter PH. A systematic review of systematic reviews of spinal manipulation. *JR Soc Med*. 2006; 99: 192-6.
[58] Sjaastad O, Fredriksen TA, Pfaffenrath V. Cervicogenic headache: diagnostic criteria. The Cervicogenic Headache International Study Group. *Headache*. 1998; 38:442-445
[59] Jadad AR, Moore RA, Carroll D, et al. Assessing the quality of reports of randomized clinical trials: is blinding necessary? *Control Clin Trials*. 1996; 17:1-12.

# 5

# Use of Physiotherapeutic Methods to Influence the Position of the Foot

Eva Buchtelová
*Department of Health Studies, University J. E. Purkyne, Usti Nad Labem, Czech Republic*

## 1. Introduction

Objective of this study was to examine a group of children who repeatedly attended physical therapy for pain in the Achilles tendons, heels, knees and back pain. Children kinesiological analysis we determined the posture and gait.

We focused on type and abrasion of footwear and we observed present deformity in the axis of foot, the whole limb and in the deviation in heel position and the instep difference in loading and unloading.

Measurements were performed on tenzometric mat - FOOTSCAN - static (standing) and dynamic (gait) analysis of selected segments of the load under the foot. With static analysis, we examined the percentage distribution of pressure between the right and left foot and the front and the posterior part of the foot. With the dynamic analysis technique, we observed unwinding of the base legs on the mat, position and movement of the COP, momentum of subtalar joint and timing of loading below the selected segments.

The results led us in pelvic area and showed functional relationships. The resullts offered various possibilities of using physiotherapeutic methods.

## 2. Issue

The structural function of the foot is enabled by an elaborate system of arches, which cause the foot to behave like an elastic spring, which compresses and flexes as necessary. For the body to be stable it must be supported at three points and the centre of gravity must be between these three points. The foot also has three supporting points: the tuberosity of the heel bone, the head of the 1st metatarsus and the head of the 5th metatarsus. Two systems of arches are formed between these points - the longitudinal and the traverse arches. However, this static tripod model is currently considered surpassed and is only accepted during anatomic description due to tradition and general comprehensibility. From the functional aspect it is more precise to compare the foot arch to a "flexible bow", where the string compressing the bow is formed of the tendons and muscles, which maintain the arches in the foot. They flex and soften the shock of impact with the ground. The arches protect the soft tissues of the sole and enable flexible transfer of the body's weight when standing and enable the foot to strike the ground flexibly when walking and running.

From the aspect of kinesiology the fact that the movements in the foot joints take place in chains is very important. In a closed chain, when the foot is under strain, it is not possible to carry out movement of only one joint. On the contrary in an open chain, when the foot is not under strain, movements can be made in one joint only

*Functional relations between the subtalar and transverse tarsal joint*

The scope of movement in these joints is significantly affected by the mutual position of the talus and the calcaneus. During supination in the subtalar joint the axes of the joint surfaces of the talus and calcaneus for connection to the os naviculare and the os cuboideum are parallel. On the anterior level they are perpendicular to projection of the diagonal axis of rotation in the transverse tarsal joint, so movement into dorsal and plantar flexion take place in their direction. As a result of their parallelism the maximum possible scope of dorsal flexion is possible in this joint, which is simultaneously less stable. The divergence of the axes of the joint surfaces in the transverse tarsal joint increases along with increasing supination in the subtalar joint, even though stability also increases the overall scope of motion decreases at the same time (Vařeka, Vařeková, 2009).

### 2.1 Developmental stages

During the new born period, if the baby is suspended under its armpits the child will support itself on its feet, however, this is a spinal reflex mechanism. During the 1st trimenon the supporting reaction of the legs disappears and a baby suspended under its armpits will not press its lower legs against a surface and the weight of its body is not supported by the lower limbs. At the end of the 2nd trimenon and the beginning of the 3rd trimenon the supporting function reappears if the child is vertical, so-called "stright" reflex. This period is typical in that the child begins to become aware of space three-dimensionally. During the 4th trimenon the postural bearing changes, the child becomes vertical and is capable standing by itself, i.e. it becomes bipedal. This concerns innate programs and innate global mobility formulae (Vojta, 1999, Vlach, 1979). The function of the sole changes, it enters a dynamic relationship between human ambulation and the floor and acts as a contact surface subject to the rules of biomechanics and also has the ability to grip, i.e. actively grip onto the floor. Here we must emphasise the great significance of proprioception and exteroception by the sole, which acts as a massive display. These are qualities that must be kept in mind, because these abilities are gradually lost as a result of wearing shoes. Shoes protect the foot, but also act as a splint (Véle, 1997).

Feet that are not sufficiently actively proprioceptive and the posture of a new born have two specifics:

1. The new born's feet are in eversion. The longitudinal axis of the calcaneus withdraws in relation to the position of the talus laterally.
2. The heel is positioned high, because the calcaneus has not yet moved below the talus. The calcaneus only acquires its position below the talus in relation to the change in posture of the whole body and erective functions. The position changes as a result of muscular function. Before being able to walk independently each child has a pes valgus. This position of the arch will only change during the 3rd year, when all foot arches develop on the basis of muscular differentiation.

An untrained adult should be able to stand on one foot for at least 10 seconds. This time is reduced as the age of the subject increases. Children are only able to stand stably on one foot during their 3rd year. Children can remain standing on one foot for around 10 seconds during their 6th year.

If an injury occurs defects may result, which combine with a functional defect. Functional defects of the musculoskeletal system are manifested by pain and increased tension. Muscular imbalance, and also hypermobility are predispositional factors.

### 2.2 Standing as a postural situation

Standing is the culmination of postural development. When standing still in a balanced position not many muscles appear to be active. The muscles in the feet, m. soleus, the hamstrings, m. rectus femoris, the hip flexors and autochthonous spinal muscles are active. However, during deviation from a stable position a gradually greater number of muscles in the lower limbs and body become involved in order to stabilise the postural situation (Véle, 1997; Nashner, Cordo, 1981).

If balance is lost it is possible to see activity in the calf muscles and the muscles on the frontal side of the lower limb even if the body otherwise remains unmoving, we identify this as "increased play of the toes" (Janda, 1984).

Equipment used for quantifying the posture at rest is a stabilographic platform (posturograph). It measures the body's deviations and their frequency using tensometric sensors, or more precisely it measures changes to the centre of pressure (CoP). The CoP changes and its route is drawn out (Spaepen et al., 1977; Mizrahi et al., 1989).

### 2.3 Function of the foot

During its development as an acral area of the lower limb the foot has adapted to an erect posture and bipedal locomotion. It is capable of adapting to terrain irregularities and absorbing the shock the foot's of impact against a surface. It also forms a solid base when walking, distributes excessive strain on the lower limb equally and reduces energy demands during movement of the body forwards (Gross, Fetto, Rosen, 2002).

Its postural function when standing bipedally consists of maintaining stability, because it transfers the body's and ground's reaction force (Vařeka, Vařeková, 2009). Continuous muscular coordination is necessary for this. We perceive this coordination and stability of the foot as a certainty when standing or walking.

When standing normally there should be no "play of ligaments", which is evidence of increased muscular tension. If standing for extended period the foot arches begin to flatten. Walking is important to maintain the shape and function of arches (Véle, 2006).

And finally the foot serves as a source of proprioceptive and exteroceptive information for the central nervous system (Vařeka, Vařeková, 2009).

### 2.4 Muscle chains affecting the lower limb

The foot does not touch the ground throughout its whole surface, but only the heel, the lateral edge of the foot and in front along the connecting line of the metatarsal heads. The burden is directed to three points: the heel, the big toe metatarsals and the little toe metatarsals. Their burden corresponds to the size of the bone structure. The medial area mostly has no contact, because the sole arches upwards and forms the main part of the longitudinal arch. The summit of the arch is formed by the sustentaculum tali on the calcaneus.

The burden is not directed towards the centre of the heel, but medially and this creates the tendency for the calcaneus to tilt medially (pronation of the calcaneus). The m. flexor hallucis connecting the fibula with the distal phalange of the toe acts against this force. Its activity lifts the sustentaculum tali and the longitudinal arch; it is active when standing, when the foot lifts from the floor and when standing on tiptoe. When standing, the distance between the heel and the bigtoe metatarsal is shorter then when lying down, because the muscles supporting the foot arch are activated. When the arch falls, if the body weight is excessive, the heel pronates and pes valgus occurs. It is stated that the arch has a developmental connection to climbing trees, where it is required to "grip" onto branches and the trunk (Véle, 2006).

*Influence of the axial skeleton on the lower limb*

The femur and the tibia form a long vertical lever, opposite which the short talus handle is located. Rotation of the femur when standing is transferred to the position of the foot and inversely the position of the foot is transferred to the pelvis through the calf. According to Kapandji, movements of the pelvic joint affect the function of the foot.

If the femur is rotated towards medially when standing, the patella moves in the direction of the big toe and rotation of the femur is transferred to the foot through the lower limb, which it forces to pronate along with reducing the longitudinal arch of the foot. If the femur is rotated laterally, the patella moves towards the little toe and the foot has a tendency to supinate and the longitudinal foot arch increases (Fig. no. 1).

The ability to lock the position of the joints in the lower limb is necessary to maintain stability when standing, this is achieved in the proximodistal direction by the following methods:

1. reducing the number of axes in the joints distally
2. the shape of the bone fork (ankle fork)
3. reinforcement of the joints by the medial and lateral ligaments

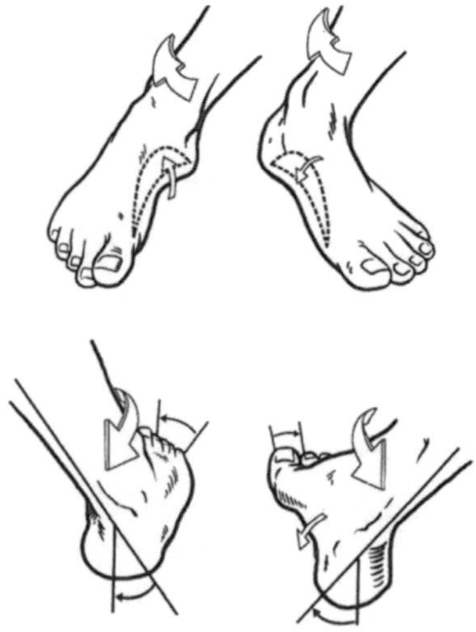

Fig. 1. Effect of femoral rotation on the foot arch.

The foot is connected to the femur through the mm. gastrocnemii, with the tibia and the fibula in the rear through the flexors and in the front through the extensors and mm. peronei. The plantar flexing muscles are important in maintaining the verticals and lifting the foot when walking. M. flexor hallucis longus stretches the longitudinal arch, m. peroneus longus stretches the traverse arch.

The stirrup holding the longitudinal arch of the foot is made up of two loops:

**The m. tibialis anterior – m. peroneus longus loop:**

**fibula** - m. peroneus longus – **metatarsus I. – os cuneiform I** – m. tibialis ant. – **tibia**

M. tibialis anterior flexes the foot, it can pronate and supinate it according to the condition of the peronei and therefore this loop has significant effect on the shape of the foot arch.

**M. tibialis posterior – m. peroneus brevis loop**

**fibula** - m. peroneus brevis – **calcaneus – os cuboideum** – m. tibialis post. - **tibia**

This loop forms a functional whole and also acts as control from the lateral and medial side on the longitudinal foot arch, which it maintains. M. quadratus plantae, which connects the heel with the front of the foot and is therefore involved in maintaining the longitudinal arch must also be included in the muscles affecting the foot arch..

The lower limb forms a complete muscle chain, the function of which can be affected from above and from below. Consequently it is important, when examining defects in the foot, to consider influence from higher areas, i.e. from the position of the pelvis, hip and knee joints. Similarly the inverse influence from the position of the plantar must also be taken into consideration.

## 2.5 Effects of stimuli on control of stabilisation

When standing the distribution of the overall burden on the sole fluctuates in relation to internal factors: the shape of the foot arch, the direction of the body's axis in relation to gravity, the projection of the centre of gravity (CoP) onto the supporting surface, the position of the femur head in the hollow of the hip joint and on the position and configuration of the axial skeleton. Distribution of the burden also depends on exterior factors: on the tilt of the supporting platform, on its profile and friction properties of the ground and shoes. Fluctuation of the burden on the soles can be scanned using equipment recording the course of the burden on individual sections of the sole when standing on pressure plates, where the course of artificial destabilisation by impact with the spine can also be seen (Otáhal). The fluctuation of pressure on individual parts of the soles statically and dynamically can be differentiated by colour on the record. Visible deviations of the CoP caused by breathing movement can also be assessed from the aspect of posture. When assessing strain on the front and rear of the foot using a device according to Kohena – Raze.

Distribution of pressure on the sole changes markedly during movement of the trunk or limbs. Information about these changes is transferred to the CNS and forms an important component during control of stabilisation of the body's position. When standing symmetrically on two scales we discover a regular difference between the sides, fluctuating between 5 – 15% of the total weight (Véle, 2006). The strain on the sole of the foot appears asymmetrically on three supporting points (figure no. 2): 1st and 5th metatarsal of the little toe and heel. The size of the burden on the supporting point is projected into the structure of the skeleton in the three aforementioned points supporting the foot arch unequally. The foot arch is made up of three arches: the practically flat lateral arch, the low transverse arch and the higher longitudinal foot arch. The shape of the foot arch changes depending on distribution of the burden on the sole in relation to whether the foot is supported on the ground when standing or is lifting when walking or is in the swinging phase of the step. When standing the longitudinal arch increases as a result of muscular activity and the distance between the heel - big toe metatarsal reduces compared to the situation without strain and when lying down. This finding means that the foot arch is maintained when standing through the activities of the postural muscles. The shape of the foot arch is influenced not only by the shape of the foot bones, but also the activity of the muscles, particularly m. peroneus longus, m. tibialis posteriori and m. abductor hallucis brevis. The position of the head of the femur against the hip joint and the position of the pelvis also affect the foot arch (see above - - *effects of the axial skeleton on the lower limb*). Information from the plantar and from the position of the hip joint affect body stabilisation and posture when standing.

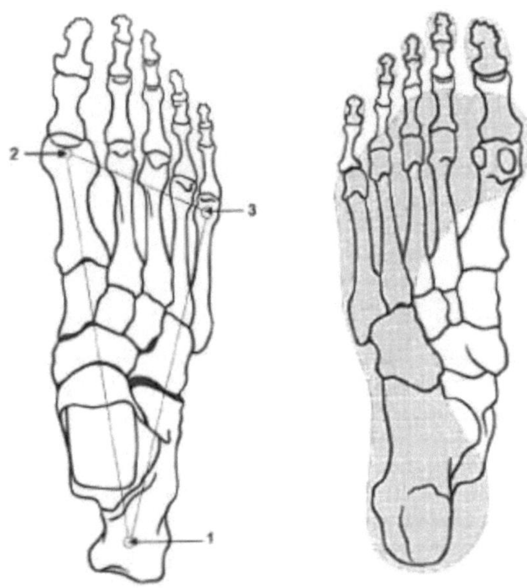


Fig. 2. Brace load distribution on the basis.

### 2.6 Test for stabilisation defects in early phases

Correction movements when standing take place in the disto-proximal direction. During increased instability plantar flexion occurs in the toes initially, thereby extending the supporting base forward. Later on, activity is expanded to the calf muscles, visible as a "play of ligaments", then the thigh muscles are activated, followed by the muscles in the trunk and finally the muscles of the upper limbs (abduction of the arms). The test for stabilisation defects during their initial phase is based on the disto-proximal direction of correction of erect posture.

The relationship between function of m. flexor digitorum brevis and the function of m. flexor digitorum longus on the lower limbs can be used to assess the beginning of arising instability. During the initial phase of stabilisation defects increased demands are placed on maintenance of stability, which appears as extension of the supporting base forward so that the foot does not contact the ground only at the heads of the metatarsals as it should under normal conditions, but also at the last phalanges of the toes - Véle's test (Fig. no. 3). During perfect stabilisation it is possible to insert a sheet of paper under the terminal phalanges. During the initial phase of a defect the stronger m. flexor digitorum longus overcomes the weaker m. flexor digitorum brevis.

This imbalance causes the distal phalanges of the toes to press into the ground throughout their surface during increased demands for stabilisation, which appears as a change in their configuration and creates a step located between the distal and slightly raised proximal toe phalange. If instability continues to increase the pressure on the end phalanges increases, their proximal ends are raised from the ground because m. flexor digitorum longus exerts a great force on the tip of the end phalange inwards and flexor digitorum brevis is not sufficient to keep the proximal phalanges on the ground, consequently the base of the distal phalanges is raised and they assume a claw like shape.

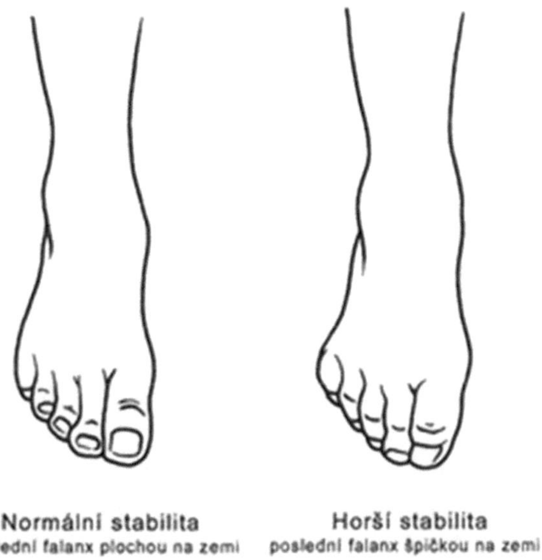


Fig. 3. Stability Test (Véle).

## 2.7 Examining stabilisation of posture when erect

Stabilisation of erect posture depends on the ability to dynamically keep standing for a long period without significant titubation. Stability of erect posture should not be significantly affected by eliminating visual control. During regularly adjusted stabilisation accompanied by the sensation of certainty when standing, it is practically impossible to see any fluctuations in standing and this finding means a good stabilisation function in all directions. If fluctuation appears when the eyes are closed (titubation), accompanied by an increased "play of ligaments" or expansion of the base, this is a symptom of worsened stabilisation when standing. Stabilisation in the frontal-rear direction is less than side stabilisation. This is given by the method of bipedal locomotion when vertical.

Defects in stabilisation appear as a subjective sense of uncertainty or even dizziness. Extension of the supporting base can be seen when standing still. The scope of the defect can be emphasised by eliminating visual control and also by examination when standing on a narrowed base. The supporting base can be narrowed by standing incorrectly, with feet together, or standing on one limb, or on the toes or heel of one limb. Examination of the ability to stay standing on one foot is necessary because standing in this manner normally appears when walking during the swing phase of the step. An adult untrained individual should remain standing on one foot with visual control for approximately 10 s. This time grows shorter along with the increasing age of the subject. Standing stably on one foot is possible approximately from 3 years of age, when postural stability when walking is guaranteed along with safety and certainty. The level of stabilisation when standing can also be tested by the effects of exterior force perpendicular to the body's axis, for instance by exerting pressure on the chest horizontally.

## 3. Methodology

Within the scope of this study we examined 24 children aged between 10–13 years, of this 12 were boys (average age = 11.5 ± 1.04 years, height = 155.58 ± 6.14 cm, weight = 48.33 ± 9.16 kg, BMI = 19.93 ± 3.18 kg.m-2) and 12 were girls (average age = 11.5 ± 0.96 years, height = 157.33 ± 4.8 cm, weight = 49.92 ± 6.08 kg, BMI = 20.15 ± 2.26 kg.m-2).

The children visited elementary school in Ústí nad Labem

We assessed incorrect movement stereotype when walking and standing. We focused on the type and wear of shoes and noticed present deformities, the axis of the foot and the whole lower limb, deviation of the heel and differences in foot arch when under strain and not.

Measurements were taken using FOOTSCAN equipment – static (standing) and dynamic (walking) analysis of the strain below selected segments of the sole.

During the static analysis we established the percentage of distribution of pressure between the right and left foot and between the front and rear of the foot. During dynamic analysis we monitored the technique of lifting the sole from the surface, the position and movement of the CoP, mobility of the subtalar joint and the progress of the strain below selected segments.

## 4. Results

According to Lewit it is known that dysfunction of the foot can be the cause of gross disruption of the movement stereotype and difficulties in the area of the pelvis, the sacral spine and even in the area of the upper cervical spine.

During examination of the position of the pelvis (table no. 1) tilting of the pelvis (lateral tilt) occurred most frequently in boys (50%) and anteversion of the pelvis (tilt forward) in girls (67%). Another indicator is occurrence of shortened muscles in the ischia crural muscles – fknee flexors (67% in boys) , m. quadratus lumborum bilaterally (58% in boys and 50% in girls) - this was established unilaterally in 20% of the children. Reduction in the upper section of m. trapezius occurred most frequently (83% in boys and 50% in boys) (tab. no. 2)

We examined the foot arch roughly by inserting an index finger below the sole from the inside when the child was standing. Flat foot was typical (61% children), toe-out (45% children), toe-in (48%) and hallux valgus (18% of children), in 30% of the children the foot sweated excessively, 15% children underwent repeated treatment for mycosis. 40% children has calluses in the area of the instep or ankle. Insufficient stimulation by movement was clear as well as the consequences of unsuitable ergonomics of the environment and shoes. 44 % children has supination of the position of one foot when standing spontaneously and pronation of the position of the other foot (48%). Children who toed-in complained of pain in the knee joint in the medial area. Stabilisation defects appeared when standing on one lower limb in 45% of children.

We registered the presence of trigger points in the area of m. quadratus plantae in 42% boys and 58% girls and the children stated excessive tiredness and experienced a sensation of heavy feet and static swelling. Trigger points on the Achilles tendon occurred in boys (50%) and an identical proportion of both genders had painful heel spurs (50%)

When examining joint mobility (table no. 3) we recorded occurrence of blocks in boys in both the Chopart joint (92%) and also in the subtalar joint (42%). In girls we found hypermobility in both joints - subtalar (42%) and Chopart (75%). The hypermobility in girls is related to the stability test according to Velé, which was positive in 58% of girls (table no. 5).

Table no. 6 shows examination of standing on two scales when the difference between sides fluctuated in the range of 5-15% of the total weight.

The distribution of the strain on the sole of the foot is described in % in table no. 7. The greatest strain occurred on the front right foot in both genders /(28.1%).

Gait was hard (70%), without shock absorbance when striking the ground the position of the pelvis also changed - it became anteflexive with hyperlordosis of the sacral spine. This was distorted by a weakened abdominal wall and seepage in the area of the sacral character of discs.

Table no. 8 and no. 9 show examination on the FOOTSCAN. The frequency of first contact appeared differently on each foot throughout the whole research group. On the left foot this was initially the lateral edge of the heel, then the medial edge of the heel, followed by the 5$^{th}$ metatarsal, the 4$^{th}$ metatarsal, the 3$^{rd}$ metatarsal, the centre of the foot, the 2$^{nd}$ metatarsal, the 1$^{st}$ metatarsal, the 2$^{nd}$ - 5$^{th}$ toe and finally the big toe.

On the right foot first contact occurred on the medial and then the lateral edge of the sole, this was followed by the 4$^{th}$ and 2$^{nd}$ metatarsal, the centre of the foot, the 5$^{th}$, 2$^{nd}$ and 1$^{st}$ metatarsal, the 2$^{nd}$ and 5$^{th}$ toe and finally the big toe.

Average max. pressures again differed depending on foot. On the left foot these occurred at the 3$^{rd}$ metatarsal for the whole group; of this girls recorded the highest pressure on the 3$^{rd}$ metatarsal and boys on the lateral edge of the heel. On the right foot the maximum pressure occurred at the medial edge of the foot for the whole group (table no. 10 and a 11).

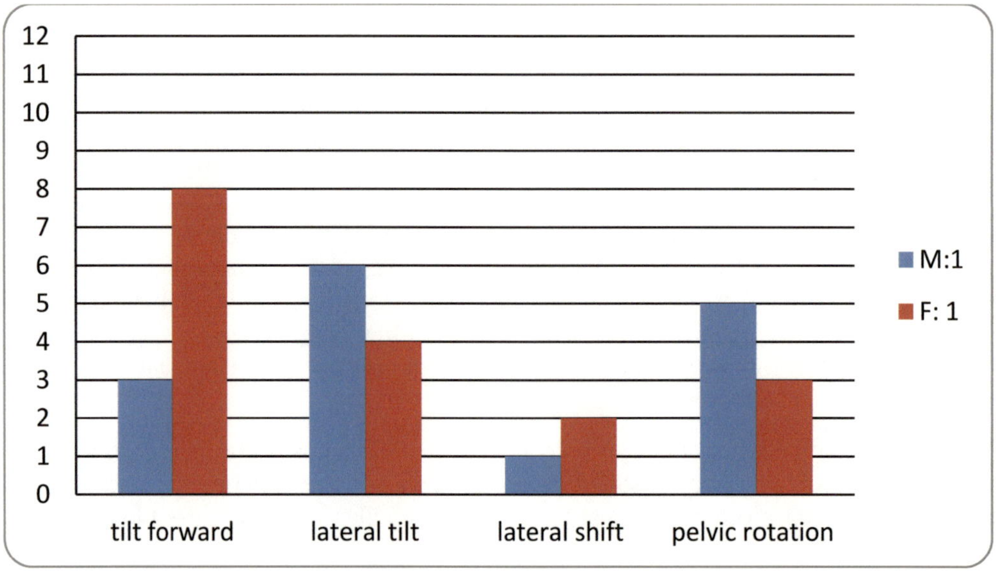


Table 1. Examination of the pelvis (M and F are differentiated in color –:M-male, F -female).

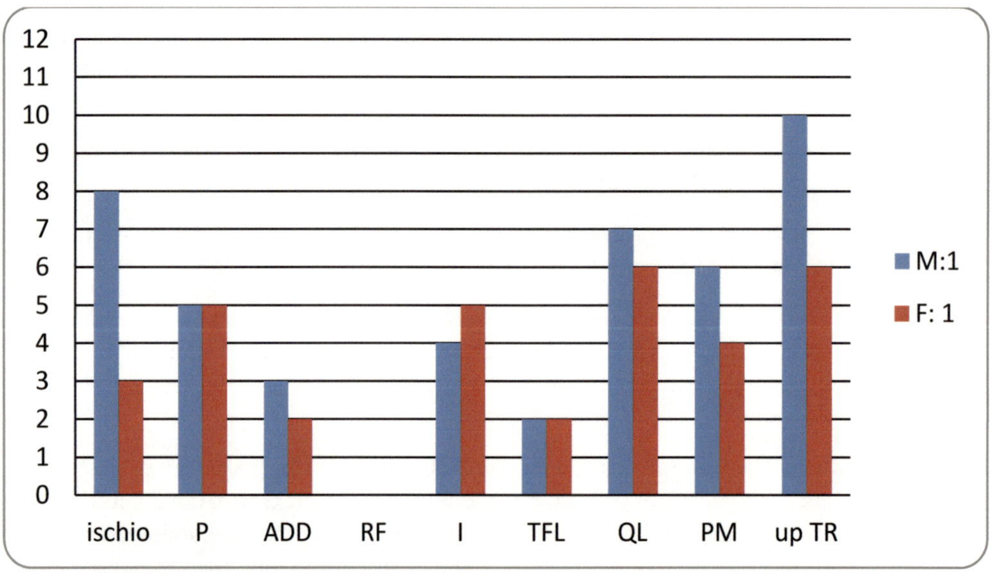


ischio - knee flexors (hamstrings+m.biceps femoris)
P – m.piriformis
ADD - hip adductors
RF – m.rectus femoris
I – m.iliopsoas
TFL –m. tensor fasciae latae
QL – m. quadratus lumborum
PM – m. pectoralis maior
upTR – up trapezius

Table 2. Shortened test structures (M and F are differentiated in color: M-male, F –female).

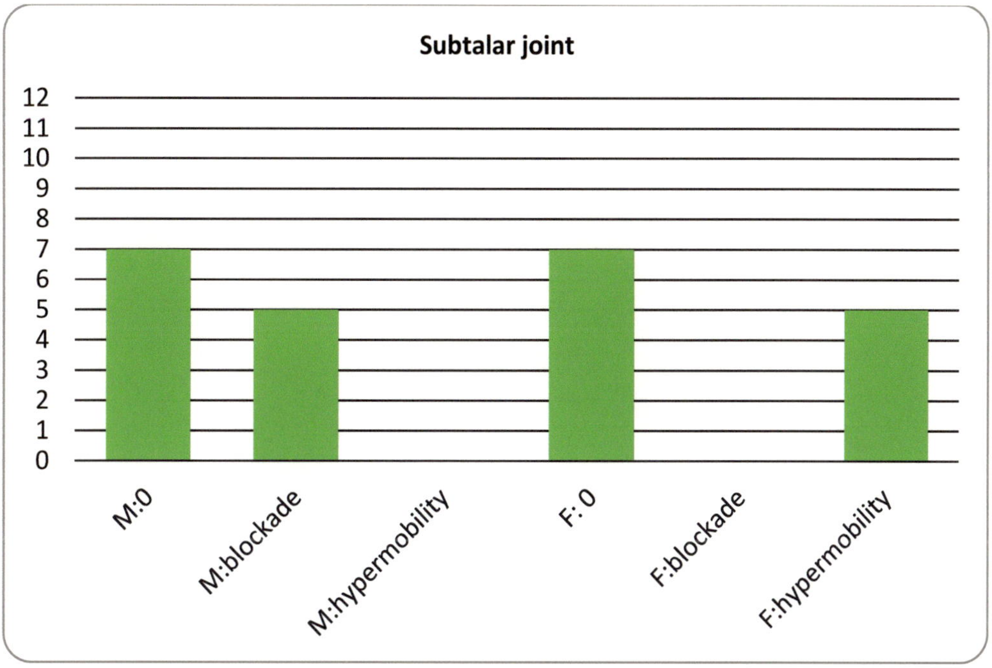


M-male, F - female

Table 3. Examination will articular subtalar joint (here M category on the left, F right).

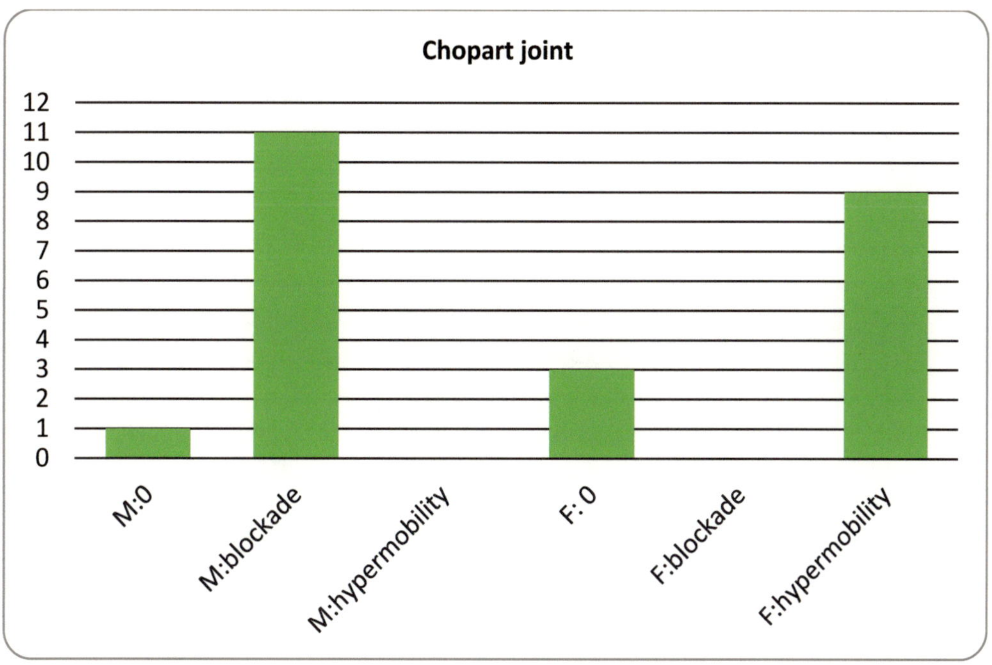


M-male, F - female

Table 4. Examination of joints in the joint will Chopart (here M category on the left, F right).

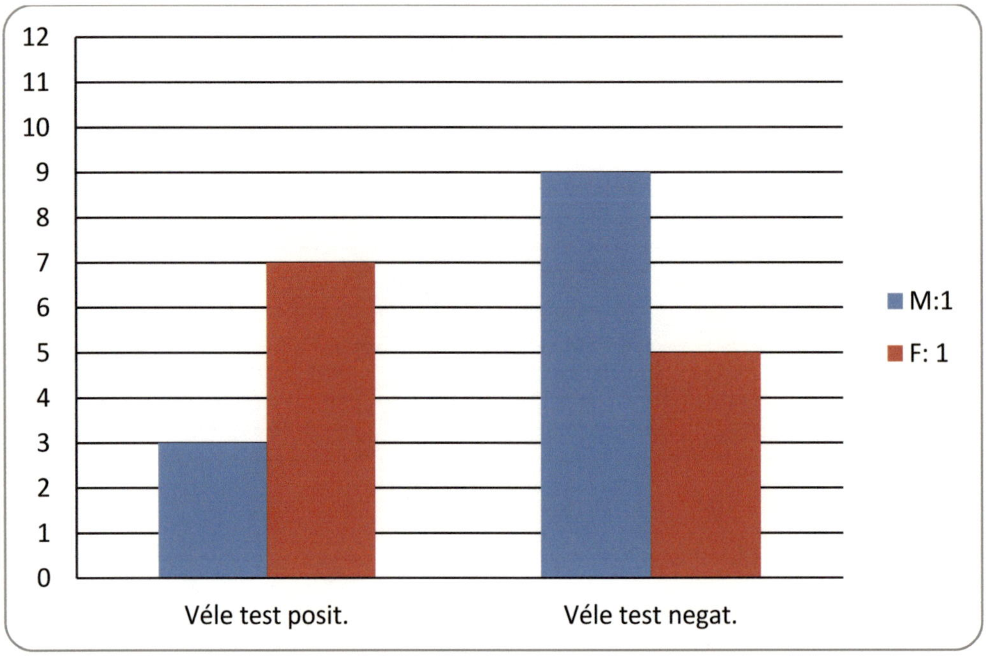


relationship between m.flexor digitorum longus a m. flexor digitorum brevis

Table 5. Stability Test Véle - (M and F are differentiated in color: M-male, F -female).

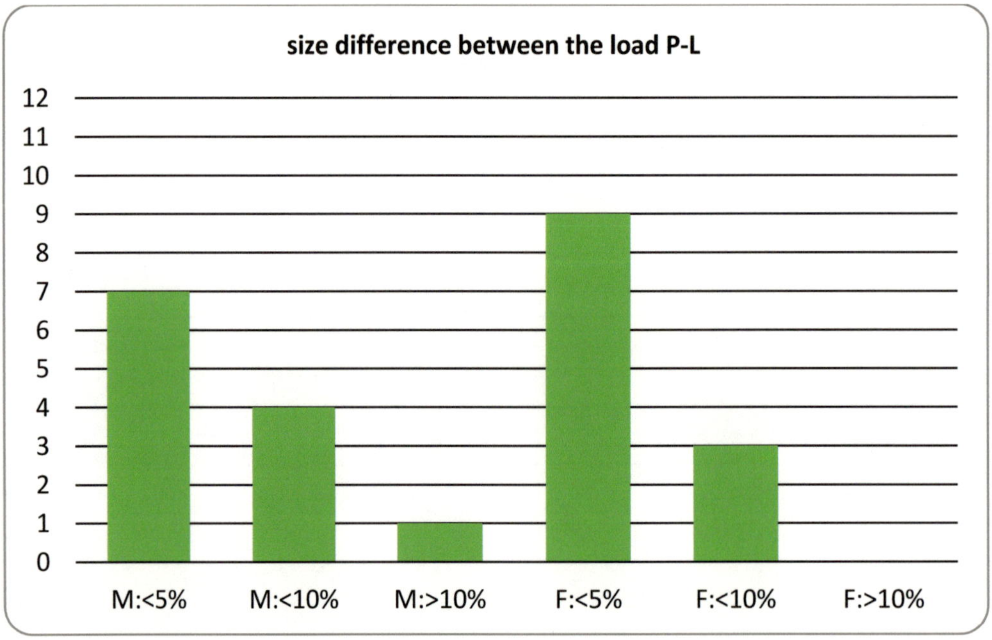


(here M category on the left, F right)

Table 6. Examination of standing on two scales the size difference max 5-15% of 5-15% of the total weight.

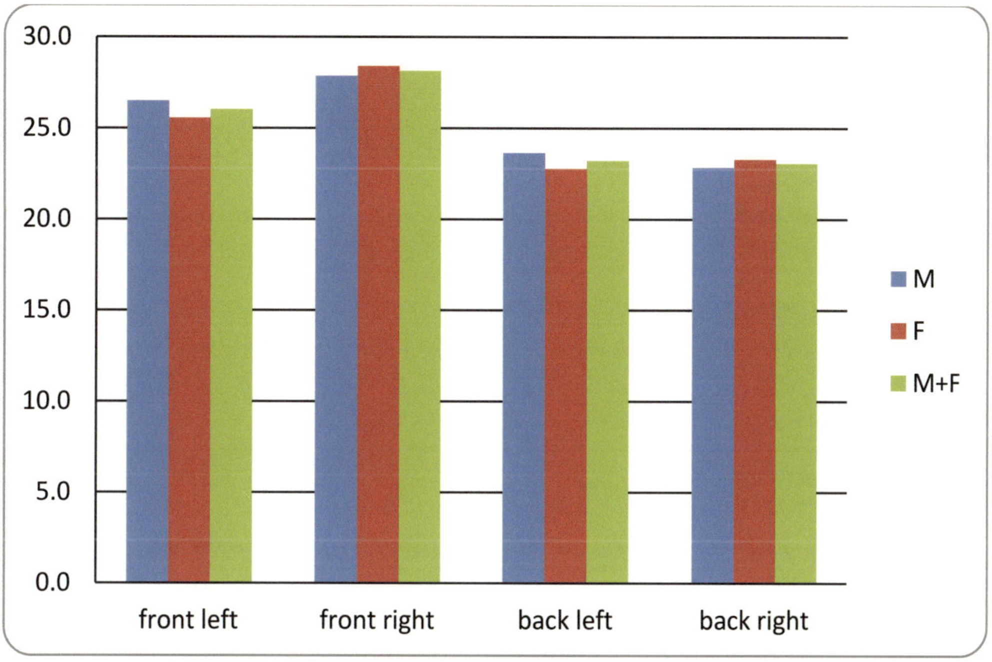


and the posterior part of the foot (%) - M-male, F -female

Table 7. Static analysis - the percentage distribution of pressure between the right and left foot and the front.

| | L-big toe | L-his toes 2-5 | L-Meta1 | L-Meta2 | L-Meta3 | L-Meta4 | L-Meta5 | L-center | L-heel1 | L - heel2 |
|---|---|---|---|---|---|---|---|---|---|---|
| M | 9 | 10 | 8 | 7 | 5 | 4 | 3 | 6 | 2 | 1 |
| F | 10 | 9 | 8 | 7 | 5 | 3 | 4 | 6 | 2 | 1 |
| M+F | 9 | 10 | 8 | 7 | 5 | 4 | 3 | 6 | 2 | 1 |

M- male,F- female
heel 1 medial edge of the heel;
heel 2 – lateral edge of the heel

Table 8. Dynamic analysis - beginning of the contact according to the sum - left leg.

| | P-big toe | P-his toes2-5 | P-Meta1 | P-Meta2 | P-Meta3 | P-Meta4 | P-Meta5 | P-center | P-heel1 | P-heel2 |
|---|---|---|---|---|---|---|---|---|---|---|
| M | 10 | 9 | 8 | 7 | 4 | 3 | 6 | 5 | 1 | 2 |
| F | 10 | 9 | 8 | 7 | 4 | 3 | 5 | 6 | 1 | 2 |
| M+F | 10 | 9 | 8 | 7 | 4 | 3 | 5,5 | 5,5 | 1 | 2 |

heel 1 medial edge of the heel;
heel 2 – lateral edge of the heel

Table 9. Dynamic analysis - beginning of the contact according to the sum - right leg.

| | L-big toe | L-his toes 2-5 | L-Meta1 | L-Meta2 | L-Meta3 | L-Meta4 | L-Meta5 | L-center | L-heel1 | L-heel 2 |
|---|---|---|---|---|---|---|---|---|---|---|
| M | 14 | 7 | 8 | 14 | 17 | 13 | 11 | 2 | 17 | 19 |
| F | 16 | 6 | 6 | 15 | 22 | 18 | 13 | 3 | 18 | 17 |
| M+F | 15 | 6 | 7 | 15 | 19 | 16 | 12 | 2 | 17 | 18 |

heel 1 medial edge of the heel;
heel 2 - lateral edge of the heel

Table 10. Dynamic analysis - average maximum pressure (N/cm2) - left leg.

| | P-big toe | P-his toes2-5 | P-Meta1 | P-Meta2 | P-Meta3 | P-Meta4 | P-Meta5 | P-center | P-heel1 | P-heel2 |
|---|---|---|---|---|---|---|---|---|---|---|
| M | 8 | 4 | 12 | 17 | 17 | 12 | 4 | 3 | 23 | 11 |
| F | 6 | 6 | 10 | 20 | 20 | 13 | 6 | 5 | 24 | 12 |
| M+F | 7 | 5 | 11 | 18 | 19 | 12 | 5 | 4 | 23 | 11 |

heel 1 medial edge of the heel;
heel 2 - lateral edge of the heel

Table 11. Dynamic analysis - average maximum pressure (N/cm2) - right leg.

## 5. Discussion

In its function the position of the sole of the foot is closely linked to the pelvic floor, the deep stabilisation system of the lumbar spine, the abdominal wall, the diaphragm and the upper thorax apparatus with the floor of the oral cavity. The functional interlinkage between hip stabilisers and the sole is of great importance. The position and function of the sole also influences the function and involvement of muscles in the pelvic floor and vice versa, the pelvic floor influences the function and position of the sole. The muscles of the pelvic floor are significantly involved in the posture of the body, they support the trunk and pelvis together with the muscles of the so-called deep stabilisation system. The position of the diaphragm changes, moving into the horizontal level and beginning to participate strongly on stabilisation of the lower and central thoracic spine. The function of the sole, pelvic girdle and the function of the abdominal wall changes fundamentally in accordance with this.

The deep muscular stabilisation system of the lumbar spine includes the muscles of the pelvic floor (m. levator or m. coccygeus) chiefly the deeper muscles in the abdominal wall (m. transversus abdominis, partially the diagonal abdominal muscles), m. quadratus lumborum, the shorter deep paraxial muscles bridging the individual segments of the spine (mm. intertransversarii, mm. interspinosi), the central muscular layer and mm. multifidi and not least the diaphragm.

During restriction of the Chopart joint through increased tension in m. biceps femoris the block in the foot arch is linked to a defect in the structure of the hip with forward body posture and the inability to relax the m. gluteus maximus when standing (Lewit, 1999). Disruption of the function of the hip flexors leads to disruption of the function of the diaphragm and the circle closes. Mobilisation of the instep and modification of afferentation in the are of the sole results in prompt full correction of the function of the foot and correction of posture in the area of the pelvis with regard to this defect.

Through stretching shortened muscle groups and exercises to adjust involvement of phasic muscles on posture we endeavour to implement higher postural models. The key sites for implementing higher global models are the sole, the pelvic area and the sacroiliac joints and the area of the nape.

There are many methodologies we can use. From yoga positions, through erecting response to Vojta's reflexion locomotion we attempt to achieve a centring of the joints and involvement of the phasic muscles after they have been toned.

A non-functional foot arch is a source of distorted afferentation and on the contrary, after it is stimulated and better adjusted under strain, changes occur to the position of the pelvis and the deeper layers of the pelvic floor are directly activated. This is carried out initially when sitting on a ball, then when standing, after this has been managed and following regular exteroceptive stimulation (from skin receptors) it is possible to exert strain by exercises on unstable surfaces according to Freeman and Janda. The basic aids simplifying sensor-motoric stimulation are round and cylindrical sections, balance sandals, rotana (turntable, twister), mini-trampoline.

We can also use the Posturomed. The Posturomed means new quality in therapy of proprioceptive afference. (Rašev, 1995).

Opening up of a frequently blocked instep also enable changes to the position of the pelvis.

We place emphasis on re-education of breathing (basal programmes according to Čápová, attitude 3rd month) or on correct positioning of the foot. We burden the foot on four key points - the 1st and 5th metatarsal and the medial and lateral edges of the tuber calcanei. During the initials phase of training it is possible to use tapping of the longitudinal arch (Flandera, 2004) (figure no. 4). It is a good idea to use tapping during correction of the whole lower limb; i.e. the femur is rotated laterally and the lower leg is rotated medially (Fig.no 5 A,B,C).

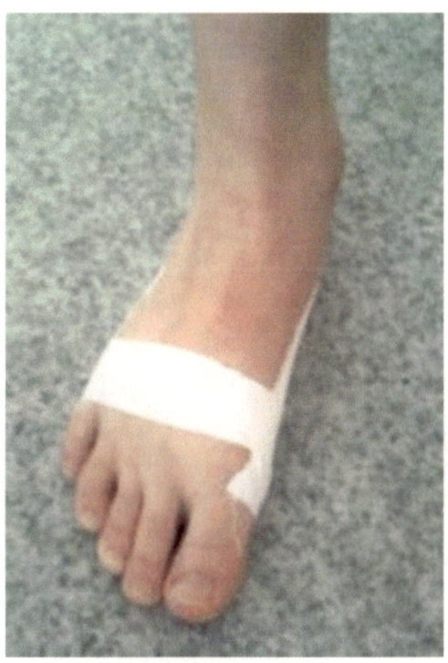

Fig. 4. Tapping of the longitudinal et transversal arch.

Use of spiral dynamics is a very suitable therapy. This leads towards correct movement - three dimensional, dynamic and systematic. It teaches "anatomically correct movement". It improves flexibility, coordination, the ability to tolerate stress and also increases the individual's performance and stamina. The success of the therapy leads to improved economy of movement. It helps prevent acute damage and chronic over-burdening of the musculoskeletal system.

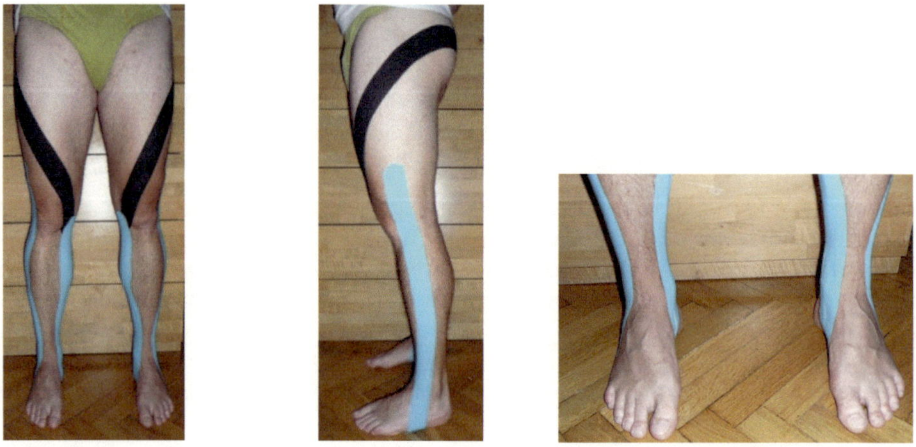

A B C

Fig. 5. Tapping the entire lower extremity - A - front view; B - side view; C - feet details

Exercising when supported is also very appropriate, where we endeavour to use the supporting phases of the foot; i.e. its maximum extension (fan) with support on the 1st metatarsal. This creates a pillar and we again place strain on the foot at four key points - the 1st and 5th metatarsal and the medial and lateral edges of the tuber calcanei. Support on the 1st metatarsal, when we help ourselves with extended and abducted toes in the MP ( metatarse phalangea) toe joints (Fig. no. 6) is considered very important here. This support is projected into the whole body - for example the os tibiale externum dissapears and muscles in the area of the pelvis and thorax are activated.

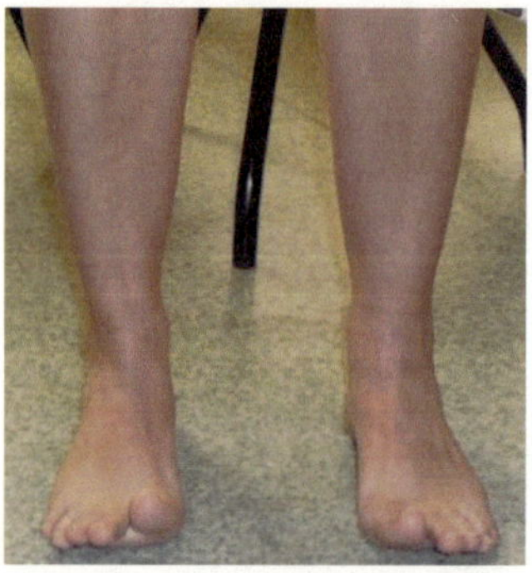

Fig. 6. Support on the 1st metatarsal, when we help ourselves with extended and abducted toes in the MP.

We can recommend the basic stances used in martial arts (Thai chi, karate), where key attention is paid to precise positioning and centring of joints during movement. The phasic system is also facilitated during these. In combination with precise positioning and the related precise afferentation we can achieve internal coordination in the muscles of the pelvic floor and improvement of the position of the pelvis if repeated often enough.

Control of own posture (kibidachi –Fig. no.7, or ashi dachi – Fig. no.8), which is the pelvis in retroversion, the thorax erect, the head straight and the chin pulled in towards the neck, during which time the energy is maintained in the lower part of the abdomen.

**toe joints**

Fig. 7. Kibadači

Fig. 8. Neko aši dači

This stance can be found in all traditional martial arts - (the head and the upper part of the body are perpendicular to the ground, the lower limbs are bent with a tendency to push the knees outward, the knees and ankles are fixed, the lower part of the pelvis is pushed slightly in front of the body, a vertical line from the centre of the front part of the knee should fall near the side of the toe, the soles are placed on the ground throughout their surface, the knees face in the direction of the 3rd metatarsal, the distance between the feet is approximately twice the width of the shoulders (Král and coll.,2004).

## 6. Conclusion

The human foot is an organ, which was formed as the result of the long-term process of adaptation by our ancestors to bipedal locomotion and to erect body posture. A foot fulfilling a static and locomotory function has gradually developed from a foot originally adapted to gripping, jumping and climbing as a result of a more and more erect posture. To assure these functions correctly the structure of the foot changed completely. The tarsus enlarged enormously while the toes and the metatarsal bones regressed. Heavy pressure on the big toe when walking led to its relative enlargement. On the contrary, the other toes underwent various degrees of noticeable reduction. In comparison to primate feet the human foot is a more rigid structure with strong ligamentous support and it formed both longitudinal and transverse arches. A typical characteristic of the human foot is loss of toe opposition and mobility. The foot thereby becomes more a supporting organ than one used for gripping items, even though it has the potential ability to develop the same ability to grip items as the hand, as evidenced by people who have lost the use of their upper limbs.

The fact that six meridians pass through the sole of the foot must also be taken into consideration. Some of these run along the medial, other along the lateral side of the foot. Apart from this the foot itself is one reflexive zone next to another. The meridians of the spleen, pancreas and liver pass along the big toe. However, we are also interested in it because it is the reflexive zone for the head with all its brain glands - the pituitary gland and the hypothalamus. The MP (metatarse phalangea) joint in the big toe is the reflexive zone for the nape. The second and third toe are subject to the meridian for the stomach, the fourth toe is the end point for the meridian of the gall bladder and the meridian of the urinary bladder ends and the meridian of the kidneys begins on the little toe.

Corns, blisters, warts, calluses, abrasions or fissures are among the varied speech of the body and mind.

Dysfunctional strain on the sole and unsuitable ergonomics of the environment and shoes are manifested throughout the body.

## 7. References


Gross, MJ Fett, J., Rosen, E.: Examination of the locomotor apparatus, 1 ed Prague: Triton, 2005, 599th ISBN 80-7254-720-8.

Flandera, P.: Taping - prevention and treatment of the musculoskeletal system., 1 ed Olomouc: Cognition, 2004., 95 p. ISBN 80-86606-47-3.

Janda, V. Fundamentals Clinic functional (neparetických) driving faults. Brno.IDV CAP 1984.139s. 57-855-84

Mizrahi, J. - Susak, Z. Bi-lateral reactive force pattern in postural sway activity of normal subjects. Biol. Cybern., 60, 1989, p. 279 - 305th

Nashner, LM - Cordo, PJ: Relation of automatic postural responses of human leg muscles. Exp. Brain Res., 43, 1981, p. 395 – 404

Spaepen, A. - Vranken, M. - Willens, EJ: Comparison of the Movements of the center of gravity of pressure in srabelometric studies. Agressologie 18, 1977, B, p. 109 - 113th

Tichý, M.: Joint dysfunction. V, Lower limb, 1 ed Praque: Miroslav Tichy Publishers, 2008. 123 p. ISBN 978-80-254-2251-9.

Vařeka, I, Vařeková R. Kinesiology feet. 1st edition. Olomouc: Palacky University in Olomouc, 2009. 189 p. ISBN 978-80-244-2432-3.

Vařeka, I, Vařeková R: Clinical typology feet., Rehab. Physical medicine., 2003, Vol. 10, No. 3, p. 94-103

Vele, F.: Kinesiology. 2nd ed Prague: Triton, 2006. ISBN 80-7254-837-9

Vlach V: Chapters of infant neurology. Prague, Avicenum 1979th

Vojta, V. - Peters, A.: Vojta principle. Springer Verlag, Berlin - Heidelberg 1992, Translation - Grada Publishing, Prague

# 6

# Effectiveness of Passive Joint Mobilisation for Shoulder Dysfunction: A Review of the Literature

Judy Chen
*Physiotherapy Department,*
*The Prince of Wales Hospital, Sydney,*
*Australia*

## 1. Introduction

Shoulder pain and stiffness is common in the general community. It is the third most frequent site of musculoskeletal pain after back and neck. Not only does shoulder pain and stiffness impact on the physical functioning, it also contributes significantly to the emotional and psychological distress of the patient. Furthermore, it imposes considerable financial burden on the affected individual and the society. The first half of this chapter will review in detail the literature on the scope of the problems associated with shoulder dysfunction, in particular the prevalence, clinical course, and specific target groups such as the elderly, occupational and sporting groups. It will also review patients with specific medical problems such as spinal cord injury, stroke and diabetes mellitus. In addition personal suffering and the financial burden of shoulder pain, occupational risk factors and prognostic factors will be discussed, and the diagnostic dilemmas associated with shoulder problems will be highlighted.

Many approaches have been employed in the treatment of shoulder disorders. Physiotherapy intervention is often the first line management of this problem, and a wide array of physical therapies has been used, in particular, passive joint mobilisation and exercise therapy. Whilst there is growing evidence for exercise therapy (Ainsworth & Lewis, 2007; Ginn et al, 1997; Ginn and Cohen, 2005; Grant et al, 2004; Trampas and Kitsios, 2660), there is a paucity of research investigating the effectiveness of this commonly-used technique on shoulder disorder. The second part of this chapter is a descriptive review of the latest evidence in support of the efficacy of passive joint mobilisation in the management of shoulder disorders.

## 2. Scope of the problem

### 2.1 Prevalence

Shoulder pain is a common problem in the general community, with prevalence ranging from 7% to 34% (van der Windt et al, 1995; van der Heijden, 1999; Vogt et al, 2003; Luime et al, 2004; Picavet & Schouten, 2003; Pope, 1997). An estimated 20% of the population will suffer shoulder pain during their lifetime (Pope, et al 1997), with a yearly incidence of 15 new episodes per 1,000 patients seen in the primary care setting (van der Windt et al, 1995). Brox (2003) suggested that half the population has at least one episode of shoulder pain yearly. Furthermore, Pope et al (1997) has found that prevalence of shoulder pain could be influenced by case definition. Hence according to this definition, prevalence could range even higher to 51%. The shoulder is the most frequent site of musculoskeletal pain after the lower back, neck and knee (Parsons et al, 2007; McCormick et al, 1996), and shoulder complaints were found to be the most common contributor to the total incidence figure for non-traumatic arm, neck and shoulder pain (Feleus et al, 2008).

Not everyone consults the general practitioner for their shoulder complaints. It has been estimated that the proportion of the population seeking treatment ranged from between 20% to 50% (Picavet and Schouten, 2003; Chard et al, 1991; Badcock et al, 2003; Walker Bone et al 2004). In particular, Chakravarty & Webley (1990) found only 47% of elderly patients had consulted their general practitioner about shoulder pain. Chard & Hazleman (1987, 1991) found underreporting of disorders of shoulder in elderly patients in the hospital and in the community. Fewer than 40% of subjects sought medical attention for shoulder disorders in the community survey, and from the hospital survey, only 3 out of 21 patients with shoulder symptoms had sought medical treatment. This was probably due to the higher acceptance of illness and disability of the elderly population. Hence the prevalence of shoulder problem is higher than what is quoted in the literature. With the population ageing, the incidence of shoulder dysfunction will continue to rise. This has been reflected in the increasing number of referrals for treatment of shoulder disorders in the current clinical settings.

### 2.2 Specific target groups

Shoulder disorders are commonly encountered in the community, but more prevalent in specific populations such as the elderly, workers involved with repetitive work, those who undertake sports/recreation involving repetitive arm movements, and those with specific medical problems. The following is a brief survey of what has been reported in the literature.

As mentioned previously, surveys have reported a high prevalence of chronic shoulder problems in the elderly in the general community, with rates ranging from 26% (Chard et al, 1987, 1991) to 34% (Chakravarty & Webley, 1990) to 74% (Vecchio et al, 1995). Shoulder pain, either alone or in conjunction with pain in other joints, has a substantial impact on the function and well-being of older adults (Chakravarty & Webley, 1990). Of those aged 85 years and over who reported shoulder joint problems, most had difficulty or were dependent on others in activities in daily living (Badley & Tennant, 1992). Measures of physical performance involving the upper extremity were also decreased in the elderly, together with increased prevalence of joint pain at other body sites (Vogt et al 2003).

Restricted shoulder movement is also common in the elderly, especially in females (Chard & Hazleman, 1987; Chard et al, 1991). It has been shown that there is a mean of 30 degrees less movement in the elderly as compared with younger subjects, with a 10 degree reduction per decade. Prevalence of symptomatic shoulder disorders in the elderly population is high. The potential disability and unmet needs caused by shoulder disorders in the elderly is considerable.

Workers involved in repetitive work and manual handling report higher prevalence rates of shoulder pain than the general community (Silverstein et al, 1998), with considerably higher rates of shoulder pain reported in occupations such as dentists (Fish & Morris-Allen, 1998; Lalumandier et al, 2001), sewing machine operators (Kaegaard & Andersen, 2000); forestry workers (Miranda et al, 2001); hotel restaurant workers (Chyuan et al, 2004); computer workers (Brandt et al, 2004); construction workers (Ludewig & Borstad, 2003), and nursing home and elderly care workers (Luime et al, 2004/2005). The six month prevalence of shoulder pain among workers exposed to repetitive work has been reported to be 45% (Leclerc et al, 2004). Performing artists also report musculoskeletal problems related to the shoulder. Musicians working in an elevated arm position (eg. violinists, violists, flautists, trumpet players) had a higher prevalence of neck-shoulder pain than those working in a more neutral position (Nyman et al, 2007).

A high incidence of shoulder problems is common in overhead sports (Gohlke et al, 1993), such a baseball (Ruotolo et al, 2006), tennis (Vad et al, 2003), and volleyball (Wang & Cochrane, 2001). In fact, painful shoulder conditions are the most common musculoskeletal complaint in swimmers, with prevalence rates of as high as 76% being reported (Pink & Tibone, 2000; Weldon & Richardson, 2001). It is also the second most affected joint in golfers (Gosheger et al, 2003), and in world-class badminton players, 52% of players reported shoulder pain and stiffness (Fahlstrom et al, 2006).

Shoulder pain in individuals with traumatic spinal cord injury is common, and frequently results in chronic debilitating pain recalcitrant to treatment. The incidence of people with spinal cord lesions reporting shoulder pain ranges from 30% to 69% (Gironda et al, 2004; Ballinger et al, 2000), with tetraplegics reporting an even higher prevalence of 80% (McCasland et al, 2006; Samuelsson et al, 2004). In addition, shoulder pain is also not self-limiting in patients with acute traumatic quadriplegia. Waring & Maynard (1991) reported 23% of this group of patients still had the same or worse shoulder pain at the 6 months discharge. Regardless of the level of spinal cord injury or the result of functional movement patterns, muscle imbalances around the shoulder joint are thought to contribute to shoulder disorders in wheelchairs users (Sinnott et al, 2000).

Like spinal cord injury, shoulder pain is also a common complication of stroke. Stroke-related shoulder problems have been categorised into frozen shoulder, shoulder-hand syndrome and subluxation (Chard & Hazleman, 1987). The prevalence of shoulder pain post stroke has been reported ranging from 30% to 67% (Sackley et al, 2008; Lindgren et al, 2007). It can impede rehabilitation and has been associated with poorer outcomes and prolonged hospital stay (Turner-Stokes & Jackson, 2002).

Lundbaek (1957) first described the association between shoulder pain and diabetes mellitus. It was noted that the incidence of adhesive capsulitis was two to four times higher in patients with diabetes than in the general population, and it has been described as the most disabling of the common musculoskeletal manifestations of this disease, affecting up to

58% of people within this population (Thomas et al, 2007; Laslett et al, 2008). In addition, older female patients with diabetes were more likely to report shoulder pain or disability (Laslett et al, 2008).

### 2.3 Personal suffering and financial burden

Studies have reported on the severity and impact of shoulder pain and restriction of movement, and the resultant disability and reduced health of the sufferers. Functional limitations in activities of daily living such as personal care, dressing, washing, cooking, housework, gardening and general activities are commonly reported (Pope et al, 1996; Largacha et al, 2006; Smith et al, 2000). As mentioned before, the impact of personal suffering is more prominent in the elderly (Chard et al, 1991; Chakravarty & Webley, 1993; Vecchio et al, 1995; Vogt et al, 2003; Smith et al, 2000). Sleep disturbance was the most common complaint in patients with shoulder pain (Andersen et al, 2002; Ostor et al, 2005). Croft et al (1994) and Smith et al (2000) reported up to 83% of patients with shoulder pain were unable to sleep on the affected side.

The presence of rotator cuff pathology is predictive of impaired physical health quality of life (MacDermid et al, 2004; Chipchase et al, 2000; Smith et al, 2000; Andersen et al, 2002; Ostor et al, 2005; Winters et al, 1999). Self assessment of general health status in patients with common shoulder conditions rank in severity with major medical conditions such as hypertension, congestive heart failure and acute myocardial infarction, and clinical depression (Gartsman et al, 1998; Kaergaard & Anderson, 2000). Ostor et al (2005) found significant difference in 6 of the 8 domains in SF-36 (self-assessment of general health status questionnaire) between population norms and those with shoulder pain, being especially marked for emotional, physical function and physical role. Badcock et al (2002) also found significant psychological distress and disability scores in subjects reporting unilateral shoulder pain.

Rotator cuff tears can have a profound effect on a person's activity level (Harryman et al, 2003). Lippitt et al, (1993) found patients with a tear could only perform 4.4 of the 12 functions on the Simple Shoulder Test. The greatest functional deficits were the inability to throw overhand and lifting 3.6kg to shoulder level (Lippitt et al, 1993; Largacha et al, 2006; Smith et al, 2000).

Apart from the personal and social costs to sufferers of shoulder dysfunction, direct and indirect health care costs pose significant economic burden on the health care system. Swedish insurance data showed that 18% of disability payments made for musculoskeletal disorders was spent on neck and shoulder problems (Nygren et al, 1995). Up to 30% of workers have reported sick leave due to shoulder pain (Kuipers et al, 2006), with sick days ranging from 25 to over 101 days (Ekberg & Wildhagen, 1996). Silverstein et al (1998) reported the average cost of a claim for an occupational shoulder disorder was nearly $US16,000, and in 2000, the direct costs for the treatment of shoulder dysfunction in the United States totalled a staggering $7 billion (Meislin et al, 2005). Many workers with chronic shoulder pain are unable to resume full-time work (Chipchase et al, 2000) and this has important socio-economic implications as the associated disability is likely to result in time lost from work (Croft et al, 1994).

### 2.4 Prognostic factors

Little is known about the aetiology of shoulder disorders, hence many studies have aimed to determine the prognostic factors associated with shoulder complaints in order to avoid a protracted clinical course. A multitude of factors such as patient demographics (Kennedy et al, 2006; Zheng et al, 2005), duration of symptoms (Croft et al, 1996; Thomas et al, 2005), pain intensity (van der Windt et al, 1996; Macfarlane et al, 1998), baseline disability (Kuipers et al, 2004; Croft et al, 1996), range of motion (van der Heijden, 1999; Croft et al, 1996), hand dominance (Thomas et al, 2005; Bartolozzi et al, 1994), concomitant neck pain (Thomas et al, 2005; van der Windt et al, 1996), trauma (van der Windt et al, 1996), early presentation (Viikari-Juntura et al, 2000), psychosocial factors (Reilingh et al, 2008; Kuipers et al, 2006), medication (Brox & Brevik, 1996), injection (van der Windt & Bouter, 2003), education (Brox & Brevik, 1996; Reilingh et al, 2008; Feleus et al, 2007), occupation (van der Windt et al, 2000; Cassou et al, 2002; Miranda et al, 2001; Viikari-Juntura et al, 2000; Kaergaard & Andersen, 2000) and even acromial morphology (Morrison et al, 1997) have been repeatedly identified as potential predictors of outcome; however these are not consistent findings in the literature and results of studies often do not agree.

Increasing age, for example, has been found to be a predictor of outcome in patients with shoulder pain in many studies (Kennedy et al, 2006; Ginn and Cohen, 2004; Linsell , 2006; Miranda, 2001; Morrison, 1997; Ostor, 2005; Pope, 1997; Kuipers et al, 2004); yet just as many studies do not support this (Feleus, 2007; Kennedy et al, 2006; Kuipers et al, 2006; Picavet, 2003; Reilingh et al, 2008; Solomon et al, 2001; Winter et al, 1997; van der Windt, 1996). Similarly gender (female) was found by Cassou et al (2002), Kennedy at al (2006), Picavet (2003), Smith et al (2000) and van der Heijden (1999) to be a predictor of poor outcome, but again other studies did not agree (Feleus, 2007; Kuipers et al, 2006; Pope, 1997; Solomon et al, 2001; van der Windt et al, 1996).

The conflicting and inconsistent outcomes of the prognostic studies could be attributed to the heterogeneity of the studies. There was a wide variety among the studies in terms of the study population, length of follow-up, prognostic factors investigated, outcome measures used and method of analysis. For example, some studies investigated shoulder pain only (Kuipers at al, 2006; Croft et al, 1996; Picavet, 2003; Reilingh et al, 2008; Winters et al, 1999), while others included shoulder pain with or without restriction (Ginn and Cohen, 2004; Kennedy et al, 2006; van der Windt et al, 1996). Treatment of shoulder complaint was included in some study models but not in others. Even when treatment was included, comparison of results was not possible due to the variability of treatment administered, such as physiotherapy with or without surgery (Brox et al, 1993,1999); physiotherapy to strengthen rotator cuff, cortisone injection (Bartolozzi et al, 1994), pendulum exercises and wall-climbing exercises (Chard et al, 1988), exercise aimed at restoring neuromuscular control of the shoulder girdle muscles (Ginn et al, 1997; Ginn and Cohen, 2005), physiotherapy stretching program (Griggs et al, 2000) and general physiotherapy (Kennedy et al, 2006; Linsell et al, 2006).

In general there is no consensus for prognostic indicators that can identify patients at high or low risk of chronicity. Kuipers et al (2004) concluded that there is strong evidence that high pain intensity predicted a poorer outcome and middle age is associated with poor outcome in occupational population. They also found moderate evidence that a long duration of symptoms and high baseline disability score predicted a poor outcome. However these results are based on small number of studies with large heterogeneity; therefore the results should be interpreted with caution. It is not known whether subjects in these studies had shoulder pain only, or had both shoulder pain and stiffness. Hence little is known about the prognostic factors associated with painful restricted shoulder dysfunction, and only one study (Zheng et al, 2005) has specifically investigated the clinical course of shoulder symptoms in patients treated conservatively for shoulder pain and stiffness.

### 2.5 Occupational risk factors

Studies of occupational diseases have shown the proportion of workers with neck/shoulder pain is high (Cassou et al, 2002). Occupational factors relating to both physical aspects of the work undertaken as well as psychological factors concerning work and the working environment have been associated with musculoskeletal symptoms in the shoulder. Results from numerous relevant studies evaluating occupational risk factors of shoulder pain are in agreement regarding the occupational physical demands associated with shoulder pain. Biomechanical factors such as heavy workload (Frost et al, 2002), duration of employment (van der Windt, 2000), duration of working above shoulder level (Svendsen et al, 2004; Silverstein et al, 2008; Pope et al, 2001; Harkness, 2003), repetitive movements (Andersen et al, 2002; Frost et al, 2002), awkward postures (Pope et al, 2001) and vibration (van der Windt, 2000) have been attributed as potential causes of shoulder problems in workers.

Psychosocial risk factors have also been reported to be associated with the development of shoulder pain. These include depression (Miranda et al 2005), age (Cassou et al, 2002; Bonde et al, 2003), poor control at work (van der Windt, 2000), job dissatisfaction (van der Windt, 2000), high job demand (Andersen et al, 2002), and poor social support (Kaergaard & Andersen 2000; Grooten et al 2004). Shoulder pain and poor work conditions have been associated with long term sickness absence amongst workers (Ekberg & Wildhagen, 1996; Viikari-Juntura et al, 2000). In order to design cost-effective measures for the prevention of shoulder pain, data on the importance of each of these risk factors and the dose-response is needed (Bongers, 2001).

### 2.6 Diagnostic dilemma

The main problem with shoulder studies is that there is no agreement on the diagnosis and classifications of shoulder disorders. This poor agreement between health care providers is due to poor reliability of diagnostic tools and clinical tests. Medical staff and physiotherapists utilise diagnostic classification of shoulder disorders to form the hypothetical framework with which management approach is adopted. Many conditions underlie shoulder pain, yet there is no generally accepted explanation for the aetiology of shoulder problems. Patients with high pain severity, chronic complaints and bilateral involvement represent a diagnostic challenge for clinicians. Furthermore, many patients seen with shoulder disorders have recurrent complaints, and the nature of these complaints varies over the course of time, leading to changes in diagnostic category (Winters et al, 1999). Koester et al (2007) highlighted the difficulty in making a diagnosis in their systematic review of the efficacy of subacromial corticosteroid injection in the treatment of rotator cuff disease. They found that even for a relatively focused topic of rotator cuff disease, there were a variety of pathologic conditions ranging from acute strains to full-thickness cuff tears described in the literature.

### 2.6.1 Interobserver agreement

Most studies have shown that there is little interobserver agreement among surgeons (Kuhn et al, 2007); physiotherapists (de Winter et al, 1999); rheumatologists (Bamji et al, 1996); and between general practitioners and physiotherapists (Liesdek, 1997) in diagnosing soft-tissue shoulder disorders. In these well-powered studies, all the practitioners were experienced and well-trained. The results showed a disappointingly low observed agreement, in particular, in one study (Liesdek, 1997), where the physiotherapists were not blinded for the diagnosis of the general practitioners, and still, the agreement between the two professions was low. Only one study reported almost perfect inter-observer agreement of the Cyriax method for the assessment of shoulder pain by trained and experienced physiotherapists (Pellecchia et al, 1996).

### 2.6.2 Reliability of clinical tests

A large number of tests are used by clinicians to help with the diagnose shoulder pain, but none of these have been standardized. The sensitivity, specificity, positive predictive value, negative predictive value, interobserver reliability and overall accuracy of commonly-used physical examination tests of the shoulder have been comprehensively evaluated in numerous studies (Calis et al, 2000; Park et al, 2005; Silva et al, 2008; Ostor et al, 2004; Nomden et al, 2009; Dinnes et al, 2003). The results of these studies demonstrated that there is a wide variation in the reliability of these tests, with poor to moderate concordance between observers. This was further confirmed in two recent high quality systematic reviews of clinical tests for shoulder pathology. Hughes et al (2008) reviewed 13 studies which evaluated 14 clinical tests commonly used to diagnose rotator cuff pathology. The authors found that most of the tests for rotator cuff pathology were inaccurate and cannot be recommended for clinical use. Hegedus et al (2008) systematically reviewed studies concerning the accuracy of clinical tests for the shoulder, and they included studies on all shoulder pathology. Of the 45 studies reviewed, half were considered to be of high quality according to the Quality Assessment of Diagnostic Accuracy Studies tool, but only 2 studies had adequate sample size (Park et al, 2005; Litaker et al, 2000). Nonetheless, the review found that no tests demonstrated significant diagnostic accuracy. Hegedus et al (2008) examined 10 of the 13 papers included in the review by Hugh et al (2008), and concurred with their assessment. Both reviews concluded that most physical examination tests used for shoulder pathology are inaccurate.

Based on the findings of the literature it is questionable whether these commonly-used clinical tests are useful at all in differentially diagnosing pathologies of the shoulder. As a result of the low reliability of clinical tests for shoulder problems, imaging techniques have been recommended to be used to better define shoulder lesions (Silva et al, 2008). The following section briefly describes some of the current literature in this area.

### 2.6.3 Diagnostic imaging techniques

Due to the diagnostic dilemma and lack of agreement between observers in their diagnosis of shoulder complaints based on clinical examination, diagnostic imaging techniques such as diagnostic ultrasound and magnetic resonance imaging (MRI) are increasingly being used to evaluate patients with painful shoulders.

#### 2.6.3.1 Ultrasound

Ultrasound is commonly used to diagnose soft tissue disorders of the shoulder (Ptasznik, 2001). Numerous studies have found ultrasonography to be highly accurate for detecting full-thickness rotator cuff tears, characterising their extent, and visualising dislocations of the biceps tendon (Teefey et al, 2000; Moosikasuwan et al, 2005; Naredo et al, 2002). In addition, ultrasound has been used in detecting acromioclavicular pathologies in 30 patients with anterior shoulder pain (Blankstein et al, 2005). When compared with 30 asymptomatic controls, degenerative changes which were undetected in plain radiographs were found in the patient group.

Milosavljevic et al (2005) evaluated the accuracy of high-resolution ultrasonography compared to arthroscopy in the detection of rotator cuff tears preoperatively in 190 shoulders. Ultrasound correctly depicted 118 of 124 rotator cuff tears, all 94 full-thickness tears, and 24 of 30 partial-thickness tears. They concluded that ultrasound is a highly accurate diagnostic method for detecting full-thickness rotator cuff tears, but is less sensitive in detecting partial-thickness rotator cuff tears. This is in agreement with the results of Norregaard et al (2002) who found ultrasound to be less sensitive for detecting partial-thickness rotator cuff tears and ruptures of the biceps tendons. Nonetheless ultrasound is quick, non-invasive, and relatively inexpensive; hence it should be used wherever possible to improve diagnosis and treatment of painful shoulder even though the effectiveness of ultrasound is dependent on the type of disorder and the skills and experience levels of the operator (O'Connor et al, 2005).

#### 2.6.3.2 Magnetic Resonance Imaging (MRI)

Magnetic resonance imaging (MRI) is used to detect joint disorders in the shoulder. It is considered an excellent non-invasive method of diagnosing shoulder problems (Toyoda et al, 2005; Mohtadi et al, 2004; Bearcroft et al, 2000). It has also been used to help in the planning of rotator cuff repairs (Ertl et al, 1998). Management plans have been shown in a literature review by Bearcroft (2000) to be subsequently changed due to the impact of MRI on the clinician's diagnosis. The impact on the clinician's diagnosis varied widely between papers: the primary diagnosis was altered in 23% to 68% of cases, and the management plans were subsequently changed in 15% to 61% of cases.

The effectiveness of MRI has been compared with ultrasound. In a systematic review of the effectiveness of diagnostic tests for the assessment of shoulder pain due to soft tissue disorders, Dinnes et al (2003) concluded that MRI and ultrasound could equally be used for detection of full-thickness rotator cuff tears, although ultrasound may be better at picking up partial tears, and MRI are accurate for detection of full-thickness rotator tears. When patients' perceptions and satisfaction with MRI and ultrasound were compared, most patients with shoulder pain preferred sonography to MRI (Middleton et al 2004). Given the large differential in the cost of the two procedures and patient satisfaction, the implication from current evidence is that ultrasound is the more cost-effective and preferred test for identification of full-thickness tears.

### 2.7 Asymptomatic rotator cuff tears

Clinical decisions of shoulder management should be based on the correlation of physical examination with investigative procedures as many patients could be clinically asymptomatic. In fact, magnetic resonance imaging and ultrasound studies have identified a high incidence of asymptomatic rotator cuff tears in the community, especially in the ageing population (Schibany et al, 2004; Milgrom et al, 1995; Miniaci et al, 1995; Yamaguchi et al, 2001; Yamaguchi et al, 2006). There is a high correlation between the onset of rotator cuff tears (either partial or full thickness) and increasing age. Surprisingly despite positive findings of rotator cuff tears on MRI, some individuals reported no pain or functional deficits, although strength was significantly lower in those with complete supraspinatus tendon tear (Schibany et al, 2004). Apart from the older population, Connor et al (2003) found 40% dominant shoulders of asymptomatic shoulders of overhead athletes had clinical findings consistent with partial or full thickness rotator cuff as compared with 0% of the non-dominant shoulders. None of these athletes reported any symptoms 5 years later. Indeed Miniaci et al (1995) also found a wide array of abnormal MRI signals in shoulders of young asymptomatic individuals.

## 3. Evidence of the effectiveness of passive joint mobilisation

Whilst many approaches have been employed in the treatment of shoulder disorders, ranging from surgical to pharmacologic interventions such as oral drug therapy and intra-articular injections, non-pharmacologic interventions such as physiotherapy intervention is often the first line management for shoulder problems, with 53% to 79% of general practitioners referring to physiotherapists (Gentle et al, 1994; Glazier et al, 1998, van der Windt et al, 1995). Peters et al (1994) found shoulder pain accounts for up to 10% of all referrals to physiotherapists in Australia which was similar to findings of Hackett et al (1993).

A wide array of physical therapies is commonly used to treat shoulder disorders, including passive joint mobilisation which is commonly used by physiotherapists in the treatment of musculoskeletal disorders. They are procedures designed to increase the joint range of movement by positioning of the joint and the application of oscillatory movements of that joint in various available ranges. Determination of dosage can be modification of duration, intensity or position of the joint (Maitland 1991). Investigators have examined the physiological effects of passive joint mobilisation (Vicenzino et al, 2007; Teys et al, 2008) and the most effective positions and techniques to perform passive joint mobilisations to optimise the effects (Hsu et al, 2002; Vermuelen et al, 2006; Yang et al, 2007). However, there was a paucity of research specifically investigating the effects of passive joint mobilisation on shoulder disorders as this treatment mode was usually embedded in a package of standardised treatment program. Hence the efficacy of this commonly-used intervention for shoulder disorders has not been established until now.

The following literature review will examine the evidence in support of the efficacy of passive mobilisation directed to shoulder region joints only, in the management of shoulder disorders. To ensure only high quality studies are included, the National Health and Medical Research Council (Australia) evidence hierarchy guidelines and the Physiotherapy Evidence Database (PEDro) rating scale have been used as a guide in this review. Where available, only Level I systematic reviews/meta-analysis And/or Level II evidence of randomised controlled trials of moderate to high quality, rating 5 or more on the PEDro scale are included.

The National Health and Medical Research Council (NHMRC) in Australia has been engaged in the development of evidence-based clinical practice guidelines to rate the strength of evidence in research trials. The strength of evidence includes the level of evidence, quality of evidence and magnitude of treatment effect. The level of evidence reflects on the study design used by the investigators to minimise bias. The quality of evidence assesses the methodological quality of the study. The highest level of evidence (Level I) is represented by a systematic review of high-quality randomised controlled clinical trials measuring relevant outcomes which demonstrates a strong, clinically important, beneficial effect of the intervention. Level II evidence is provided by at least one properly designed randomised controlled trial.

The Physiotherapy Evidence Database (PEDro) contains abstracts of systematic reviews, randomised controlled trials and evidence-based clinical practice guidelines in physiotherapy. The trials have been rated for quality using a rating scale which has sufficient reliability for use in systematic reviews of physical therapy randomised controlled trials (Sherrington et al, 2000; Maher et al, 2003). Studies are rated against a checklist which identifies the internal validity and statistical interpretability of the trials. To assess internal validity the following aspects are included: random allocation, concealment of allocation, comparability of groups at baseline, blinding of patients, therapists and assessors, analysis by intention to treat and adequacy of follow-up. To assess interpretability, between-group statistical comparisons and reports of point estimates and measure of variability are included. The PEDro score is determined by counting the number of checklist criteria that are satisfied in the trial report. The total maximum score is ten. Of note, PEDro does not rate the external validity of the trial.

For this review, a search of the literature was conducted using the Cumulative Index of Nursing Allied Health Literature; MEDLINE; EMBASE from January 1980 to April 2011. Search limits were set to include English language, abstract and human studies. Key search terms included shoulder, shoulder pain, shoulder stiffness, shoulder impingement, physiotherapy, exercise, manual therapy, mobilisation, manipulation and rehabilitation. Inclusion criteria required randomised controlled trials where some form of shoulder "diagnosis" or dysfunction symptoms and some form of manual/manipulative therapy with or without adjunctive or multimodal therapy were used. Articles were excluded when the pain was referred from a spinal source; if there was surgical intervention; sinister pathology such as infection, malignancy, widespread neurological symptoms etc; and contra-indications for manual/manipulative therapy (eg RA, ligamentous instability). The articles were not included if a specific diagnosis of adhesive capsulitis or frozen shoulder was used, unless the article was identified in a systematic review. In addition, conference proceedings, non-peer reviewed literature and case reports were not included.

All relevant articles were read, synthesised and assessed with the PEDro scale. Only those articles which have scored 5 or more on the PEDro rating scale have been included in this review. Essential requirements were randomisation and blinding, with allocation concealment, intention-to-treat and adequate power desirable. A total of 178 citations were retrieved of which 11 met the inclusion/exclusion criteria. Very few peer-reviewed articles restricted passive joint mobilisation to the shoulder region joints alone. Manual therapy/passive joint mobilisation was often part of a multimodal treatment package included with exercise, electrical modalities, massage, stretching etc. The findings of these studies are summarised in Table 1.

| Author | Population Details | Intervention | Control | Outcome Measures | Results | Conclusion |
|---|---|---|---|---|---|---|
| Nicholson 1985 | Pain, restriction GHJ<br>n = 20<br>Age = 51-55 | Passive joint mobilisation (PJM)-glenohumeral joint (GHJ); active exercises<br>Frequency: 2-3x/wk<br>4 weeks | Active exercises within restriction; resistive exercise to increase strength | Pain<br>Range of motion (ROM)<br>Follow-up: 4 weeks | Passive abduction improved in mobilisation group;<br>No difference in pain score between groups | PJM does not add extra benefit to exercises in treatment of painful stiff shoulders |
| Conroy & Hayes 1998 | Primary impingement syndrome; pain, limited functional movement pattern<br>n = 14<br>Age = 50-55 | PJM of GHJ, and multiple physical therapy modalities (hot packs, active ROM, stretching, strengthening, soft tissue mobilisation, patient education)<br>Frequency: 3x/wk<br>3 weeks | Multiple physical therapy modalities | Pain<br>Active ROM<br>Function<br>Follow-up: 24 hours, 3 weeks | Mobilisation group: decreased 24hr pain and pain with subacromial compression test<br>Control group: increase mobility and function<br>No difference between groups in mobility, functional gains | PJM relieved 24 hours pain, but very small sample size in study, therefore unable to detect significant differences |
| Van der Windt et al 1998 | Shoulder pain and stiffness<br>n = 109<br>Age = 57-60 | Corticosteroid injection plus analgesic<br>2 x 30 minutes/wk<br>7 weeks | PJM ; exercises; ice &/or heat &/or electrotherapy (no ultrasound); analgesics | Self-perceived improvement<br>Pain (VAS)<br>Functional disability<br>ROM<br>Follow-up: 7, 26, 52 weeks | Corticosteroid group: greater decrease in pain, increase in function, and passive external ROM<br>No difference between groups in passive abduction ROM | Corticosteroid injection more efficacious than PJM in the short term |
| Winters et al 1999 | 1) Painful stiff shoulders, "synovial " in origin<br>n= 114<br>Age = 46-53<br>2) Painful stiff shoulders, "shoulder girdle" in origin<br>n = 58<br>Age = 43-46 | 1) "Synovial"<br>Corticosteroid injection: 1-3<br>2) "Shoulder girdle"<br>Massage, exercise, electrotherapy | 1)" Synovial"<br>Massage, exercises, electrotherapy (2x/wk)<br>Mobilisation and manipulation of shoulder region and vertebral column joints (1/wk- 6 weeks)<br>2) Mobilisation and manipulation of shoulder region and vertebral column joints | Pain (numerical rating scale)<br>Active and passive ROM<br>Follow-up: 2,6, 11 weeks | 1) "Synovial"<br>Improvement in pain intensity in all groups<br>More rapid improvement in injection group<br>Lower drop out rate due to treatment failure in injection group<br>2) "Shoulder girdle"<br>Improvement in both groups in pain intensity<br>Greater and more rapid improvement in manipulation group<br>Lower dropout due to treatment failure in manipulation group | For synovial disorders, corticosteroid injection seems to be more effective<br>For shoulder girdle disorders, manipulation produces larger, more rapid results |

Table 1. Summary of research on randomised controlled trials for passive joint mobilisation for shoulder dysfunction.

| | | | | | | |
|---|---|---|---|---|---|---|
| Bang & Deyle 2000 | Shoulder impingement<br>n = 52<br>Age = 42-45 | Exercise and PJM to shoulder region joints and shoulder girdle, and/or upper vertebral column joints<br>2 x/wk for 3 weeks | Flexibility and strengthening exercises | Pain (VAS)<br>Function (self-report using Likert scale)<br>Muscle strength<br>Follow-up: 1, 2 months | Improvements in both groups in terms of pain, strength and function<br>Greater improvement in mobilisation group | PJM has added benefit to exercises in the short term for treatment of shoulder pain |
| Bergman et al 2004 | Shoulder pain and dysfunction<br>n = 150<br>Age = 48 | Manipulative therapy to vertebral column joints plus usual medical care<br>6 sessions over 12 wks | Usual medical care | Self-perceived improvement scale<br>Pain<br>Functional disability<br>General health<br>Follow-up: 12, 26, 52 weeks | Similar improvement in both groups up to 6 weeks<br>At 12 weeks, significant difference between groups in pain, disability and general health in favour of manipulation<br>At 52 weeks similar improvement in both groups | Addition of manipulative therapy to vertebral column joints to usual medical care accelerates recovery |
| Dickens et al 2005 | Subacromial impingement syndrome; failed conservative treatment, awaiting surgery<br>n = 85<br>Age = 55 | PJM to shoulder region and vertebral column joints<br>Exercise therapy, postural advice, strapping, electrotherapy<br>Frequency not reported | No intervention | Constant score<br>Follow-up: 6 months | Intervention group: all improved means of 20 in Constant score; 11 did not require surgery<br>Control group: mean improvement in Constant score of 0.65; all required surgery | Physiotherapy is effective in patients with subacromial impingement where surgery could be avoided in some |
| Chen et al 2009 | Shoulder pain and stiffness (less than 140 degrees of active flexion and abduction; or greater than 10cm hand-behind-back deficit compared to unaffected side)<br>n = 90<br>Age = 65 | PJM to shoulder region joints; exercise to restore neuromuscular control and advice (avoid painful activities; pain free exercises)<br>Up to 10 sessions over 8 weeks | Same exercise and advice as intervention group | Shoulder Pain and Disability Index (SPADI); global perceived effect; active shoulder ROM<br>Follow-up: 1, 6 months | Similar improvement in all outcome measures in both groups; differences between groups were small and statistically non-significant | The addition of PJM applied to shoulder region joints to advice and exercise is not more effective than exercise and advice along for painful restricted shoulders |

Table 1. Summary of research on randomised controlled trials for passive joint mobilisation for shoulder dysfunction. (Continuation)

| | | | | | | |
|---|---|---|---|---|---|---|
| Bergman et al 2010 | Shoulder pain, restriction of movement; cervicothoracic spine and rib dysfunction<br>n = 150<br>Age = 48 | Manipulative therapy (MT) to spine only; none to shoulder region joints; advice<br>6 sessions over 12 weeks | Usual care | Patient perceived recovery<br>Shoulder pain<br>Functional disability<br>General health<br>Economic evaluation<br>Follow-up: 6, 12, 26 weeks | At 6 weeks: no difference between groups<br>At 12 weeks: significant improvement with MT group for shoulder and neck pain<br>At 26 weeks: MT favoured for shoulder pain and mobility and neck mobility | MT with usual care decreases shoulder and neck pain and increases shoulder and neck mobility<br>Spinal mobilisation was more cost-effective for neck pain than usual care |
| Bennell et al 2010 | Chronic rotator cuff disease<br>Shoulder pain and restriction<br>n = 112<br>Age = 60 | Soft tissue massage; PJM of glenohumeral joint; scapular retraining exercise; taping; home exercise program<br>10 sessions over 8 weeks | Sham ultrasound; nontherapeutic gel | SPADI<br>Self-perceived global improvement<br>SF-36<br>AQoL<br>Isometric shoulder muscle strength; compliance to protocol<br>Follow-up: 11, 22 weeks | At 11 weeks: no significant difference between groups for SPADI; intervention group significantly better in self-reported measures and strength<br>At 22 weeks: MT significantly better in SPADI, muscle strength, interference with activity and AQoL | No immediate additional benefit with treatment over placebo, but benefits of manual therapy ane exercise man accrue over time.<br>Physiotherapy intervention more valuable for improving function than pain |
| Yiasemides et al 2011 | Shoulder pain with minimal movement restriction<br>n = 98<br>Age = 60 | Passive mobilisation to shoulder region joints; exercise and advice<br>1 or 2 sessions/week for first month; where necessary, additional treatment over following 4 weeks to maximum of 12 sessions | Exercise and advice | SPADI; self-rated change; active ROM (flexion and abduction)<br>Follow-up: 1, 3, 6 months | No statistically significant differences in any outcome measurements at each follow-up | Addition of passive mobilisation to shoulder regions joints to exercise and advice is not more effective than exercise and advice alone in shoulder pain and minimal movement restriction |

Table 1. Summary of research on randomised controlled trials for passive joint mobilisation for shoulder dysfunction. (Continuation)

As can be seen, prior to 2009, only one study had specifically studied the efficacy of the commonly-used physiotherapeutic technique of passive joint mobilisation in the treatment of painful stiff shoulders. Nicholson (1985) investigated the effects of passive joint mobilisation to the shoulder joints and active exercises in patients with painful restricted shoulders. Twenty patients with shoulder pain and limited passive motion at the glenohumeral joint were randomised into an experimental group receiving mobilisation and active exercises or the control group receiving only active exercises. Outcome measures included pain questionnaire and range of glenohumeral motion at baseline and weekly intervals for four weeks. The results demonstrated a mean reduction in pain for the experimental group of -5.1 out of 10 (SD 4.6) compared with -2.9 out of 10 for the control group (SD 4.4). This represented a non-significance difference of – 2.2 (95% CI -6.4 to 2.0). Only passive abduction increased significantly in the experimental group than in the control group. However, due to the small sample size, the study lacked the statistical power to detect small but clinical-meaningful effects of passive joint mobilisation to shoulder region joints.

In 2009, Chen et al conducted a single-blinded randomised clinical trial to assess whether the addition of passive mobilisation of shoulder region joints to advice and exercise for patients with shoulder pain and stiffness was more effective in reducing pain and disability than advice and exercise alone. The experimental group received passive mobilisation directed to the shoulder region joints only. Both experimental and control groups received exercises with proven efficacy which aimed at improving neuromuscular control of the shoulder muscles in order to restore the dynamic stability and muscle force couple co-ordination of the shoulder region. They also received advice on how to use pain-free methods to perform activities of daily living. No other electrotherapeutic modalities were used. Primary outcome measures included pain and disability measured with the 13-point Shoulder Pain and Disability Index. Secondary outcome measures were self-perceived global improvement measured on a 6-point Likert scale and active ranges of motion. All subjects received a maximum of 10 sessions of therapy, and outcome measurements were performed at baseline, one month and six month post randomisation. Results showed that one month after randomisation, participants in both groups had improved in all outcome measure. Further improvements were seen at six months. However between group differences in all outcome measures were small and statistically non-significant. Hence this study demonstrated that the addition of passive mobilisation of shoulder region joints is not more effective than advice and exercise alone for shoulder pain and stiffness.

Results of the previous study were further supported by Yiasemides et al (2011) who investigated the efficacy of passive mobilisation of shoulder region joints for people with shoulder pain and minimal movement restriction. Participant were randomly allocated to the experimental group where they received passive mobilisation of shoulder region joints, exercise and advice, or to the control group where they received exercise and advice alone. Outcome measures similar to those of Chen et al (2009) were taken at baseline and repeated at 3 and 6 months.

Similarly the results demonstrated that the addition of passive mobilisation of shoulder region joints to exercise and advice is not more effective than exercise and advice alone in decreasing pain and painful range of motion and improving function and self-rated change in symptoms in their cohort, with no significant differences in any of the outcome measurements between the 2 groups at short-, medium-, or longer-term follow-up.

Other studies have included passive joint mobilisation as a treatment package together with exercises, modalities, corticosteroid injections, massage, taping etc (Conroy and Hayes, 1998; Winter et al, 1999; van der Windt et al, 1998; Dickens et al, 2005; Bennell et al, 2010). Hence it is not possible to tease out the benefits of passive mobilisation directed to the shoulder region joints alone in these studies. Other studies have directed passive joint mobilisation to the cervicothoracic spine and ribs only (Bergman et al, 2004, 2010) or to a combination of spinal and glenohumeral joints (Bang and Deyle, 2000; Winters et al, 1999; Dickens et al, 2005). In these studies, the cervicothoracic spine and adjacent ribs were deemed to be symptomatic. In contrast, McClatchie et al (2009) investigated mobilising asymptomatic cervical spine (C5, 6, 7) for shoulder pain with painful arc. They found significant decrease of shoulder pain but no difference in muscle strength following spinal mobilisation.

Systematic reviews have found some evidence in support of manual techniques (van der Heijden et al, 1997; Green et al, 1998; Desmeules et al, 2003; Michener et al, 2004; Grant et al, 2004; Green et al, 2005; Faber et al, 2006; Trampas and Kitsios ,2006; Kuhn, 2009; Kromer et al, 2009; Dorrestijn et al, 2009), but these studies have included the above-mentioned trials (except Bennell et al, 2011), which directed passive joint mobilisation at both shoulder regions joints and the vertebral column joints. Ho et al (2009) conducted a systematic review of randomised controlled trials to determine specifically the effectiveness of manual therapy (MT) for the management of musculoskeletal disorders of the shoulder. The review included studies where at least one application of manual therapy, defined as manipulation, passive joint or soft tissue mobilisation techniques or massage, was applied to either the shoulder girdle, cervical or thoracic spine. Fourteen RCTs were included in this well-designed and comprehensive review. Interventions included joint mobilisations (Bulgen et al, 1984; Conroy and Hayes, 1998; Maricar and Chok, 1999; Vermuelen et al, 2006), mobilisation of the upper quarter (Winters et al, 1997; Bang and Deyle, 2000; Bergman et al, 2004), manipulation (Winters et al, 1997; Bergman et al, 2004), Cyriax' manipulation and deep transverse frictions (Guler-Ulysal and Kozanoglu, 2004), "mobilisation-with-movement" (MWM) (Teys et al, 2008) and soft tissue massage (Van den Dolder and Roberts, 2003). Manual therapy was used in isolation (Winter et al, 1997; Winters et al, 1999; Van den Dolder and Roberts, 2003; Vermuelen et al, 2006; Teys et al, 2008) or in combination with exercises (Nicholson, 1985; Conroy and Hayes, 1998; Maricar and Chok, 1999; Bang and Deyle, 2000; Guler-Ulysal and Kozanoglu, 2004; Citaker et al, 2005), hot packs (Conroy and Hayes, 1998; Citaker et al, 2005) or medical care (Bergman et al, 2004).

The authors concluded that:

- For patients with adhesive capsulitis, MT was not more effective than other interventions in the short term for decreasing pain, improving range of motion (ROM) and function
- For patients with subacromial impingement syndrome, there was conflicting evidence for the use of MT for decreasing pain and improving function in the short term, and moderate evidence that MT was no more effective for improving ROM in comparison to other interventions in the short term
- Combination of soft tissue and joint mobilisation techniques, and therapeutic exercise may be more effective than an exercise program alone
- Conflicting evidence for MT in the management of nonspecific shoulder pain in the short term compared to other interventions

- Massage and MWM are useful in patients with musculoskeletal disorders of the shoulder in the short term compared with no treatment

However these conclusions do not specify the site at which the manual therapy was directed.

Even though systematic reviews provide the highest level of evidence for clinical trials, due to the heterogeneity of the studies, these reviews are not without their difficulties. Due to the low number of articles meeting inclusion criteria of the reviews, conclusions are often product of critical assessments of trials only, or based on single studies which might be of low methodological quality. The ability to generalise the findings of such studies to inform clinical practice is questionable. This present review specifically investigated the evidence for passive joint mobilisation to the shoulder region joints only and can now provide conclusive evidence that the addition of passive mobilisation to shoulder region joints alone do not confer additional benefit in the short or medium term to exercise and advice alone in the treatment of painful shoulders, with or without restriction (Chen et al, 2009; Yiasemides et al, 2011). It provides evidence that there is immediate benefit in manual therapy and home exercise program in people with chronic rotator cuff disease. There is also evidence to suggest that passive mobilisation to glenohumeral joints, cervicothoracic joints and adjacent ribs, together with exercise and advice may be of value in improving shoulder disorders and that these benefits may accrue over time, especially in terms of improving shoulder function rather than pain.

## 4. Conclusion

Shoulder pain and stiffness is a common and widespread problem affecting many individuals in the community, with even higher prevalence in particular target groups. Not only does shoulder pain and stiffness impact on the physical functioning, it also contributes significantly to the emotional and psychological distress of the patient. Furthermore it also imposes considerable financial burden on the individual and society. The search for a cost-effective, evidence-based management for this complex problem is of utmost importance. Prognostic information can help to distinguish between patients with a favourable outcome and those with a high risk of persistent pain and disability, thus facilitating decisions regarding treatment and referral of patients. Little is known about the cause of shoulder disorders and despite the numerous studies identifying the various factors associated with treatment outcome, there is no agreement in the literature on these putative factors, thus rendering the present clinical prediction guidelines inadequate. The main problem is the lack of consensus on the diagnosis of shoulder disorders amongst health care providers. The myriad of clinical tests commonly used by clinicians to help with diagnosis of shoulder problems are of questionable reliability. As a result imaging techniques have been recommended to better define shoulder lesions, yet these are not without their problems and constraints. In addition the validity and reproducibility of the diagnosis and classification system and commonly-used clinical tests for shoulder complaints is inadequate. This lack of agreement regarding the diagnosis of shoulder disorder raises serious doubts about its usefulness. This also underpins the reason that despite the large number of studies conducted, there is little evidence for the treatment of shoulder problems.

The present review has been able to fill the gap in the current literature on the evidence-based management of shoulder dysfunction. Effective treatment should be administered according to signs and symptoms rather than diagnosis. Given the competition for the forever decreasing health dollar, it is important that evidence-based and cost-effective treatments are used for shoulder disorders. Further research is needed for a better understanding of the scope of the problem and effectiveness of the different physiotherapy treatment regimes which will help inform and guide clinical practice in the prevention and management of patients with shoulder dysfunction.

## 5. References

[1] Ainsworth R, Lewis JS. (2007). Exercise therapy for the conservative management of full thickness tears of the rotator cuff: a systematic review. *Br J Sports Med*, 41:200-210

[2] Andersen J, Kaergaard A, Frost P, Thomsen JF, Bonde J, Fallentin N, Borg V, Mikkelsen S. (2002). Physical, psychosocial, and individual risk factors for neck/shoulder pain with pressure tenderness in the muscles among workers performing monotonous, repetitive work. *Spine*, 27:660-7

[3] Badcock L, Lewis M, Hay E, McCarney R, Croft PR.(2002). Chronic shoulder pain in the community: a syndrome of disability or distress? *Ann Rheum Dis*, 61:128-31

[4] Badcock L, Lewis M, Hay E, Croft PR. (2003). Consultation and the outcome of shoulder-neck pain: a cohort study in the population. *J Rheumatol*, 30:2694-9

[5] Badley E, Tennant A. (1992). Changing profile of joint disorders with age: findings from a postal survey of the population of Calderdale, West Yorkshire, United Kingdom. *Ann Rheum Dis*, 51:366–71

[6] Ballinger D, Rintala D, Hart K. (2000). The Relation of shoulder pain and range-of-motion problems to functional limitations, disability, and perceived health of men with spinal cord injury: a multifaceted longitudinal study *Arch Phys Med Rehabil*, 81:1575-81

[7] Bamji A, Erhardt C, Price T, Williams P. (1996). The painful shoulder: Can consultants agree? *Br J Rheumatol* 35:1172-1174

[8] Bang M, Deyle G. (2000). Comparison of supervised exercise with and without manual physical therapy for patients with shoulder impingement syndrome. *J Orthop Sports Phys Ther*, 30(3):126-137

[9] Bartolozzi A, Andreychik D, Ahmad S. (1994). Determinants of Outcome in the Treatment of Rotator Cuff Disease. *Clinical Orthop*, 308:90-97

[10] Bearcroft P, Blanchard T, Dixon A, Constant C. (2000). An assessment of the effectiveness of magnetic resonance imaging of the shoulder: a literature review. *Skeletal Radiol*, 29:673-679

[11] Bennell K, Wee E, Coburn S, Green S, Harris A, Staples M, Forbes A, Buchbinder R. (2010). Efficacy of standardised manual therapy and home exerxcise programme for chronic rotator cuff disease: randomised placebo controlled trial. *BMJ*, 340:c2756

[12] Bergman GJ, Winters JC, Groenier KH, Pool JJ, Meyboom de-Jong B, Postema K, Van der Heijden GJ. (2004). Manipulative therapy in addition to usual medical care for patients with shoulder dysfunction and pain. *Ann Int Med*, 141:432-439

[13] Bingefors K, Isacson D. (2004). Epidemiology, co-morbidity, and impact on health-related quality of life of self-reported headache and musculoskeletal pain - a gender perspective. *Eur J Pain*, 8(5):435-50

[14] Blankstein A, Ganel A, Givon U, Dudkiewicz I, Perry M, Diamant L, Chechik A. (2005). Ultrasonography as a diagnostic modality in acromioclavvicular joint pathologies. *Isr Med Assoc J*, 7:28-30

[15] Bonde JP, Mikkelsen S, Andersen JH, Fallentin N, Baelum J, Svendsen SW, Thomsen JF, Frost P, Thomsen G, Overgaard E, Kaergaard A and the PRIM Health Study Group. (2003). Prognosis of shoulder tendonitis in repetitive work: a follow-up study in a cohort of Danish industrial and service workers. *Occup Environ Med*, 60(9):E8

[16] Bongers PM. (2001). The cost of shoulder pain at work. *Br Med J*, 322:64–5

[17] Brandt LP, Andersen JH, Lassen CF, Kryger A, Overgaard E, Vilstrup I, Mikkelsen S. (2004). Neck and shoulder symptoms and disorders among Danish computer workers. *Scand J Work Environ Health*, 30(5):399-409

[18] Brox JI. Regional musculoskeletal conditions: shoulder pain. (2003). *Best Pract Res Clin Rheumatol* , 17(1):33056

[19] Brox JI, Brevik JI. (1996). Prognostic factors in patients with rotator tendinosis (stage II impingement syndrome) of the shoulder. *Scand J Prim Health Care*,14:100–5

[20] Bulgen DY, Binder Aim Hazleman BL, Dutton J, Roberts S. (1984). Frozen shoulder: perspective clinical study with an evaluation of three treatment regimens. *Ann Rheum Dis*, 43:353-60

[21] Calis M, Agkun K, Birtane M, Karacan I, Calis H, Tuzun F. (2000). Diagnostic accuracy of the Hawkins and Neer subacromial impingement syndrome. *Ann Rheum Dis*, 59:44-47

[22] Cassou B, Derriennic F, Monfort C, Norton J, Touranchet A. (2002). Chronic neck and shoulder pain, age and working onditions: longitudinal results from a large random sample in France. *Occup Environ Med*, 59:537-544

[23] Chakravarty KK, Webley M. (1990). Disorders of the shoulder: an often unrecognised cause of disability in elderly people. *BMJ*, 300(6728):848-9

[24] Chakravarty KK, Webley M. (1993). Shoulder joint movement and its relationship to disability in the elderly. *J Rheumatol*, 20:1359-61

[25] Chard MD, Hazleman BL. (1987). Shoulder disorders in the elderly (a hospital study). *Ann Rheum Dis*, 46:684-7

[26] Chard MD, Hazleman R, Hazleman BL, King RH, Reiss BB. (1991). Shoulder disorders in the elderly: a community survey. *Arthritis Rheum*,34:766-9

[27] Chen JF, Ginn KA, Herbert RD. (2009). Passive mobilisation of the shoulder region joints plus advice and exercise does not reduce pain and disability more than advice and exercise alone: a randomised trial. *Aust J Physiother*, 55:17-23

[28] Chipchase L, O'Connor D, Costi J & Krishnan J. (2000). Shoulder impingement syndrome: Preoperative health status. *J Shoulder Elbow Surg*, 9: 12-15

[29] Chyuan JY, Du CL, Yeh WY, Li CY. (2004). Musculoskeletal disorders in hotel restaurant workers. *Occup Med (Oxford)* 54:55-7

[30] Citaker S. Taskiran H, Akdur H, Arabaci UO, Ekici G. (2005). Comparison of the mobilization and proprioceptive neuromuscular facilitation methods in the treatment of shoulder impingement syndrome. *The Pain Clinic*, 27:197-202

[31] Connor PM, Banks DM, Tyson AB, Coumas JS, D"Allessandro DF. (2003). Magnetic resonance imaging of the asymptomatic shoulder of overhead athletes. *Am J Sports Med,* 31:724-727

[32] Conroy DE & Hayes KW. (1998). The Effect of Joint Mobilization as a Component of Comprehensive Treatment for primary Shoulder Impingement Syndrome. *J Orthop Sports Phys Ther,* 1: 3-14

[33] Croft P, Pope D, Zonca M, O'Neill T, Silman A. (1994). Measurement of shoulder related disability: results of a validation study. *Ann Rheum Dis,* 53:525-8

[34] Croft P, Pope D, Silman A. (1996). The clinical course of shoulder pain: prospective cohort study in primary care. Primary Care Rheumatology Society Shoulder Study Group 96. *BMJ,* 313(7057):601-2

[35] Desmeules F, Cote CH, Fremont P. (2003). Therapeutic exercise and orthopaedic manual therapy for impingement syndrome: a systematic review. *Clin J Sport Med,* 13:176-82

[36] De Winter AF, Jans MP, Scholten RJ, Deville W, van Schaardenburg D, Bouter LM. (1999). Diagnostic classification of shoulder disorders: Interobserver agreement and determinants of disagreement. *Ann Rheum Dis,* 58:272-277

[37] Dickens VA, Williams JL, Bahmra MS. (2005). Role of physiotherapy in the treatment of subacromial impingement syndrome: a prospective study. *Physiotherapy,* 91: 159-164

[38] Dinnes J, Loveman E, McIntyre L, Waugh N. (2003). The effectiveness of diagnostic tests for the assessment of shoulder pain due to soft tissue disorders: a systematic review. *Health Technol Assess (Winch Eng),* 7:iii, 1-166

[39] Dorrestijn O, Stevens M, Winters J, van der Meer K, Diercks R. (2009). Conservative or surgical treatment for subacromial impingement syndrome? A systematic review. *J Shoulder Elbow Surg,*18:652-660

[40] Ekberg K, Wildhagen I. (1996). Long-term sickness absence due to musculoskeletal disorders: the necessary intervention of work conditions. *Scand J Rehab Med,* 28:39-47

[41] Ertl JP, Kovacs G, Burger RS. (1998). Magnetic resonance imaging of the shoulder in the primary care setting. *Med Sci Sports Exer,* 30:S7-S11

[42] Faber E, Kuiper JI, Burdofr A, Miedema HS, Verhaar JA. (2006). Treatment of impingement syndrome: a systematic review of the effects on functional limitations and return to work. *J Occup Rehabil,* 16:7-25

[43] Fahlstrom M, Yeap JS, Alfredson H, Soderman K. (2006). Shoulder pain - a common problem in world-class badminton players. *Scand J Med Sci Sports,* 16:168-173

[44] Feleus A, Bierma-Zeinstra SMA, Miedema HS, Bernsen RMD, Verhaar JAN, Koes BW. (2008). Incidence of non-traumatic complaints of arm, neck and shoulder in general practice. *Manual Therapy,* 13:426-433

[45] Feleus A, Bierma-Zeinstra S, Miedema H, Verhagen A, Nauta A, Burdofr A, Varhaar J, Koes B. (2007). Prognostic indicators for non-recovery of non-traumatic complaints at arm, neck and shoulder in general practice - 6 months follow-up. *Rheumatology,* 46:169-176

[46] Fish DR, Morris-Allen DM. (1998). Musculoskeletal disorders in dentists*New York State Dental Journal*, 64:44-8

[47] Frost P, Bonde JP, Mikkelsen S, Andersen JH, Fallentin M, Kaergaard A, Thomsen JF. (2002). Risk of shoulder tendinitis in relation to the shoulder loads in monotonous repetitive work. *Am J Ind Med*, 41:11-8

[48] Gartsman GM, Brinker MR, Khan M, Karahan M. (1998). Self-assessment of general health status in patients with five common shoulder conditions. *J Shoulder Elbow Surg*, 7:228-237

[49] Gentle P, Herlihy P, Roxburgh I. (1994). Controlled trial of open-access physiotherapy service. *Br J Gen Pract*, 34:371-6

[50] Ginn KA, Herbert RD, Khouw W, Lee R. (1997). A randomized, controlled clinical trial of a treatment for shoulder pain. *Phys Ther*, 77:802-9

[51] Ginn KA, Cohen ML. (2004). Conservative Treatment for Shoulder Pain. Prognostic Indicators of Outcome. Arch Phys *Med Rehabil*, 85 :1231-5

[52] Ginn KA, Cohen M. (2005). Exercise therapy for shoulder pain aimed at restoring neuromuscular control: a randomized comparative clinical trial. *J Rehabil Med*, 37:115-22

[53] Glazier RH, Dalby DM, Badley EM, Hawker GA, Bell MJ, Buchbinder R, Linekar SC. (1998). Management of common musculoskeletal problems: a survey of Ontario primary care physicians. *CMAJ*, 158:1037-1040

[54] Gironda RJ, Clark ME, Neugaard B, Nelson A. (2004). Upper limb pain in a national sample of veterans with paraplegia. *J Spinal Cord Med, 27:120-7*

[55] Gohlke F, Lippert MJ, Keck O. (1993). Instability and impingement of the shoulder of the high performance athlete in overhead stress. *Sportverletzung Sportschaden*, 7:115-21

[56] Gosheger G, Liem D, Ludwig K, Greshake O, Winkelmann W. (2003). Injuries and overuse syndromes in golf. *Am J Sport Med*, 31:438-443

[57] Grant HJ, Arthur A, Pichora DR. (2004). Evaluation of interventions for rotator cuff pathology: a systematic review. *J Hand Ther*, 17:274-99

[58] Green S, Buchbinder R, Glazier R, Forbes A. (1998). Systematic review of randomised controlled trials of interventions for painful shoulder: selection criteria, outcome assessment, and efficacy. *BMJ*, 316:354–360

[59] Green SE, Buchbinder R, Forbes A, Glazier R. (2005). Interventions for shoulder pain. *The Cochrane Database of Systematic Reviews*, Vol (1)

[60] Griggs SM, Ahn A, Green A. (2000). Idiopathic adhesive capsulitis: a prospective functional outcome study of nonoperative treatment. *J Bone Joint Surg*, 82-A:1398-1407

[61] Grooten WJ, Wiktorin C, Norrman L, Josephson M, Tornqvist EW, Alfredsson L. (2004). For the MUSIC-Norrtalje Study Group. Seeking care for neck/shoulder pain: a prospective study of work-related risk factors in a healthy population. *J Occup Environ Med*, 46:138-146

[62] Guler-Uysal F, Kozanoglu E. (2004). Comparison of the early response to two methods of rehabilitation in adhesive capsulitis. *Swiss Med Weekly*, 134:353-8

[63] Hackett GI, Bundred P, Hutton JL, O'Brien J, Stanley IM. (1993). Management of joint and soft tissue injuries in three general practices: value of onsite physiotherapy. *Br J Gen Prac,* 43:61-64

[64] Harkness EF, Macfarlane G.J, Nahit E.S, Silman AJ, McBeth J. (2003). Mechanical and psychosocial factors predict new onset shoulder pain: a prospective cohort study of newly employed workers. *Occup Environ Med,* 60:850-857

[65] Harryman DT, Hettrich CM, Smith KL, Campbell B, Sidles JA, Matsen FA. (2003). A prospective multipractice investigation of patients with full-thickness rotator cuff tears. The importance of comorbidities, practice, and other covariables on self-assessed shoulder function and health status. *J Bone Joint Surg (Am),* 85:690-696

[66] Hegedus EJ, Goode A, Campbell S, Morin A, Tamaddoni M, Moorman CT, Cook C. (2008). Physical examination tests of the shoulder: a systematic review with meta-analysis of individual tests. *Br J Sports Med,* 42:80-92

[67] Ho CCY, Sole G, Munn J. (2009). The effectiveness of manual therapy in the management of musculoskeletal disorders of the shoulder: a systematic review. *Manual Therapy,* 14:463-474

[68] Hsu AT, Hedman T, Chang JH, Vo C, Ho L, Ho S, Chang L. (2002). Changes in abduction and rotation range of motion in response to simulated dorsal and ventral translational mobilisation of the glenohumeral joint. *Phys Ther,* 82:544-556

[69] Hughes PC, Taylor NF, Green RA. (2008). Most clinical tests cannot accurately diagnose rotator cuff pathology: a systematic review. *Aust J Physiother,* 54:159-70

[70] Kaergaard A, Andersen JH. (2000). Musculoskeletal disorders of the neck and shoulders in female sewing machine operators: prevalence, incidence, and prognosis. *Occup Environ Med,* 57:528-534

[71] Kennedy CA, Manno M, Hogg-Johnson S, Haines T, Hurley L, McKenzie D, Beaton DE. (2006). Prognosis in soft tissue disorders of the shoulder: predicting both change in disability and level of disability after treatment. *Phys Ther,* 86:1013-1032

[72] Koester MC, Dunn WR, Kuhn J, Spindler KP. (2007). The efficacy of subacromial corticosteroid injection in the treatment of rotator cuff disease: a systematic review. *J Am Acad Orthop Surg,* 1:3-11

[73] Kromer TO, Tautenhahn UG, de Bie RA, Bart Staal J, Bastiaenen C. (2009). Effects of physiotherapy in patients with shoulder impingement syndrome: a systematic review of the literature. *J Rehabil Med,* 41:870-880

[74] Kuhn JE. (2009). Exercise in the treatment of rotator cuff impingement. A systematic review and a synthesised evidence-based rehabilitation protocol. *J Shoulder Elbow Surg,* 18:138-160

[75] Kuhn JE, Dunn WR, Ma B, Wright RW, Jones G, Spencer EE, et al. (2007). Multicenter Orthopaedic Outcomes Network-Shoulder (MOON Shoulder Group). Interobserver agreement in the classification of rotator cuff tears. *Am J Sports Med,* 35:437-41

[76] Kuijpers T, van der Windt DA, van der Heijden GJ, Bouter LM. (2004). Systematic review of prognostic cohort studies of shoulder disorders. *Pain,* 109:420-31

[77] Kuijpers T, van der Windt DAWM, Boeke AJP, Twisk JW, Vergouwe Y, Bouter LM, Van der Heijden GJ. (2006). Clinical prediction rules for the prognosis of shoulder pain in general practice. *Pain,* 120:276-85

[78] Lalumandier JA, McPHee SD, Parrott CB, Vendemia M. (2001). Musculoskeletal pain: prevalence, prevalence, prevention, and differences among dental office personnel. *Gen Dentistry,* 49:160-6

[79] Largacha M, Parsons IM 4th, Campbell B, Titelman RM, Smith KL, Matsen F 3rd. (2006). Deficits in shoulder function and general health associated with sixteen common shoulder diagnoses: a study of 2674 patients. *J Shoulder Elbow Surg* 15:30-9

[80] Laslett L, Burnet L, Redmond SP, McNeil CL. (2008). Predictors of shoulder pain and shoulder disability after one year in diabetic outpatients. *Rheumatology,* 47 :1583-6

[81] Leclerc A, Chastang J-F, Niedhammer I, Landre M-F, Roquelaure Y. (2004). Incidence of shoulder pain in repetitive work. *Occup Environ Med,* 61:39-44

[82] Liesdek C, van der Windt DAWM, Koes B, Bouter LM. (1997). Soft-tissue disorders of the shoulder. A study of inter-observer agreement between general practitioners and physiotherapists and an overview of physiotherapeutic treatment. *Physiotherapy,* 83:12-17

[83] Lindgren I, Jonsson A, Norrving B, Lindgren A. (2007). Shoulder pain after stroke: a prospective population-based study. *Stroke,* 38:343-348

[84] Linsell l, Dawson J, Zondervan K, Rose P, Randall T, Fitzpatrick R, Carr A. (2006). Prevalence and incidence of adults consulting for shoulder conditions in UK primary care; patterns of diagnosis and referral. *Rheumatology,* 45:215-221

[85] Lippitt SB, Harryman DT, Matsen FA III. (1993). A practical tool for evaluating function: The simple shoulder test. In Matsen PA III, Fu FH, Hawkins RJ (editors) The Shoulder: A Balance of Mobility and Stability. Rosemont IL: *American Academy of Orthopaedic Surgeons,* 501-518

[86] Litaker D, Pioro M, El Bilbeisi H, et al. (2000). Returning to the bedside: using the history and physical examination to identify rotator cuff tears. *J Am Geriatr So,* 48:1633–7

[87] Ludewig P & Borstad J. (2003). Effects of a home exercise programme on shoulder pain and functional status in construction workers. *Occup Environ Med,* 60: 841-849

[88] Luime JJ, Koes BW, Hendriksen IJ, Burdorf A, Verhagen AP, Miedma HS, Verhaar JA. (2004). Prevalence and incidence of shoulder pain in the general population: a systematic review. *Scand J Rheumatol,* 33:73-81

[89] Luime JJ, Kuiper JI, Koes BW, Verhaar JA, Miedema, HS, Burdorf A. (2004). Work-related risk factors for the incidence and recurrence of shoulder and neck complaints among nursing-home and elderly-care workers. *Scand J Work Environ Health,* 30:279-86

[90] Luime J, Koes BW, Miedem HS, Verhaar JA, Burdorf A. (2005). High incidence and recurrence of shoulder and neck pain in nursing home employees was demonstrated during a 2-year follow-up. *J Clin Epidemiol,* 58:407-13

[91] Lundbaek K. Stiff hands in long term diabetics. (1957). *Acta Med Scand,* 158:447-51

[92] MacDermid JC, Ramos J, Drosdowech D, Faber K, Patterson S. (2004). The impact of rotator cuff pathology on isometric and isokinetic strength, function, and quality of life. *J Shoulder Elbow Surg,* 13:593-8

[93] MacDonald PB, Clark P, Sutherland K. (2000). An analysis of the diagnostic accuracy of the Hawkins and Neer subacromial impingement signs. *J Shoulder Elbow Surg,* 9:299-301

[94] MacFarlane GJ, Hunt IM, Silman AJ. (1998). Predictors of chronic shoulder pain: a population based prospective study. *J Rheumatology,* 25:1612-5

[95] Maher CG, Sherrington C, Herbert RD, Moseley AM, Elkins M. (2003). Reliability of the PEDro scale for rating quality of randomised controlled trials. *Phy Ther,* 83:713-21

[96] Maitland GD.(1991) *Peripheral Mobilisation* (3rd Ed Oxford: Butterworth-Heinemann Chs 4-5

[97] Maricar NN, Chok B. (1999). A comparison of the effect of manual therapy with exercise therapy and exercise therapy alone for stiff shoulders. *Physiotherapy Singapore,* 2:99-104

[98] McCasland LD, Budiman-Mak E, Weaver FM, Adams E, Miskevics S. (2006). Shoulder pain in the traumatically injured spinal cord patient: evaluation of risk factors and function. *J Clin Rheumatol,* 12:179-86

[99] McClatchie L, Laprede J, Martin S, Jaglal SB, Richardson D, Agur A. (2009). Mobilisations of the asymptomatic cervical spine can reduce signs of shoulder dysfunction in adults. *Man Ther,* 14:369-74

[100] Meislin RJ, Sperling JW, Stitik TP. (2005). Persistent shoulder pain: epidemiology, pathophysiology , and diagnosis. *Am J Orthop (Belle Mead NJ),* 34(12 suppl):5-9

[101] Michener LA, Walsworth MK, Burnet EN. (2004). Effectiveness of rehabilitation for patients with subacromial impingement syndrome: a systematic review. *J Hand Ther,* 17:152-64

[102] Middleton WD, Payne WT, Teefy SA, Hildebolt CF, Rubin DA, Yamaguchi K. (2004). Sonography and MRI of the shoulder: comparison of patient satisfaction. *Am J Roentgenol,* 183:1449-52

[103] Milgrom C, Schaffler M, Gilbert S, van Holsbeeck M. (1995). Rotator-cuff changes in asymptomatic adults. The effect of age, hand dominance and gender. *J Bone Joint Surg (Br),* 77-B:296-298

[104] Milosavljevic J, Elvin A, Rahme H. (2005). Ultrasonography of the rotator cuff: a comparison with arthroscopy in one-hundred-and-ninety consecutive cases. *Acta Radiologica,* 46:858-865

[105] Mohtadi NG, Vellet AD, Clark ML, Hollinshead RM, Sasyniuk TM, Fick GH, Burton PJ. (2004). A prospective, double-blind comparison of magnetic resonance imaging and arthroscopy in the evaluation of patients presenting with shoulder pain. *J Shoulder Elbow Surg,* 13:258-65

[106] Moosikasuwan JB, Miller TT, Burke B. (2005). Rotator cuff tears: clinical radiographic, and US findings. *RadioGraphics,* 5:1591-1607

[107] Naredo E, Aguado P, De Miguel E, Ulson J, Mayorodomo L, Gijon-Banos J, Martin-Mola E. (2002). Painful shoulder: comparison of physical examination and ultrasonographic findings. *Ann Rheum Diseases,* 61:132-136

[108] Nicholson GG. (1985). The effects of passive joint mobilization on pain and hypomobility associated with adhesive capsulitis of the shoulder. *J Orthop Sports Phys Ther,* 6:238-246

[109] Nomden JG, Slagers AJ, Bergman GJ, Winters JC, Kropmans TJ, Dijkstra PU. (2009). Interobserver reliability of physical examination of shoulder girdle. *Manual Therap,* 14:152-9

[110] Norregaard J, Krogsgaard MR, Lorenzen T, Jensen EM. (2002). Diagnosing patients with longstanding shoulder joint pain. *Ann Rheum Dis,* 61:646-9

[111] Nygren A, Berglund A, Von Koch M. (1995). Neck and shoulder pain: an increasing problem. Strategies for using insurance material to follow trends. *Scand J Rehabil Med Suppl,* 32:107-112

[112] Nyman T, Wiktorin C, Mulder M, Johansson YL. (2007). Work postures and neck-shoulder pain among orchestra musicians. *Am J Ind Med,* 50:370-376

[113] O'Connor PJ, Rankine J, Gibbon WW, Richardson A, Winter F, Miller JH. (2005). Interobserver variation in sonography of the painful shoulder. *J Clin Ultrasound,* 33:53-6

[114] Ostor AJK, Richards CA, Prevost AT, Speed CA, Hazleman BL. (2005). Diagnosis and relation to general health of shoulder disorders presenting to primary care. *Rheumatology,* 44:800-805

[115] Park HB, Yokota A, Gill HS, El Rassi G, McFarland EG. (2005). Diagnostic accuracy of clinical tests for the different degrees of subacromial impingement syndrome. *J Bone Joint Surg Am,* 87:1446-55

[116] Parsons S, Breen A, Foster NF, Letley L, Pincus T, Vogel S Underwood M. (2007). Prevalence and comparative troublesomeness by age of musculoskeletal pain in different body locations. *Family Practice,* 24:308-316

[117] Pellecchia GL, Paolino J, Connell J. (1996). Inter-tester reliability of the Cyriax evaluation in assessing patients with shoulder pain. *J Orthop Sports Phys Ther,* 23:34-38

[118] Peters D, Davies P, Pietroni P. (1994). Musculoskeletal clinic in general practiced: study of one year's referrals. *Br J Gen Pract,* 44:25-29

[119] Picavet HS, Schouten JS. (2003). Musculoskeletal pain in the Netherlands: prevalences, consequences and risk groups, the DMC (3)-study. *Pain,* 102:167-78

[120] Pink MM, Tibone JE. (2000). The painful shoulder in the swimming athlete. *Orthop Clin North Am,* 31:247-61

[121] Pope D, Croft P, Pritchard C, Macfarlane G, Silman A. (1996). The frequency of restricted range of movement in individuals with self-reported shoulder pain: results from a population-based survey. *Br J Rheumatol,* 35:1137-41

[122] Pope DP, Croft PR, Pritchard CM, Silman AJ. (1997). Prevalence of shoulder pain in the community: the influence of case definition. *Ann Rheum Dis,* 56:308–12

[123] Pope DP, Silman AJ, Cherry NM, Pritchard C, Macfarlane GJ. (2001). Association of occupational physical demands and psychosocial working environment with disabling shoulder pain. *Ann Rheum Dis,* 60:852-8

[124] Ptasznik R. (2001). *Sonography of the shoulder.* Van Holsbeeck MT, Introcaso JH, editors. (2001). Musculoskeletal Ultrasound. 2nd Ed. St Louis: Mosby; p.463-516

[125] Reilingh ML, Kuijpers T, Tanja-Harfterkamp AM, van der Windet DA. (2008). Course and prognosis of shoulder symptoms in general practice. *Rheumatology,* 47:724-730

[126] Ruotolo C, Price E, Panchal A. (2006). Loss of total arc of motion in college baseball players. *J Shoulder Elbow Surg,* 15:67-71

[127] Sackley C, Brittle N, Patel S, Ellins J, Scott M, Wright C, Dewey ME. (2008). The prevalence of joint contractures, pressure sores, painful shoulder, other pain, falls and depression in the year after a severely disabling stroke. *Stroke,* 39:3329-34
[128] Samuelsson KA, Tropp H, Gerdle B. (2004). Shoulder pain and its consequences in paraplegic spinal cord-injured, wheelchair users. *Spinal Cord,* 42:41-46
[129] Schibany N, Zehetgruber H, Kainberger F, Wurnig C, Basslamah A, Herneth AM, Lang T, Gruber D, Breitenseher MJ. (2004). Rotator cuff tears in asymptomatic individuals: a clinical and ultrasonographic screening study. *Eur J Radiology,* 51:263-268
[130] Sherrington C, Herbert RD, Maher CG, Moseley AM. (2000). PEDro: A database of randomised trials and systematic reviews in physiotherapy. *Manual Therapy,* 5:223-226
[131] Silva L, Andreu JL, Munoz P, Pastrana M, Millan I, Sanz J, Barbadillo C, Fernandez-Castro M. (2008). Accuracy of physical examination in subacromial impingement syndrome. *Rheumatology,* 47:679-683
[132] Silverstein B, Welp E, Nelson N, Kalat J. (1998). Claims incidence of work-related disorders of the upper extremities: Washington State, 1987 through 1995. *Am J Public Health,* 88:1827-1833
[133] Silverstein BA, Bao SS, Fan ZJ, Howard N, Smith C, Spielholz P, Bonauto D, Viikari-Juntura E. (2008). Rotator cuff syndrome: personal, work-related psychosocial and physical load factors. *J Occup Environ Med,* 50:1062-76
[134] Sinnott KA, Milburn P, McNaughton H. (2000). Factors associated with thoracic spinal cord injury, lesion level and rotator cuff disorders. *Spinal Cord,* 38:748-753
[135] Smith KL, Harryman II DT, Antoniou J, Campbell B, Sigles JA, Matsen III FA. (2000). A prospective, multipractice study of shoulder function and health status in patients with documented rotator cuff tears. *J Shoulder Elbow Surg,* 9:395-402
[136] Solomon DH, Bates DW, Schaffer JL, Horsky J, Burdick E, Katz JN. (2001). Referrals for musculoskeletal disorders: patterns, predictors, and outcomes. *J Rheumatol,* 28:2090–5
[137] Svendsen SW, Bonde JP, Mathiassen SE, Stengaard-Pedersen K, Frich LH. (2004). Work related shoulder disorders: quantitative exposure-response relations with reference to arm posture. *Occup Environ Med,* 61:844-53
[138] Teefey SA, Hasan AS, Middleton WD, Patel M, Wright RW, Yamaguchi K. (2000). Ultrasonography of the rotator cuff: A comparison of ultrasonographic and arthroscopic findings in one hundred consecutive cases. *J Bone Joint Surg (Am),* 82:498-504
[139] Teys P, Bisset L, Vicenzino B. (2008). The initial effects of a Mulligan's mobilisation with movement technique on range of movement and pressure pain threshold in pain-limited shoulders. *Manual Therapy,* 13:37-42
[140] Thomas E, van der Windt, DAWM, Han EM, Smidt N, Dziedzic K, Bouter LM, Croft PR. (2005). Two pragmatic trials of treatment of shoulder disorders in primary care: generalisability, course and prognostic indicators. *Ann Rheum Dis,* 64:1056-61
[141] Thomas SJ, McDougall C, Brown ID, Jaberoo MC, Stearns A, Ashraf R, Fisher M, Kelly IG. (2007). Prevalence of symptoms and signs of shoulder problems in people with diabetes mellitus. *J Shoulder Elbow Surg,* 16(6):748-51

[142] Toyoda H, Ito Y, Tomo H, Nakao Y, Koike T, Takaoka K. (2005). Evaluation of rotator cuff tears with magnetic resonance arthrography. *Clin Orth Related Res,* 439:109-115

[143] Trampas A, Kitsios A. (2006). Exercise and manual therapy for the treatment of impingement syndrome of the shoulder: a systematic review. *Phys Ther Reviews,* 11:125-142

[144] Turner-Stokes L, Jackson D. (2002). Shoulder pain after stroke: a review of the evidence base to inform the development of an integrated care pathway. *Clin Rehabil,* 16:276-98

[145] Vad VB, Gebeh A, Dines D, Altchek D, Norris B. (2003). Hip and shoulder internal rotation range of motion deficits in professional tennis players. *J Sci Med Sport,* 6:71-75.

[146] Van den Dolder PA, Roberts DL. (2003). A trial into the effectiveness of soft tissue massage in the treatment of shoulder pain. *Aust J Physiother,* 49:183-8

[147] Van der Heijden GJ (1999). Shoulder disorders: a state-of-the-art review. Best Pract Res Clin Rheumatol, 13:287-309

[148] van der Heijden GJ, van der Windt, DA, de Winter AF. (1997). Physiotherapy for patients with soft tissue shoulder disorders: a systematic review of randomised clinical trials. *Br Med Journal,* 31:25-30

[149] van der Windt, Koes BW, de Jong BA, Bouter LM. (1995). Shoulder disorders in general practice: incidence, patient characteristics, and management. *Ann Rheum Dis,* 54:959-964

[150] van der Windt DA, Koes BW, Boeke AJ, Deville W, De Jong BA, Bouter LM. (1996). Shoulder disorders in general practice: prognostic indicators of outcomes. *Br J Gen Pract,* 46(410):519-23

[151] van der Windt DA, Koes BW, Deville W, Boeke AJP, de Jong BA, Bouter LM. (1998). Effectiveness of corticosteroid injections versus physiotherapy for treatment of painful stiff shoulder in primary care: randomised trial. *BMJ,* 317:1292-1296

[152] Van der Windt DAWM, Koes BW, Deville W, Boeke AJP, deJong BA, Bouter LM (1998) Corticosteroid injections versus physiotherapy for painful stiff shoulder in primary care: randomised trial. BMJ;317:1292-1296

[153] van der Windt DA, Thomas E, Pope DP, de Winter AF, Macfarlane GJ, Bouter LM, Silman AJ. (2000). Occupational risk factors for shoulder pain: a systematic review. *Occup Environ Med,* 57:433-42

[154] van der Windt DA, Bouter LM. (2003). Physiotherapy or corticosteroid injection for shoulder pain? *Ann Rheum Dis,* 62:385-387

[155] Vecchio PC, Kavanagh RT, Hazleman BL, King RH. (1995). Community survey of shoulder disorders in the elderly to asses the natural history and effects of treatment. *Ann Rheum Dis,* 54:152-154

[156] Vermuelen HM, Rozing PM, Oberman WR, le Cessie S, Vliet V, Thea PM. (2006). Comparison of high-grade and low- grade mobilization techniques in the management of adhesive capsulitis of the shoulder: randomized controlled trial. *Phys Ther,* 86:355--368

[157] Vicenzino B, Paungmali A, Teys P. (2007). Mulligan's mobilization-with-movement, positional faults and pain relief: current concept from a critical review of literature. *Manual tTherapy,* 12(2):98-108

[158] Viikari-Juntura E, Takala E, Riihimaki H, Martikainen R, Jappinen P. (2000). Predictive validity of symptoms and signs in the neck and shoulders. *J Clin Epidemiol,* 53:800–8

[159] Vogt MT, Simonsick EM, Harris TB, Nevitt MC, Kand JD, Rubin SM, Kritchevsky SB, Newman AB. (2003). Neck and shoulder pain in 70- to 79-year-old men and women: findings from the Health, Aging and Body Composition Study. *The Spine Journal,* 3:435-441

[160] Walker Bone K, Palmer KT, Reading I, Coggon D, Cooper C. (2004). Prevalence and impact of musculoskeletal disorders of the upper limb in the general population. Arthritis Rheum, 51:642-51

[161] Wang CJ, Ko JY, Chen HS. (2001). Treatment of calcifying tendonitis of the shoulder with shock wave therapy. *Clin Orthop,* 387):83–9

[162] Wang HK, Cochrane T. (2001). Mobility impairment, muscle imbalance, muscle weakness, scapular asymmetry and shoulder injury in elite volleyball athletes. *J Sports Med Phys Fitness,* 41:403-410

[163] Waring WP, Maynard FM. (1991). Shoulder pain in acute traumatic quadriplegia. *Paraplegia,* 29:37-42

[164] Weldon EJ, Richardson AB. (2001). Upper extremity overuse injuries in swimming. A discussion of swimmer's shoulder. *Clin Sports Med,* 20:423-438

[165] Winters JC, Groenier KH, Sobel JS, Arendzen HH, Meyboom-de-Jongh B. (1997). Classification of shoulder complaints in general practice by means of cluster analysis. *Arch Phys Med Rehab,* 78:1369-74

[166] Winters JC, Sobel JS, Groenier KH, Arendzen JH, Meyboom-de Jong B. (1997). The course of pain and the restriction of mobility in patients with shoulder complaints in general practice. *Rheumatology Inter,* 16:219-25

[167] Winters JC, Sobel JS, Groenier KH, Arendzen JH, Mayboom-de Jong B. (1997). Comparison of physiotherapy, manipulation, and corticosteroid injection for treating shoulder complaints in general practice: randomised, single blind study. *BMJ,* 314(7090):1320-1325

[168] Winters JC, Jorritsma W, Groenier KH, Sobel JS, Meyboom-de Jong B, Arendzen HJ. (1999). Treatment of shoulder complaints in general practice: long term results of a randomised, single blind study comparing physiotherapy, manipulation, and corticosteroid injection. *BMJ,* 318(7195):1395-1396

[169] Winters JC, Sobel JS, Groenier KH, et al. (1997). Comparison of physiotherapy, manipulation and corticosteroid injection for treating shoulder complaints in general practice. *J Fam Pract (Journal Club),* 45(2):103-4

[170] Winters JC, Sobel JS, Groenier KH, Arendzen JH, Mayboom-de Jong B. (1999). The long-term course of shoulder complaints: a prospective study in general practice. *Rheumatology,* 38:160-3

[171] Yamaguchi K, Tetro M, Blam O, Evanoff BA, Teefey SA, Middleton WD. (2001). Natural history of asymptomatic rotator cuff tears: A longitudinal analysis of asymptomatic tears detected sonographically. *J Shoulder Elbow Surg,* 10(3):199-297

[172] Yamaguchi K, Konstantinos D, Middleton WD, Hildebolt CF, Galatz LM, Teefey SA. (2006). The demographic and morphological features of rotator dcuff disease. A comparison of symptomatic and symptomatic shoulders. *J Bone Joint Surg (Am),* 88:1699-1704

[173] Yang JL, Chang CW, Chen SY, Wang SF, Lin JJ. (2007). Mobilisation techniques in subjects with frozen shoulder syndrome: randomised multiple-treatment trial. *Phys Ther,* 87:1307-15

[174] Yiasemides R, Halaki M, Cathers I, Ginn K. (2011). Does passive mobilization of shoulder region joints provide additional benefit over advice and exercise alone for people who have shoulder pain and minimal movement restriction? A randomized controlled trial *Phy Ther,* 91(2):1-11

[175] Zheng, XH, Simpson JA, van der Windt, DAWM, Elliot AM. (2005). Data from a study of effectiveness suggested potential prognostic factors related to the patterns of shoulder pain. *J Clin Epidemiol,* 58(8):823-30

# Part 3

# Use of Cryotherapy and Thermotherapy in Physical Therapy

7

# Cryotherapy: Physiological Considerations and Applications to Physical Therapy

Anna Lubkowska
*Department of Physiology, Faculty of Biology, Szczecin University,*
*Chair and Department of Biochemistry and Medical Chemistry,*
*Pomeranian Medical University,*
*Poland*

## 1. Introduction

Contemporary cryogenics has been developing from the end of the 19th century since the liquefaction of oxygen, nitrogen, carbon dioxide and hydrogen, and also the industrial production and storage of liquid coolants, enabled the development of cryobiology and the use of extremely low temperatures in medicine. The term "cryotherapy" was first used in 1908 by A.W. Pusey to describe the treatment of skin lesions with very low temperatures [Zagrobelny et al. 1999; Jezierski 2006;]. Currently, cryotherapy refers to various treatments aimed at lowering the body surface temperature without tissue destruction, whereas in cryosurgery diseased tissues are destroyed through freezing. The world's first cryogenic-temperature chamber was set up in Japan, in 1978. Thanks to Yamauchi and his team, cryotherapy began to be widely used in medicine [Zagrobelny 2003; Skrzek 2009].

The clinical application of low temperatures is recommended for inflammatory conditions, such as swelling and acute localised pain. It is essential to clarify the distinction and clear labelling of methods based on low temperature usage and the intended purpose, as the body's response to low temperature depends on the temperature, method of application, exposure time, method and rate of heat loss, humidity of the cooled air, and the characteristics and age of the subjects.

*Cryostimulation* requires making use of a suitable croyogenic liquid as a coolant source. It is based on the use of very low temperatures (-100°C and lower) in order to induce a physiological reaction to cold.

A different category of treatment in modern medicine and rehabilitation, the *cooling of tissues*, is based on different methods and rates of tissue heat loss using various temperatures and methods of application achieved by the use of bags of ice, frozen silicone gel, salt solution, wet cold (which is not tolerated well by many people), partial bathing in cold water or whole body bath (temperature below 10°C). Cooling results in heat loss, the treated body area experiences vasoconstriction, a long-duration decrease in the temperature of the exposed tissues, decrease in inflammatory reaction and inhibition of strong symptoms of inflammatory reaction, but without any stimulatory effect [Rawecka & Rokita, 2006].

Depending on the size of the body area to be treated with cryogenic temperatures, the following types of cryotherapy may be distinguished:

1. *Local cryotherapy* (involving a small body area): ventilation with nitrogen vapours (gas temperature at the outlet of the nozzle can range from -160° C to -196° C), ventilation with a mixture of nitrogen vapours and cooled air (temperature of the gas mixture varies from -100° C to -178° C), and ventilation with cold air (temperature ranges from -30° C to -34° C).
2. *Whole-body cryotherapy* (involving the entire body, including the head): cryosauna, two-stage chambers, cryochamber with cool retention effect; chambers cooled with a mixture of liquid nitrogen vapour and air, or with a compression cooling system. During low-temperature treatment in a chamber, the cold air cools the whole body.

After the first cryochamber in Japan (1978), another one was created in Germany in 1982, and the third in Wrocław, Poland, in 1989. At present a few dozen cryochambers may be found around the world. A growing number of centres conduct scientific research on the effects of whole body cryostimulation on the human body, yet knowledge on the subject is still lacking. While whole-body cryostimulation (WBC) is not harmful or detrimental to healthy subjects, the effects are being studied in relation to: 1. Motor activity and physical efficiency; 2. Cardiovascular response; 3. Lipid profile; 4. Hematology; 5. Hormones; 6. Antioxidant Defense System; 7. Immunology and inflammation 8. Recovery from exercise-induced muscle damage.

## 2. Cryogenic liquids

Cryogens (cryogenic liquids) are those liquids with a boiling point not exceeding 120 K under normal atmospheric pressure. The following are various types of cryogens and respective temperature ranges:

1. wet cold - water with ice, 0 ° C
2. damp cold - cold air, temperature from -15 ° C to -30 ° C
3. dry cold- vapours of liquefied gases:
   a. liquid nitrogen vapour: -196 °C
   b. carbon dioxide vapour: -75 °C

The most rapid effect can be achieved with water (wet cold). Subjecting a patient to a cold bath, however, is unpleasant and can cause shock, and therefore is used only in selected cases. Cryostimulation using dry cold (vapours of liquefied gases: nitrogen, air, carbon dioxide) is also quick, convenient and much more patient-friendly [Bojek 2006; Rawecka & Rokita 2006].

Cryogenic liquids, vapours and cold gases used in the wrong way are able to cause serious damage related to frostbite. However, liquid nitrogen in short contact is not harmful, because at the time of contact with warm tissue, it evaporates intensely. It is important to avoid contact with the eyeball as it may cause serious and irreversible damage to the cornea. The danger with these substances is in their low temperature and high density. They are kept in specially insulated containers that are designed to withstand rapid temperature changes.

## 3. Cryostimulation procedures

Whole-body cryotherapy is often used as preparation of the patient to physiotherapy. However, cryotherapy is also increasingly often performed for the purposes of biological regeneration and prevention of musculoskeletal overload and injury (athletes), and as a way of increasing immunity. In some cases with symptoms of inflammation (acute and chronic musculoskeletal injuries), the purpose of cryostimulation is to decrease the temperature of the inflamed tissue as cold is known to limit inflammation symptoms. This protects against oedema of periarticular tissues and any further damage caused by pressure and movement of tissues. In such cases very good results are obtained by local cryotherapy.

### 3.1 Local cryostimulation

Local cryotherapy treatment uses usually the following cryogens: liquid nitrogen, carbon dioxide, cooled air [Sieroń & Cieslar 2003]. Prior to cryostimulation, the patient should thoroughly dry the body surface to be subjected to treatment. It is most desirable to keep the patient in a standing position or if not possible, in a sitting or lying position. Procedure duration is defined individually for each patient, depending also on the body structure, muscle mass and fat thickness. Cryotherapy is not used for more than about 5 joints at the same time which should not be longer than 12-15 minutes. The hand, foot and spine are considered as a single set of small joints [Księżopolska - Pietrzak 1998].

The treatment time takes usually from 30 sec to several minutes (3-4 min), depending on the patient's skin response. The treatment may be repeated twice a day but with a break of at least 6 hours. The recommended duration of the break is determined by the sustained stimulatory effect after a single session. In children, treatment should be used very cautiously and of a shorter duration than adults due to the higher sensitivity of the younger skin and usually lower amount of subcutaneous fat. The distance of a cryoapplicator from the surface of the body depends on the quality of the device, and the temperature obtained at the nozzle exit, usually about 10 cm from the body surface (in older devices, the distance can be up to 15 cm). The treatment area of the body should be progressively swept over to avoid cooling of just one spot. Initially the patients experience a burning sensation. The patient must stay in communication with the physiotherapist and inform him about the pain or the burning sensation. After reported discomfort the applicator should be positioned further from the tissue, or the movement of the nozzle should be sped up. Rapid skin blanching is an indicator for immediate cessation of the treatment [Jezierski 2006].

During cryotherapy, one must remember about the bi-phasic vascular response to extremely low temperatures, and bear in mind the desired effect. Initially, in micro-circulation, a constriction of vessels as a result of closing of pre-capillary sphincter and activation of arteriovenous anastomoses, which takes from a few to a ten or so seconds, after which the vessels dilate and arteriovenous anastomoses close, during which blood flow increases. A lower temperature of the coolant (liquid nitrogen vapour) results in a stronger and faster first stage, i.e. vasoconstriction causing tissue ischemia. In this way one can reduce the effects of trauma, reduce the rate of metabolic processes (which may be helpful for example in the physical therapy of arthritis), while hypoxia reduces perceived pain. In the second phase, local hyperaemia occurs (increase in local blood flow due to the effect of internal and external factors, known as the rebound effect), during which metabolic processes are intensified and muscle tension is reduced.

Treatment duration and temperature are diverse and depend on whether the goal is mainly to generate secondary hyperaemia and reduce muscle tension, e.g. in preparing the patient for physiotherapy or massage, or whether the first phase is the primary goal, slowing metabolism and reducing the effects of trauma. Cryotherapy is often used as an analgesic agent, and this effect can last for several hours. The manner of cryostimulation is also influenced by the patient's age, physical condition, sensitivity to low temperature, or existing disease and its severity [Kasprzak & Markowska 2008].

In some cases, excessive hyperaemia may worsen pain in the long-term (with a strong inflammatory component). In this case cold therapy instead of cryotherapy is recommended.

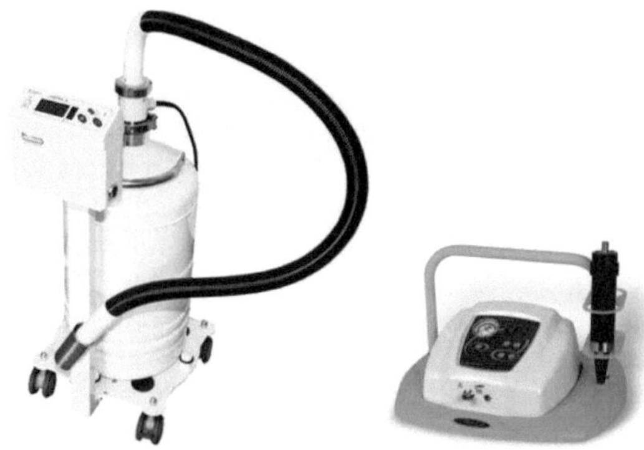

Fig. 1. Examples of stationary and portable cryotherapy apparatus

### 3.1.1 Diseases in which local cryotherapy is used

Patients with locomotor disorders are now the largest group of patients eligible for cryotherapy. An efficient muscle-joint and skeletal system is fundamental for the functioning of the human body. The inability to move affects an individual both socially as well as psycho-socially. As musculoskeletal disorders affect people of all ages and from all social groups, researchers worldwide have been looking for the most beneficial methods of treatment. Cryotherapy has gained the greatest recognition as supporting treatment of the motor system [Bieńkowska et al. 2006; Hopkins 2006; Łuczak & Michalik 2006].

**Ankylosing spondylitis**

In ankylosing spondylitis pain forces patients to reduce their motor activity. The progress of the disease and reduced physical activity leads to a reduction in joint mobility, periarticular ossification and degenerative changes. The spine assumes deep thoracic kyphosis, deep compensatory cervical lordosis, thoracic stiffness and stiffness in the peripheral joints affected by inflammation. The advancing total thoracic rigidity causes a situation where lung ventilation only takes place through diaphragm movement. Kinesitherapy and cryotherapy are currently the preferred treatments for ankylosing spondylitis. Cryostimulation decreases the concentration of acute phase proteins and immunoglobulins in serum and increases the concentration of β-endorphin.

The cryotherapy-induced reduction of inflammation and the higher threshold of pain perception allows the use of physiotherapy to a greater degree. The patient benefits from a reduced need for analgesics and anti-inflammatory drugs. Local cryostimulation should be followed by kinesitherapy based on an individual program, allowing for the period of the disease (acute or remission), the severity of the disease, its extensiveness, the patient's age and the incidence of other diseases. In the acute phase of the disease, cryostimulation plays an important role in reducing pain as well as inflammation, allowing for sufficient activity of the patient during exercises. During remission, the most important aim is to gradually regain motor functionality [Sieroń & Cieslar 2003; Jezierski 2007].

**Rheumatoid arthritis (RA)**

RA is a chronic immunologically dependent connective tissue disease that is characterised by non-specific, symmetric arthritis, sometimes accompanied by non-joint changes and systemic complications. The course of RA is characterised by alternating periods of joint inflammation and remissions. This disease can occur in three forms: mild, medium and acute. Diagnosing the disease is synonymous with pharmacological treatment that should be supported by physical therapy. Cryostimulation has beneficial effects in RA - it increases active muscle power and lowers passive muscle power. Pain and increased oedema of joints is reduced, resulting in increased range of motion in the inflamed joints. A single cryostimulation of a specific part of the body takes about 3 minutes and has both analgesic and anti-inflammatory effects. Therefore cryostimulation should be immediately followed by physical exercise [Sieroń & Cieslar 2003; Straburzyńska - Lupa et al. 2005; Krawczyk - Wasilewska et al. 2007; Jezierski 2007].

**Osteoarthritis (OA)**

Symptoms of osteoarthritis include morphological, biochemical, molecular and biochemical changes in cells of the cartilage, which lead to softening, fibromatosis, ulcers and weight loss of articular cartilage, sclerosis and thickening of bone, osteophytes and subchondral cysts. Patients with hip degeneration experience destruction of cartilage and subchondral bone, and pain in the hip. Local cryostimulation aims at maximum cooling of the hip. Cold should be applied from the front and medially to the groin and the space between the pubic symphysis and thigh adductor and from the back to the gluteal fold and ischial tuberosity. In addition, cryostimulation should involve the entire thigh, gluteus maximus and iliopsoas. The duration of treatment ranges from 3 to 6 minutes. In degeneration of the knee, cryostimulation is performed on the knee joint and thigh muscles. Cooling of the joint is done best in a sitting position with a bent knee (cold should be first applied on sides, then around the knee cap). Osteoarthritis may involve the joints of the spine or vertebral stem. Cryostimulation is performed on a relevant segment of the spine area and muscles situated in the area. The procedure is followed by exercises intended to strengthen back muscles and abdominal muscles, stretching and increasing the mobility of the spine exercises, and correction exercises [Sieroń & Cieslar 2003; Jezierski 2008; Pasek et al. 2009].

**Shoulder impingement syndrome**

Treatment of shoulder impingement syndrome is not straightforward. Pharmacological methods are often accompanied by physiotherapy and anti-inflammatory analgesic electrotherapy, ultrasound, massage, kinetic therapy and increasingly popular cryotherapy. The aim of physiotherapy treatment is improvement in the functional efficiency of the limb through the elimination or reduction of pain. Literature reports describe both cold therapy, resulting in a slowdown of inflammatory reactions and reduced metabolism in cooled tissues, and cryostimulation which aims at improving conditions in blood circulation in a given area and to exert a strong analgesic effect. Cryostimulation for patients with shoulder impingement syndrome increases the range of motion in the transverse plane, bending and movement of external rotation, and a slight effect on straightening and bending in the sagittal plane, abduction and internal rotation [Sieroń & Cieslar 2003; Lisinski et al. 2005; Boerner et al. 2007].

**Gouty diathesis**

This inflammation of joints is caused by crystallisation of sodium urate in synovial fluid and crystal deposition in tissues. Another form of this disease is chondrocalcinosis caused by calcium pyrophosphate dihydrate crystallisation in synovial fluid and crystal deposition in articular cartilage. In patients with acute gout, cryostimulation can be applied only after the attacks. Local cryotherapy causes a short-term reduction in the intensity of local inflammation, but does not stop an attack of gout. Kinesitherapy is introduced gradually and with caution, due to the persistent deposits of sharp crystals in cartilage and other connective tissues [Sieroń & Cieslar 2003; Jezierski 2008].

**Osteoporosis**

This is a bone disease with a gradual decrease in bone mass with muscles and ligaments losing elasticity, and resistance to physical effort. The decline in mechanical bone strength is accompanied by degenerative changes. The disease affects the neuromuscular system and motor pattern, and results in the painful overload of muscles, ligaments, bones and joints. Treatment with cold, as in previous cases, leads to a reduction in pain, relaxes skeletal muscles and improves their strength, and increases range of motion within the treated joints [Księżopolska - Pietrzak 1998].

**Spinal pain syndromes**

These syndromes represent a serious medical problem as they affect between 60% to 90% of the European population. The lumbosacral spine is most affected, due to the sedentary lifestyle. Chronic pain leads to lower psychophysical performance. Cryotherapy decreases muscle tension and reduces the speed of nerve conduction, which are beneficial in relieving pain and allows further treatment with physiotherapy [Woźny et al. 2006].

### 3.2 Whole-body cryotherapy

This cryotherapy involves rapid cooling of the whole body at extremely low temperatures (cryogenic, below -100°C) to trigger physiological responses to cold. Such cryotherapy is performed in a cryogenic chamber, ensuring appropriate treatment conditions - temperature and humidity. While in the chamber, the patient must have regular contact with the physiotherapist, usually through glass doors and speakers. A special alarm device allows the patient to call for help in an emergency, and the door can be easily opened by the person inside. The inside of the chamber is lined with a material that does not absorb moisture. The walls do not have any sharp protruding metal parts to avoid contact with the patient, or to cause very unpleasant feelings [Biały et al. 1998; Adamowicz 2005; Brojek & Warzocha 2006].

#### 3.2.1 Types of cryogenic chambers

There are many types of chambers available in the market, and the choice depends on the design of the chamber, one room or two, the number of persons undergoing treatment at the same time, type of coolant. The choice of chamber is often also related to economic considerations.

**Two-stage cryogenic chamber**

This type of chamber consists of one or two vestibules where the temperature is approximately -60°C, and a main chamber in which the temperature is maintained within the range -110°C to -160°C (depending on the coolant). In order to maintain such a sufficiently low temperature, liquid nitrogen is pumped directly into heat exchangers which consist of 3 systems: cooling, air preparation and control. Maintaining the temperature is possible by heat exchangers installed in the vestibule and the chamber proper. Compressor, filters and dryer constitute the system of breathing air preparation. By this arrangement it is possible to introduce purified dry air, reducing the feeling of cold, and ensuring transparency in the chamber. The control system consists of a power supply and control cabinet, controller and computer, providing the desired data parameters. Extra security is provided by a glazed door, opening under the pressure of the body and an alarm button. A chamber of this type may be used by 5 people at one time [Cholewka & Drzazga 2005].

**Cryochamber with cool retention effect**

This chamber is cooled with a mixture of liquid nitrogen and liquid oxygen in the proportions of atmospheric air. The air is pumped from the external tank and sprayed by specially prepared nozzles. This design is based on the phenomenon of cold retention in a basin, which reduces the importance of thermal insulation and elimination of the vestibule, reducing the cost of materials. Three to four people may stay in this cryochamber at the same time, although there is restricted access to wheelchair users. This type of a cryochamber may be without a vestibule, or with one or two vestibules [Cholewka & Drzazga 2005].

**Cryobarrel (cryosauna)**

This is a cabin for one person. The patient is immersed in the cold vapours up to the arms, breathing air from the room with the head is located outside the reach of extremely low temperatures. It has a swinging door which opens under the pressure of the body, and contact with the patient is very good. Liquid nitrogen or liquid air is used to generate the coolant. The cryosauna has many advantage: low cost of use, good contact with the patient, easy assembly. It is contentious whether omission of the head is an effective treatment (with the head having a significant number of cold receptors), but it allows the use of cryostimulation by a person suffering from claustrophobia [Księżopolska-Pietrzak 1998].

**Cryochamber with a compressor cooling system**

This chamber with a compressor system uses air as a coolant. The chamber consists of two vestibules with a temperature of -10°C and -60°C, and the chamber proper with a temperature of about -100°C. It can be mounted either indoors or outdoors (the chamber is isolated from the ground). Suitable temperatures in the inside are provided by three compressors, and the cooling system uses only non-combustible and non-toxic gases. This type of chamber has all the elements necessary for the safe performance of cryostimulation - additional emergency exit, visual and audio contact with the operator. All parameters are monitored by a computer. The chamber can be used in group therapy [Migaj 2005].

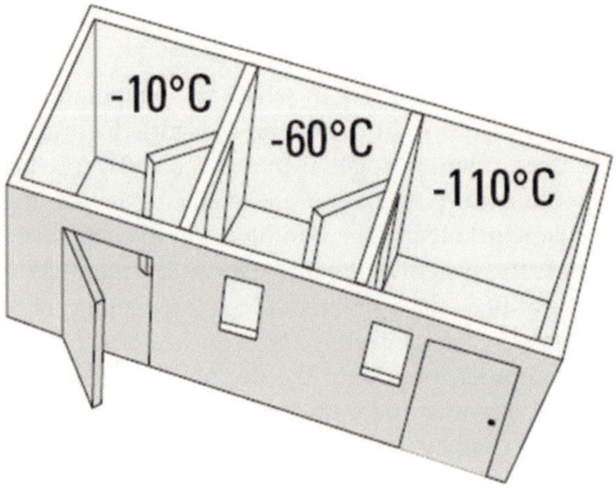


Fig. 2. The schema of crychamber with two vestibules (Zimmer MedizinSysteme GmbH, Germany)

Fig. 3. Two stage cryochamber (CREATOR Sp. z o.o., Poland)

Fig. 4. Two stage cryochamber (KRIOSYSTEM Sp. z o.o. Poland)

### 3.2.2 Preparation of patients for whole-body cryostimulation

Patients who are eligible for treatment in a cryogenic chamber must undergo medical examination prior to treatment. An interview is carried out with each patient in order to detect any contraindications to cryostimulation, including questions relating to cold tolerance of the body. In all of the candidates for cryostimulation, blood pressure and heart rate are measured [Zagrobelny 2003].

In practice, whole-body cryostimulation is not permitted for patients with cardiac disease or unstable blood pressure. In Finland, cryostimulation is recommended only for people with a blood pressure below 160/100 mmHg [Westerlund et al. 2004] and in Poland usually below 150/90 mmHg [Lubkowska & Szygula 2010; Lubkowska & Suska 2011].

Bathing before cryostimulation is not recommended, and prior to entering the cryogenic chamber the patient should be wiped completely with a towel to remove sweat - drops of sweat left on the skin could turn into ice crystals, which causes a painful and unpleasant sensation of cold. Additionally, all the patients are informed of the necessity of slow and shallow breathing while in the cryochamber. Clothes during the treatment should cover those areas of the body which are most vulnerable to frostbite. These are primarily the hands, feet, outer ears and popliteal pits.

Taking into account protection of the external ear against frostbite, it is recommended to use earmuffs or a headband instead of a hat. Hands should be protected by gloves, which should be turned inside out to dry after use in the chamber. Undried gloves may cause frostbite in the fingers during subsequent treatments. It should also be remembered that a surgical mask lined with a layer of swabs is put on the face just before entering the chamber, as waiting in this mask may cause the accumulation of moisture in the mask [Brojek &

Warzocha 2006]. The mask is necessary for the protection of the lungs - if the extremely cold air reaches lung tissues in greater amounts, then after heating in the lungs its volume may even double which may result in respiratory oppression [Zagrobelny 2003].

The distribution of temperatures in the cryochamber results in the coldest air being located in the lowest area of the device, therefore clogs are the most suitable shoes during cryostimulation. A thick sole is good isolation of the feet from the ground. Socks should be short and protect only the area of the ankles and feet. The popliteal pit is very vulnerable to surface frostbite, as it has a tendency for intense sweating which may be prevented by instructing the patient to avoid sitting or putting a leg on a leg. It is best to wait for the treatment in the standing position, and wipe the area of the popliteal pit and put a knee pad on immediately before entering the chamber [Brojek & Warzocha 2006].

Clothes used during cryostimulation should be made solely from natural components, such as wool or cotton, and should ensure full psychological comfort. It is best if pants are short and tight. Loose pants, such as boxers results in unpleasant rubbing of the frozen material against the body. Wearing any type of shirt is pointless as it inhibits the process of heat exchange during cryostimulation [Raczkowski & 2007].

### 3.2.3 Procedure of whole body cryostimulation

Duration and temperature of cryostimulation are established for each patient individually before each entry to the cryogenic chamber. Usually, when the patient enters the chamber for the first time, the treatment starts with a 1 minute exposure, gradually extending in duration for subsequent sessions to 2.5 - 3 minutes. Prepared patients are let into the vestibule by the operator where short-term adaptation to the low temperature occurs. The participants then move into the chamber proper.

Time of stay in the vestibule is not fixed, but usually it is associated with the time required to close the external door and open the door to the chamber proper. It is thus short, about 30 secs. During the entire session, the intensity of movement is limited to only a slow walk in order to eliminated the effect of skeletal muscle activity on the dynamics of cooling and heating of the body. In the middle of the session, the direction of movement is reversed. Contact with other people and the walls of the chamber should be avoided. During the entire stay in the chamber proper, patients should remain in eye and voice contact with the operators outside the cryochamber.

The staff at the site should include the physician that qualified the patients to the treatment, and a physician that could react in an emergency. After leaving the cryochamber proper, patients move to a place at room temperature. Usually kinesitherapy is applied after cryostimulation, usually for 20 to 30 minutes. It sometimes happens that after leaving the chamber patients do not perform any exercises. Undesirable effects that may occur after the session are shallow and quickly receding frostbites, usually in the legs [Suszko 2003; Zagrobelny 2003].

## 4. Indication and contraindication for cryostimulation

In clinical practice cryostimulation is used primarily in sports medicine. Professional athletes and healthy patients use cryostimulation for athletic recovery associated with a desirable effect for defensive and protective mechanisms in the human body.

In the case of ill patients, cryotherapy is used in the following diseases:

- inflammatory musculoskeletal disease: rheumatoid arthritis, ankylosing spondylitis, rheumatic fever
- degenerative diseases and secondary degenerative changes in peripheral joints and the spine
- joint disease of metabolic origin such as gout
- mixed connective tissue disease
- rheumatic and soft tissue disorders (polymyositis and dermatomyositis)
- periarticular, tendon and joint capsule inflammation
- some skin diseases involving joints: psoriatic arthritis
- autoimmune diseases
- post-traumatic changes or overload of joints and soft tissues
- chronic inflammation of the cervical spine
- discopathies
- osteoporosis
- muscle overload
- neurological disorders (spastic paresis, multiple sclerosis, radicular neuralgia)
- depression syndromes, vegetative neurosis

The treatments in cryochambers are usually well tolerated by patients and complications related with the treatment are very rare. However, one must not neglect states in which the effect of low temperatures can be adverse for the patient. In qualifying patients for cryochamber treatment, one should pay attention to the following:

- patient's age
- the existing diseases
- nutrition
- performance of vascular vessels
- duration of cold and its intensity
- intaken drugs
- consumption of alcoholic beverages
- individual characteristics of response to cold

All these factors determine the suitability of using cryogenic treatment or can disqualify patients. Additionally, qualification to cryostimulation should be based on the results of basic spirometric and cardiological tests.

The following contraindications are known for the application of cryogenic treatment:

absolute contraindications to cryostimulation

- cold intolerance
- cryoglobulinaemia
- cryofibrinogenemia
- Raynaud disease
- cold utricaria
- open wounds and ulcers
- gangrenous lesions
- thromboembolic changes and inflammation in the venous system

- agammaglobulinemia
- central nervous system disease
- sympathetic neuropathy
- hypothyroidism
- local blood flow disturbances
- significant anaemia
- claustrophobia
- mental disorders, which may prevent adequate co-operation with the patient in the chamber
- emaciation and hypothermia
- cancer
- Printzmetal syndrome
- unstable coronary artery disease
- valve disorders: aortic valve stenosis and mitral valve stenosis
- diseases of the myocardium or the mitral apparatus
- arrhythmia at a rate higher than 100 strokes per minute
- severe forms of exertional angina pectoris and spontaneous angina
- venous blood leaks in the lungs
- acute respiratory diseases of various origin
- effects of drugs, especially antipsychotics and alcohol.

Some relative contraindications also exist concerning participation in the cryogenic treatment. These include the following:

- age above 65
- venous thrombosis and a history of peripheral arterial embolism
- excessive emotional lability, expressed, inter alia, in excessive excitability

Currently whole-body cryostimulation is increasingly popular among older people, also those after 65. The age is not a serious contraindication, if other more important contraindications do not occur [Jagodziński et al. 2001; Zagrobelny 2003; Sieroń et al. 2007].

## 5. The effect of cryogenic temperatures on the human body

Exposure of the human body to cryogenic temperatures results in many favourable physiological phenomena [Jagodzinski et al. 2001; Kiljański et al 2005; Sieroń et al. 2007] such as:

- overall improvement in well-being (relaxation, physical relaxation)
- analgesic effect
- neuromuscular effect (increase of muscle strength)
- profuse flow of blood
- increased systemic immunity
- increase in serum beta endorphins, norepinephrine, adrenaline, testosterone (especially in men)
- antioxidant effect of cryostimulation

The analgesic effect of cryostimulation is the most appreciated by practitioners. Hypothermia is related to a reduction in conduction velocity in sensory fibres and in impulsation in slowly conducted C fibres disabling sensory receptors and their connections with proprioreceptors, β-endorphin secretion [Leppauloto et al. 2008] and the selection of stimuli reaching the central nervous system ( the "control gate" mechanisms) [Gregorowicz et al. 1998; Sieroń et al. 2003].

Sensory fibres experience slower conduction and blocking of afferent hyperalgesia impulsation to the spinal cord. These mechanisms raise the pain threshold. At the same time, it has been found that response to pain involves almost all the glands of internal secretion, with a special role of endorphins in the chain of endocrine responses to pain [Suszko 2003].

The impact on the psyche is a very positive effect of cryostimulation. Through the activity of the central nervous system, patients who underwent cryostimulation sessions in a cryochamber, felt lower fatigue and a radical improvement in mood. Other observed effects include elimination of clinical depression syndromes, easier sleep and improvements in sleep quality [Zagrobelny & Zimmer 1999; Rymaszewska et al. 2007; Sieroń et al. 2007].

Exposure to extremely low temperatures also reduces the intensity of nerve conduction and reduces the reactivity of peripheral sensory-motor endings, which in turn reduces muscle tension. Cryogenic temperatures increase muscle power through the favourable release of motor impulses in the fibres and increased number of motor units involved in work in different muscle groups.

Here are present the main directions of research and their results in terms of the effects of whole-body cryostimulation on the body.

**Motor activities and physical efficiency**

Research by Łuczak, carried out on a large group of athletes (300 people), was meant to find an optimal operating temperature during the whole-body cryostimulation for the improvement of motor skills. The effects of exposure were compared between two 10 minute whole-body cryostimulations (at -100°C, -130°C and -160°C) and physical efficiency, based on an assessment of agility, balance, speed and dynamic strength of abdominal muscles. The analysis of the results showed no effect of cryostimulation on the level of agility. Balance improved significantly in groups exposed to temperatures below -100°C, while no significant effects were observed for -100°C. Parameters evaluating speed and dynamic strength of abdominal muscles improved most after the application of -100°C. It was proposed that whole body cryotherapy exerts positive effects on human motor characteristics, although the lowest cryogenic temperatures should be used in only specific cases [Łuczak et al. 2006]. Subsequent studies showed that a series of 20 stimulations with an average temperature -130°C performed on martial arts competitors, resulted in an extended duration of exercise and lower subjective feeling of fatigue at increasing mean speed and angle of treadmill inclination during an exercise according to the Bruce protocol [Hagner et al. 2009]. A recent study on the effects of whole-body cryostimulation on aerobic and anaerobic capacities showed that three 10 minute sessions (average temperature -130°C) increased maximal anaerobic power in males but not in females, and did not influence aerobic capacity in either gender [Klimek et al. 2011]. There are also reports of improved exercise tolerance, expressed by a lower level of lactates, heart rate and increased threshold capacity during a rowing ergometer test by Olympic team athletes (rowers) after 23 cryostimulation sessions (3-minutes at a temperature of -150°C, 2 x day) [Chwalbińska-Moneta 2003].

**Cardiovascular response**

It is known that cold exposure is a risk factor for hypertension. In physiotherapeutic practice, it is standard procedure to test participants before cryostimulation where a blood pressure control is measured, but contraindications to the use of cryotherapy or whole-body cryostimulation do not include unstable blood pressure or hypertension. Literature data on changes in key cardiovascular indicators in humans exposed to cryogenic temperatures are ambiguous. Some of them report a significant but short-term increase in systolic and diastolic blood pressure after WBC in both normotensive and mildly hypertensive individuals [Westerlund et al. 2004; Fricke 1989; Taghawinejad et al. 1989]. Similarly, Komulainen et al. [2004] observed a rapid increase in blood pressure in mildly hypertensive subjects exposed to -15°C. Other authors reported that thermal stress (-110°C) did not cause changes in systolic or diastolic blood pressure but only a decrease in the heart rate [Zalewski 2009].

In our experiments we observed no changes in blood pressure or heart rate influenced by the potential stress resulting from the planned participation in cryostimulation. However, we observed a statistically significant increase in systolic ($p \leq 0.001$) and diastolic ($p \leq 0.05$) blood pressure immediately after a 3 minute long whole-body cryostimulation. These changes did not last more than 10 minutes after the cryostimulation when the values returned to initial levels [Lubkowska & Suska 2011]. In another study, in which we used 15 daily cryostimulations (-130°C/3min), the average increases in SBP and DBP on the first day were 20 mmHg and 6 mmHg. All the observed changes in the circulatory system subsided after 10 minutes of resting in a sitting position. Changes in blood pressure were accompanied by a decrease in heart rate by 8±4 bpm. The mentioned changes were not different on the 1st, 5th, 10th and 15th days of the experiment and it can therefore be argued that no adaptation changes occurred in response to the repeated stress associated with the cold [Lubkowska & Szyguła2010].

**Lipid profiles**

There are only a few reports in literature on the influence of whole-body cryostimulation on lipid levels in blood serum. The first information comes from experimental animal models [Skrzep-Poloczek et al. 2002]. Our last report concerned the results of lipid profiles in response to different procedures of whole-body cryostimulation in a cryochamber (-130 °C) for five, ten and twenty sessions. We ascertained that five sessions of whole-body cryostimulation in a cryogenic chamber did not change the lipid profile; while in a group subjected to a series of 10 sessions, the level of TG values statistically significantly decreased, and the changes were more pronounced in the group subjected to 20 sessions: a significant reduction in LDL, reduction in total cholesterol, while a significant increase was observed for HDL fraction. Comparing the ratios of individual lipid fractions, a statistically significant decrease in the TG fraction was observed in relation to total cholesterol, while the HDL fraction increased in comparison with total cholesterol and LDL cholesterol after 20 cryostimulations. No changes in the proportions between lipid fractions were observed after 5 and 10 cryostimulations [Lubkowska et al. 2010]

### Hematology

Literature on changes in haematological indices induced by cryostimulation are often inconsistent and insufficient. Blatteis [1998] reported a decrease in leukocytes and erythrocytes in healthy subjects after a series of cryostimulations. No significant increase in leukocytes was reported by Stanek et al. [2006], although they also observed a significantly increased percentage of monocytes in healthy individuals after a series of 10 two-minute long cryostimulations at -120ºC. Similarly, in our studies the increase concerned the number of lymphocytes and monocytes, and to a lesser extent, neutrophils and eosinophiles [Lubkowska et al. 2009]. With regards to information on the effect of cryotherapy or cryostimulation on the red blood cell system, data is still very scarce. The only available paper is Banfi et al. [2008] which investigated the effects of 5 session of whole-body cryotherapy treatment on haematological values in athletes. The paper suggested that cryotherapy does not have detrimental effects on the mentioned parameters, but a small significant decrease in haemoglobin concentration, mean corpuscular haemoglobin and mean reticulocyte volume were observed.

In terms of the participation of white and red blood cells in effort ability and tolerance, further studies are required on potential changes in haematological indices in response to varied number of cryostimulation in a series.

### Hormones

A single cryostimulation at -130°C causes increased concentration of a proopiomelanocortin-related hormone (ACTH adrenocorticotropic hormone), β-endorphin, adrenaline and noradrenaline in men and women, and a significant increase in testosterone in men [Zagrobelny 1993]. Soccer players undertaking ten sessions of cryostimulation followed by 60 minutes of kinesitherapy, had a significant decrease in the concentration of testosterone and estradiol. There were no changes in the level of luteinizing hormone (LH) and dehydroepiandrosterone sulphate [Korzonek-Szlacheta 2007]. Reports of changes in the level of cortisol (defined as the stress hormone) are divergent. Smolander et al.[2009] concluded that WBC treatments (-110°C), for 2 min, three times a week for 12 weeks, do not lead to disorders related to secretions of the growth hormone, prolactin, thyrotropin or thyroid hormones in healthy females.

Literature data indicate that in elderly women, cryotherapy at temperatures of -110ºC to -150ºC influence the level of bone conversion markers (a decline in osteocalcin, increased alpha type I collagen in blood serum) [Skrzek et al. 2003], and that rheumatoid arthritis patients were reported to have a decreased histamine level [Wojtecka-Łukasik et al. 2010].

### Antioxidant defence system

Even one session of whole-body cryostimulation causes changes in the prooxidant-antioxidant balance - the level of total oxidative status in plasma was statistically significantly decreased at 30 minutes after leaving the cryochamber and remained lower the following day, whereas the level of total antioxidative status decreased after cold exposure and increased the next day [Lubkowska et al. 2008]. Additionally, a 36% increase was observed in the activity of superoxide dismutase (SOD), glutathione peroxidase (GPx) and conjugated dienes (CD) in healthy individuals after a single stimulation [Woźniak et al. 2007]. Duge et al. [2005] observed a significant increase in total peroxyl radical trapping antioxidant capacity of plasma (TRAP) in healthy women 2 minutes after the cold stress but only after the first 4 weeks of the 12 weekend long study. Additionally those authors concluded that prolonged, regular cryostimulation or winter swimming for 12 weeks did not appear to be harmful regarding antioxidative capacity. Further studies are needed for the confirmation of potential adaptational advantages occurring in antioxidative response to cryostimulation.

**Immunology and inflammation**

Cryotherapy is used in the early treatment of acute injuries (sprains, strains, fractures) but only a few papers discuss any possible influence of whole-body cryostimulation on inflammation mechanisms or immunology. In studies by Jackowska et al. [2006] it was ascertained that IgA, IgG, IgM and C3, C4 complement protein levels increase during cryostimulation, yet after cryotreatment, the levels of the mentioned markers were similar to initial values. Leppäluoto et al. [2008] did not observe changes in plasma level of IL-1β, Il-6 and TNFα during prolonged treatment (12 weeks) with cryostimulation (-110°C; 2min; 3 times a week). In our studies, we observed an increase in the level of white blood cells in response to a series of 10 cryostimulations, and at the same time we showed that a single 3 minute long whole-body cryostimulation (-130°C) leads to an increase in the level of interleukin 6, which is maintained for the next 10 stimulations [Lubkowska et al 2010 ]. This was later confirmed in the next experiment, which additionally showed the more advantageous effect of 20 sessions compared to 10 or 5 cryostimulations. This advantageous effect - an increased level of anti-inflammatory cytokines (IL-6, IL-10, IL-12) - was maintained during the whole series of cryostimulations, and receded no earlier than after two weeks after the completion of the cryostimulations, regardless of the number of treatments. However, although the decreased level of pro-inflammatory IL-1α was observed during the series of 5 and 10 treatment, in an examination two weeks after the last stimulation the IL-1α decrease was maintained only after the series of 20 cryostimulation treatments [Lubkowska et al. 2011].

The anti-inflammatory effect of cryotreatment on rugby players lead to an increase in anti-inflammatory cytokinase IL-10 with a decrease in pro-inflammatory IL-2 and chemokine IL-8. It was postulated that cold exposure had an immunostimulating effect related to enhanced noradrenaline response and can be connected with paracrine effects. Similar tendencies in decrease in pro-inflammatory cytokine (IL-2, Il-8) and increase in anti-inflammatory one (IL-10) after cryostimulation were observed by Banfi et al. [2009].

**Recovery from exercise-induced muscle damage**

Some of the latest papers on cryotherapy relate to the use of cryostimulation in athletic recovery, acceleration of recovery to full physical ability, and removing the results of muscle fibre damage caused by intense exercise. It is promoted as a treatment method for muscle injuries, syndromes of overuse and to enhance recovery between training sessions [Banfi et al. 2010].

Difficulty with the evaluation of the advantageous effect of cryostimulation is associated, as in earlier aspects, with the very small number of papers in this field. Costello et al. [2011] performed very interesting studies to evaluate the immediate effect of cryostimulation on proprioception and tried to evaluated the effectiveness of this treatment in muscle soreness and function following eccentric exercise damage. Based on obtained results, the authors suggest that although cryostimulation does not increase the risk of proprioceptive related injury, it is ineffective in improving recovery if administered 24h after exercise. In another study [Pournot et al. 2011] researchers compared the effect of two different recovery modalities: cryostimulation (3 min at -110°C) and passive recovery, on markers of exercise-induced muscle damage and inflammation obtained after a simulated trail running race. The recovery session was applied immediately after, 24h, 48h and 72h after subjects performed a 48min running treadmill exercise. The author observed that a unique session of cryostimulation performed immediately after exercise enhanced muscular recovery by restricting the inflammatory process. They suggested that soluble receptor antagonist IL-1ra increases after single whole-body cryostimulation and restricts the inflammatory response to exercise by a decrease in the magnitude if IL-1β and protein C-reactive. This research indicates that this physiotherapeutic treatment reduces the time of recovery, although depending on the time lag from the intense effort to cryostimulation. Further studies are required for a detailed understanding of mechanisms of response to cryogenic temperatures to find their most effective application in athletic training and recovery.

## 6. Conclusions

Whole-body cryostimulation is a factor that can influence many physiological and biochemical indicators in the human body. Triggering physiological defensive mechanisms in response to a repeated external stressor (cold) may be widely used in medicine, physiotherapy, sport and athlete recovery. However, it is important to select an appropriate duration, form and number of cryostimulations, according to the needs of each patient.

The aforementioned literature data very often relates to different procedures and methodologies. Hence the results are often inconsistent and do not always give a clear view of the discussed problem. It is hard to conclude on the real effectiveness of cryostimulation when it is accompanied by daily physical training which is in itself a stressogenic factor for the human body. In order to achieve a precisely determined effect of cryostimulation on athletes or patients, one should use appropriate methods. In addition, there exists the problem of immersion in cold water which is accompanied by another stressor - i.e. hydrostatic pressure, depending on the depth of immersion.

There exists no single agreed recommendation concerning the number of cryostimulations in a series. Usually, 10 daily sessions are applied, although this is less related to therapeutic reasons than economic ones. It is also not widely agreed how many cryostimulations give real and sustained beneficial effects and whether the adaptive changes in subjects depend on the number of sessions, although this correlation seems very likely.

Because the interest in using cold temperatures in medicine, physiotherapy, sport and athlete recovery is constantly growing, and knowledge in this field is still far from satisfactory, any new research on the use of cryogenic temperatures is a welcome and precious source of information from the point of view of physiotherapeutic practice.

## 7. References

Adamowicz, B. (2005). Cryogenic chambers - a matter of choice. *Acta Bio - Optica et Informatica Medica* Vol. 11, No. 1-2, pp. 44-46

Banfi, G.; Lombardi, G.; Colombini A. & Melegati G. (2010). Whole-Body Cryotherapy in Athletes. *Sports Medicine*, Vol. 40, No. 6, pp. 509-517

Banfi, G.; Krajewska, M.; Melegati, G. & Patacchini, M. (2008). Effects of whole-body cryotherapy on haematological values in athletes. *British Journal of Sports Medicine* , No. 42, pp. 558-559

Banfi, G.; Melegati, G.; Barassi, A.; Dogliotti G.; d'Eril, G.M.; Dugué B. & Corsi M.M. (2009). Effects of whole-body Cryotherapy on serum mediators of inflammation and serum muscle enzymes in athletes. *Journal of Thermal Biology* , Vol. 34, No. 2, pp. 55-59

Biały, D.; Zimmer, K. & Zagrobelny, Z. (1998). Cryochamber - advantages of its use in rehabilitation - own experiences. Adamowicz, B. (2005). Cryogenic chambers - a matter of choice. *Acta Bio - Optica et Informatica Medica,* Vol. 4, No.4, pp. 169-172

Bieńkowska, A.; Molski, P.; Dzierżanowski, M.; Bułatowicz, I. & Hagner W. (2006). Pain and cryotherapy in the treatment of complex medical treatment of the human motor system. *Kwartalnik Ortopedyczny,* No. 4, pp. 311 - 314

Blatteis, C.M. (1998). Physiology and pathophysiology of temperature regulation. World Scientific Co. Pte. Ltd. ISBN 981-02-3172-5

Boerner, E.; Brzyk, R. & Bienias - Jędrzejewska, M. (2007). Evaluation of the effectiveness of local cryotherapy in treating the shoulder impingement syndrome. *Acta Bio - Optica et Informatica Medica,* Vol. 13, No 1, pp. 54 - 56

Bojek, W. (2006). Cryotherapy - general comments. *Balneologia Polska No.* 1,pp. 64 - 67

Brojek ,W. & Warzocha, A. (2006). Cryochamber - most frequently asked questions.(Part II). *Acta Bio -Optica et Informatica Medica,* Vol. 12, No.2, pp. 205-206

Cholewka, A. & Drzazga, Z. (2005). Whole-body cryotherapy in a two-stage cryochamber and in a cryobarrel . *Acta Bio - Optica et Informatica Medica,* Vol. 11, No 1-2, pp. 49-54

Chwalbińska-Moneta J. (2003). Influence of whole-body cryotherapy of results of selected exercise response. *Sport Wyczynowy,* No. 5-6, pp. 461-462

Different amount of sessions in healthy men. *Scandinavian Journal of Clinical & Laboratory Investigation,* Vol. 71, No. 5, pp. 419-425

Dugué, B.; Smolander, J.; Westerlund, T.; Oksa, J.; Nieminen, R.; Moilanen, E. & Mikkelsson, M. (2005). Acute and long-term effects of winter swimming and whole-body cryotherapy on plasma antioxidative capacity in healthy women. *Scandinavian Journal of Clinical & Laboratory Investigation* Vol. 65, No. 5, pp. 395-402

Fricke R. (1989). Ganzkoperkaltetherapie in einer Kaltekammer mit Temperaturen -110°C. Z Phys Med Baln Med Klim No. 18, pp. 1-10

Gregorowicz, H. & Zagrobelny, Z. (1998). Cryotherapy in Multiple Sclerosis treatment. *Acta Bio - Optica et Informatica Medica* , Vol. , No. 4, pp. 173-174.

Hagner, W.; Smolka, A. & Różańska J. (2009). Influence of whole-body cryotherapy of results of exercise test. *Balneologia Polska,* Vol.115, No.1, pp. 35-39

Hopkins, J.T. (2006). Knee Joint Effusion and Cryotherapy Alter Lower Chain Kinetics and Muscle Activity. *Journal of Athletic Training;* Vol.41, No.2, pp. 177–184

Jackowska, E.; Pisula, A.; Drulis-Kawa, Z.; Guz, K.; Bugla-Płoskońska, G.; Doroszkiewicz, W. & Stręk, W. (2006). Changes of level of immunoglobulins and C3 and C4 proteins in serum during whole-body cryotherapy. *Acta bio-Optica et Informatica Medica,* Vo. 12, No.2, pp. 101-103

Jagodziński, L.; Kubacka, M.; Wiśniowska, B.; Puszer, M. & Stanek A. (2001). Whole-body cryostimulation. Part II. *Gabinet Prywatny*, Vol. 92, No. 4, pp. 10 - 11

Jezierski C.: Cryostimulation in rheumatology, traumatology, orthopedics and rehabilitation (partII. *Acta Bio - Optica et Informatica Medica*, Vol. 4, No. 13, pp. 336 - 337

Jezierski, C. (2006). Methodology and principles of local cryostimulation techniques. *Acta Bio – Optica et Informatica Medica*, Vol. 3, No.12, pp.200 - 201

Jezierski, C. (2007). Cryostimulation in rheumatology, traumatology, orthopedics and rehabilitation treatments (part I). *Acta Bio - Optica et Informatica Medica*, Vol. 3, No. 13, pp. 240 - 242

Kasprzak, W. & Markowska, A. (2008). Physiotherapy. PZWL Wydawnictwo Lekarskie, Warsaw 2008, pp. 18 – 22

Kiljański, M.; Clayton, M.; Karpiński, J.; Szczepaniak, R.; Kiebzak, W. & Kałuża, J (2005). Evaluation of an individual cryochamber in a comprehensive physiotherapy based on own observation. *Fizjoterapia Polska* , Vol. 5, No. 2, pp. 207 - 210

Klimek, A.T.; Lubkowska, A.; Szygula, Z.; Chudecka, M. & Frączek B. (2010). The influence of the ten sessions of the whole body cryostimulation on aerobic and anaerobic capacity. *International Journal of Occupational and Environmental Health*, Vol. 23, No. 2, pp. 181-189, ISSN: 1896-494X

Klimek, A.T; Lubkowska, A.; Szygula, Z.; Frączek, B. & Chudecka M. (2011). The influence of single whole body cryostimulation treatment on the dynamics and the level of maximal anaerobic power. *International Journal of Occupational and Environmental Health*, Vol.24, No.2, pp. 184 - 191, ISSN: 1896-494X

Komulainen, S.; Oja, T.; Rintamaki, H.; Virokannas, H. & Keinanen - Kiukaanniemi, S. (2004). Blood pressure and thermal responses to whole body cold exposure in mildly hypertensive subjects. *Journal of Thermal Biology*, Vol. 29 , No. 7-8, pp. 851-856

Korzonek-Szlacheta, L.; Wielkoszyński, T.; Stanek A.; Świętochowska, E.; Karpe, J. & Sieroń, A. (2007). Influence of whole-body cryotherapy on the levels of some hormones in Professional footballers. *Polish Journal of Endocrinology*, Vol.58, No. 1, pp. 27-32

Krawczyk - Wasilewska, A.; Kuncewicz, E.; Sobieska, M. & Samborski, W. (2007). Evaluation of the effectiveness of physical therapy in relieving pain accompanying rheumatoid arthritis. *Nowa Medycyna*, No. 4, pp. 74 - 79

Księżopolska - Pietrzak, K. (1998). Cryotherapy in osteoporosis. *Polski Merkuriusz Lekarski*, Vol. 28 No.5, pp. 222 - 224, ISSN: 1426-9686

Leppäluoto, J.; Westerlund, T.; Huttunen, P.; Oksa, J.; Smolander, J.; Dugué, B. & Mikkelsson, M. (2008). Effects of long-term whole-body cold exposures on plasma concentrations of ACTH, beta-endorphin, cortisol, catecholamines and cytokines in healthy females. *Scandinavian Journal of Clinical & Laboratory Investigation* , Vol. 68, Nzo.2, pp. 145-153, ISSN: 0036-5513

Lisinski, P.; Jozwiak, D. & Samborski, W. (2005). Cold therapy and cryotherapy in treating patients with pains in the area of the shoulder joint. *Chirurgia Narządów Ruchu i Ortopedia Polska*, Vol. 70, No. 6, pp. 435 – 438

Lubkowska, A.; Dolegowska, B.; Szygula, Z. & Klimek, A.T. (2009). The activity of selected enzymes in erythrocytes and the level of plasma antioxidants in response to single whole-body cryostimulation in humans. *Scandinavian Journal of Clinical & Laboratory Investigation* , Vol. 69, No. 3, pp. 387 -394

Lubkowska, A. & Suska, M. (2011). The increase in systolic and diastolic blood pressure after exposure to cryogenic temperatures in normotensive men as a contraindication for whole-body cryostimulation. *Journal of Thermal Biology,* Vol. 36, No. 5, pp. 264-268

Lubkowska, A. & Szygula Z. (2010). Changes in blood pressure with compensatory heart rate decrease and level of aerobic capacity in response to repeated whole-body cryostimulation in normotensive, young and physically active men. *International Journal of Occupational and Environmental Health,* Vol. 23, No. 4, pp. 367-375, ISSN: 1896-494X

Lubkowska, A.; Chudecka, M.; Klimek, A.T.; Szyguła, Z. & Frączek B. (2008). Acute effect of a single whole-body cryostimulation on prooxidant-antioxidant balance in blood of healthy young men. *Journal of Thermal Biology* , Vol. 33, No.8, pp. 464-467

Lubkowska, A.; Banfi, G.; Dołęgowska, B.; d'Eril, G.M.; Łuczak, J. & Barassi, A. (2010). Changes in lipid profile in response to three different protocols of whole-body cryostimulation treatments. *Cryobiology,* No. 61, pp. 22-26, ISSN 0011-2240

Lubkowska, A.; Szyguła, Z.; Chlubek, D. & Banfi, G. (2011). Serum mediators of inflammation level: IL-1α, IL-1β, IL-6, IL-10, IL-12 and TNFα during prolonged whole-body cryostimulation treatment with different amount of sessions in healthy men. Scandinavian Journal of Clinical & Laboratory Investigation , Vol. 71, No.5, pp. 419-25

Lubkowska, A.; Szyguła, Z.; Klimek, A. & Torii, M. (2009). Do sessions of cryostimulation have influence on white blood cells count, level of IL6 and the total oxidative and antioxidative status in healthy men. *European Journal of Applied Physiology* , Vol. 109, No. 1, pp.67-72

Łuczak, J. & Michalik, J. (2006). Wpływ skrajnie niskich temperatur na wybrane cechy motoryczne człowieka. *Fizjoterapia Polska,* Vol. 3, No. 4, pp. 206-211.

Migaj, R. (2005). Treatment with cold generated by the compressor cooling system. *Acta Bio - Optica et Informatica* Medica, Vol. 11, No. 1-2, pp. 55

Pasek, J.; Pasek , T. & Sieroń, A. (2009). Local and systemic cryotherapy in patients with degenerative arthritis. *Rehabilitacja w Praktyce,* No. 2, pp. 32 - 33

Pournot, H.; Bieuzen, F.; Louis, J.; Fillard, J.R.; Barbiche, E. & Hausswirth, C. (2011). Time-Course of Changes in Inflammatory Response after Whole-Body Cryotherapy Multi Exposures following Severe Exercise. *PLoS ONE,* Vol.6, No. 7,e22748. Doi:10.1371/journal.pone.0022748

Raczkowski, J. & Daniszewska, B. (2007). Cryotherapy – use of extremely low temperatures in medical treatment. *Acta Bio - Optica et Informatica Medica* , Vol. 13, No.1, pp. 58 – 60

Rawecka, D. & Rokita E. (2006). Local cryotherapy - has everything been said? *Acta Bio – Optica Informatica Medica,* Vol. 2, No.12, pp. 111-115

Rymaszewska, J.; Ramsey ,D.; Chładzińska-Kiejna, S. & Kiejna, A. Can short-term exposure to extremely low temperatures can be helpful in the treatment of depressive disorders and fears? *Psychiatria Polska,* Vol. 41, No. 5, pp. 625-636, ISSN: 0033-2674

Sieroń, A. & Cieslar, G. (2003). Cryotherapy - treatment with cold. *a - medica Press* 2003, ISBN 83-88778 59-5

Sieroń, A.; Stanek, A.; Cieślar, G. & Pasek, J. (2007). Cryorehabilitation - Role of cryotherapy in the contemporary rehabilitation. *Fizjoterapia, Vol.* 15, No. 2, pp. 3 - 8

Skrzek A. (2009). The history of whole-body cryotherapy in Poland. *Acta Bio – Optica et Informatica Medica,* Vol. 4, No.15, pp.309 - 313

Skrzek, A.; Bolanowski, M.; Zagrobelny, Z. & Lubczyńska-Kowalska, W. (2003). The influence of whole body cryotherapy followed by relaxing gymnastics on bone turnover and muscular function in elderly women. *Acta Bio - Optica et Informatica Medica,* Vol. 9, No. 1-2, pp. 45-51

Skrzep-Poloczek, B.; Romuk, E. & Birkner, E. (2002). The effect of whole-body cryotherapy on lipids parameters in experimental rat model. *Balneologia Polska* 2002; No. 1-4, pp. 7-13

Smolander, J.; Leppäluoto, J.; Westerlund, T.; Oksa, J.; Dugue, B.; Mikkelsson, M. & Ruokonen, A. (2009). Effects of repeated whole-body cold exposures on serum concentrations of growth hormone, thyrotropin, prolactin amd thyroid hormones in healthy women. *Cryobiology,* Vol. 58, No. 3, pp. 275-278

Stanek, A.; Cieslar, G.; Rosmus-Kuczia, I, Matyszkiewicz, B.; Romuk, E.; Skrzep-Poloczek, B.; Birkner, E. & Sieroń A. (2006). Influence of whole body cryotherapy on blood morphology parameters in patients with ankylosing spondylitis and in healthy volunteers. *Acta Bio - Optica et Informatica Medica,* Vol. 12, No. 3, pp. 207-210

Straburzyńska - Lupa, A.; Czubaszewski, L.; Romanowski, W. & Świetlik, Z. (2005). Comparative studies of single treatment with cold air and liquid nitrogen vapour in patients with rheumatoid arthritis. *Fizjoterapia Polska* , Vol. 3, No. 5, pp. 323 - 328

Suszko, R. (2003). Whole-body cryotherapy. *Rehabilitacja Medyczna* , Vol. 7, No. 2, pp. 63- 71

Suszko, R. (2003). Whole-body cryotherapy. *Rehabilitacja Medyczna* Vol. 7, No. 2, pp. 63, 65 – 71

Taghawinejad, M.; Birwe, G.; Fricke, R. & Hartman, R. (1989). Ganzkoperkaltetherapie Beeinflussung von Kreislauf - un Stoffwechselparametern. Z Phys Med Baln Med Klim, No. 18, pp. 23-30

Westerlund, T.; Smolander, J.; Uusitalo-Koskinen, A. & Mikkelsson, M. (2004). The blood pressure responses to an acute and long-term whole-body cryotherapy (-110°C) in men and women. *Journal of Thermal Biology* Vol. 29, No 6, pp. 285-290, ISSN: 0306-4565

Westerlund, T.; Smolander, J.; Uusitalo-Koskinen, A. & Mikkelsson M. (2004). The blood pressure responses to an acute and long-term whole-body cryotherapy (-110°C) in men and women. *Journal of Thermal Biology,* Vol. 29, No. 6, pp. 285-290.

Wojtecka-Łukasik, K.; Księżopolska-Orłowska, E.; Gaszewska, O.;, Krasowicz-Towalska, P.; Rzodkiewicz-Maślińska, D.; Szukiewicz, S. & Maśliński, E. Cryotherapy decreases histamine levels in the blood of patients with rheumatoid arthritis. *Inflammatory Research,* Vol. 59, No. 2, pp. 253-255

Woźniak, A.; Woźniak, B; Drewa, G.; Mila-Kierzenkowska, C. & Rakowski, A. (2007). The effect of whole-body cryostimulation on the prooxidant-antioxidant balance in blood of elite kayakers after training. *European Journal of Applied Physiology* , Vol. 101, pp. 533-537

Woźny A., Kujawa J., Pieszyński I., Gworys K. & Puzder A. (2006). Evaluation of the analgesic effectiveness of McKenzie physiotherapy method in combination with local cryotherapy in patients with lumbosacral pain syndromes. *Kwartalnik Ortopedyczny* , No. 1, pp. 63 – 69

Zagrobelny Z. & Zimmer, K. (1999). The use of cryogenic temperatures in sports medicine and physiotherapy. *Rehabilitacja Medyczna* , Vol. 94, No. 15, pp. 8 - 13

Zagrobelny Z. & Zimmer, K. (1999). The use of cryogenic temperatures in medicine and sport physiotherapy. *Medycyna Sportowa* Vol.94,No. 5,pp. 8 – 13

Zagrobelny, Z. (2003). Local and whole-body cryotherapy. *Elsevier Urban & Partner;* Wrocław 2003

Zalewski, P.; Tafil-Klawe, M.; Klawe, J.; Buszko, K.; Lewandowski, A. & Panowicz, I. (2009). Influence of the whole-body cryotherapy on the hemodynamic parameters in healthy subjects. *Acta Bio - Optica et Informatica Medica,* Vol.15, No. 3, pp.209-214

# 8

# Natural Wool Fabrics in Physiotherapy

Krzysztof Gieremek[1,2] and Wojciech Cieśla[2]
[1]*Department of Physiotherapy, Academy of Physical Education in Katowice,*
[2]*Department of Tourism and Health Promotion, Faculty of Physiotherapy, Katowice School of Economics, Poland*

## 1. Introduction

Prevention of heat loss, thermal insulation and heat treatment play an important role in physiotherapy. For example, peripheral vascular disease associated with metabolic diseases (e.g., diabetes, arteriosclerosis), arthritis, paresis and paralysis of peripheral nerves all require interventions involving heat therapy. Wool products such as garment linings, shoulder-straps, knee pads and lumber pads exert a very positive effect on the above mentioned circulatory disturbances. Moreover, due to its unique physical and chemical properties, knitwear increases the pain threshold and reduces muscle tension; knitwear can be used both before and after exercise to prolong the therapeutic effect. Physiotherapy sessions for little children on wool mattresses also evoke additional sensory impulses thus improving sensorimotor outcomes. Extending the heat effect on the skin resulting from physical procedures such as massage, heat therapy and ultrasounds is regarded as one of the main theraputic advantages. Last but not least, there are also some easily perceivable economic benefits.

In addition, the influence of wool beddings on sleep patterns (associated with thermoisolation and thermoregulation) will be discussed as well as their regenerative effects in patients with somatic disease and depressive disorders.

## 2. The effect of wool products on human and animal physiology and health

### 2.1 Physicochemical properties of wool affecting medical value and health benefits of wool fabrics

Wool, a material used to manufacture the fabrics under assessment, is a product of animal origin. The basic type of wool used was merino wool sometimes enriched with camel or kashmir goat hair. Wool fibres of different origins may have varying diameters but still display similar properties.

Wool is primarily made of keratin, an insoluble protein. A wool fibre is composed of three layers, i.e., epidermis (outer layer), tubular cortex (middle layer) and medulla (inner layer).

The medulla makes up the bulk of the fibre and consists of a latticework of spindle cells with air-filled spaces in between; it determines the physicochemical properties of wool. Wool fibres are not straight but more or less bent forming crimps (measured in crimps per inch or crimps per centimetre) of different regularity. The feature distinguishes wool from other fabrics and, to a large extent, determines its uniqueness.

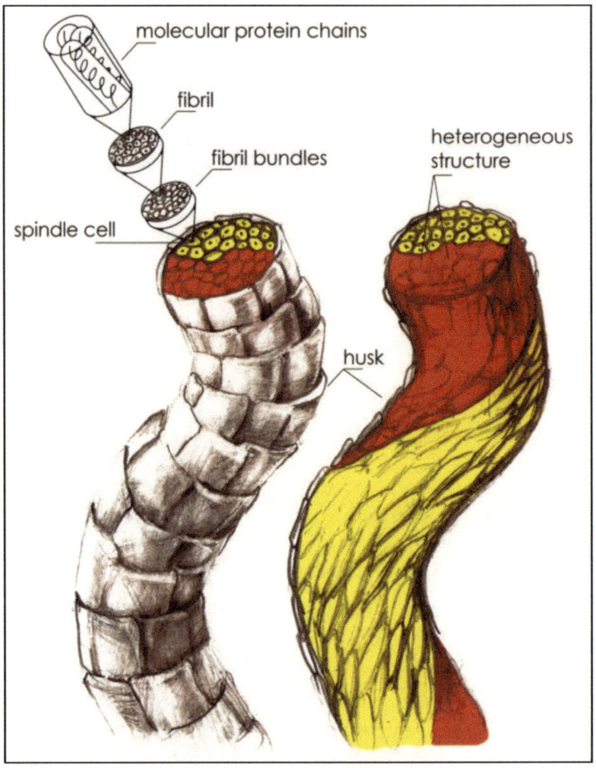


Fig. 1. The structure of a wool fibre.

The properties of wool fabrics that are deemed the most specific and *important for clinical application* are springiness, resilience, **low thermal conductivity**, elasticity and moisture-retaining and thermal insulation properties.

*Springiness* is the ability of a bent, stretched or compressed object to return to its original form when the acting force is removed.

*Resiliency* is the property of a material which causes it to resist deformation and return to its original shape when wrinkled or distorted in any way. Natural wool displays these properties; its springiness results from high sulphur content in wool keratin. Both springiness and resiliency are of importance as they prevent wrinkle formation in wool fabrics ensuring a product's durability. Along with the above mentioned wool crimp and scale structure on the fibre, springiness and resiliency cause large air content to be enmeshed in wool; the **amount** of **air** trapped between the yarns and fibers does not change while the fabric is in use. Hence, good thermal insulation and the feeling of warmth. Due to springiness, wool products do not adhere directly to human skin whereby the isolating air layer is increased; the dynamics of mechanical deformation resulting from temperature changes causes sufficient air flow to occur.

*Thermal conductivity***.** Due to its structure and chemical composition of keratin, wool fibre is characterized by low thermal conductivity, a property crucial for a product to be used in heat therapy.

*Moisture content.* Wool is the most hygroscopic among the so far described fabrics. Its fibres absorb and retain considerable amounts of moisture from the atmosphere distributed in **capillary spaces between fibres.** Through moisture absorption, the fibres prevent rapid temperature changes. Highly **hygroscopic** fibres have a tendency to *maintain temperature;* hence, the unique *insulating properties* of wool. During absorption and condensation of water vapour a considerable amount of heat is produced defined as *absorption heat* (the amount of calories released during absorption of 1g water).

The heat reduces possible effects of low external temperature and facilitates evaporation of the absorbed moisture. Atmospheric moisture is then again absorbed and, through a chemical reaction with wool fibre molecules, helps generate heat to prevent temperature changes. Animal fibres absorb the largest amounts of moisture without physical changes thus displaying considerable insulating properties.

| FABRIC TYPE | CAL/G |
|---|---|
| Wool (64g) | 27.00 |
| Argon (viscose fabric) | 25.00 |
| Natural silk | 19.35 |
| Cotton | 11.00 |
| Nylon | 7.60 |

Table 1. Immersion heat generated during the transition from total dryness to the point of complete saturation (Grycewicz & Staniszkis, 1959).

Investigations have been recently carried out on superhydrophilic wool fibres coated with silica nanoparticles which alter surface roughness of the wool fibre and make it more water absorbent. Modified wool has been proven to have excellent properties under different environmental conditions (Chen, 2010).

*Elasticity.* Water-vapour absorption-desorption cycle influences fibre flexural stiffness in contact with the skin resulting in gentle and nonspecific *mechanical stimulation* of the superficial nerve receptors. Wool products contain lanolin, natural wool wax secreted by the **sebaceous glands** of wool-bearing animals, especially merino sheep. Lanolin and moisture-retaining properties ensure wool elasticity thus preventing fibre rigidity. It is also worth mentioning that lanolin does not allow dirt to penetrate into the fibres which facilitates maintenance process and assures proper hygiene standards for product users.

*Thermal insulation.* **Woolen garments** outperform other fabrics in maintaining a fairly constant temperature of the human body. The above mentioned structure and properties including fibre crimp, low thermal conductivity, springiness and resiliency increase air entrapment and ensure outstanding thermal insulation. Transferred from dry and warm to wet and cold conditions, wool absorbs atmospheric moisture which reacts with molecules in the wool generating heat (Beuth, 1968; Grycewicz & Staniszkis, 1959; Żyliński, 1958). A wool suit weighing approximately 1.5 kg, when transferred from a dry and warm to wet and cold room will almost immediately produce the amount of heat released by the human body during one hour. Due to **high hygroscopic** susceptibility of fibres, wool garments absorb sweat and prevent overheating in summer.

## 2.2 Influence of wool fabrics on skin physiology and the efficiency of several body mechanisms

The physicochemical properties of wool fabrics clearly suggest that their primary effects on the human organism would be associated with heat therapy with wool's insulating properties being of crucial importance (Matusiak, 2010). In many patients with peripheral vascular disease, the course of treatment is affected by the equilibrium between the amount of heat generated through metabolic processes and heat release to the ambient atmosphere, commonly referred to as thermal comfort (Sudoł - Szopińska & Chojnacka, 2007). Heat transfer by wool-based fabrics depends, among others, on the product surface, thermal conductivity coefficient of the material used and its effect on the skin. Therefore sufficient knowledge of the influence of different textiles on skin physiology is indispensable.

Wollina et al. (Wollina et al, 2006). believe that all clothing fabrics including wool interact with **skin** functions in a **dynamic pattern**. Thermoregulatory processes mediated by local blood flow, perspiration and sweat evaporation are important elements of the relationship between the skin and textile fibers. Wool products may also influence skin immunity through affecting the major constituents of **skin microflora**. Using laser Doppler flowmetry, Yiming Gan et al. (Yiming et al, 2010). investigated the effects of various natural and artificial fibres on subcutaneous circulation in healthy subjects. They found out that fibre thickness, coarseness, structure and moisture influenced skin temperature and blood flow. Fabric surface characteristics play an important role, especially during the transient heat exchange at the beginning of its contact with the skin. The higher of the moisture level, the longer the duration of the fabric impact on the vascular structures of the skin. The significance of moist heat source on the skin in eliciting blood flow responses in male and female patients was confirmed by the study of Petrofsky et al. (Petrofsky et al, 2009). Moist heat packs caused a significantly higher skin blood flow (about 500% greater) than dry heat. The authors believe that moist heat renders skin thermoreceptors more sensitive due to calcium channels opening in endothelial cells.

Interesting observations have been made by Li Yong from Hong Kong Polytechnic University, who investigated the psychophysical mechanisms of the sensation of temperature and moisture while wearing jumpers made of wool or acrylic fibres (Li, 2005). It has been demonstrated that the perception of warmth follows both Stevens' power law and the Weber-Fechner law[1], and shows positive relationship with skin temperature. Since the perception of comfort is also positively related to skin temperature, wearing wool made the study subjects feel warmer, dryer and more comfortable. Similar conclusions regarding the effects of natural (wool) and artificial (100% acrylic) fabrics on human physiology prior to, during and after physical exercise have been drawn by Ciesielska et al. Each fabric, including those made of natural fibres, influences cardiorespiratory parameters and psychophysical performance (Ciesielska et al, 2009).

[1] Stevens' power law is a proposed relationship between the magnitude of a physical stimulus and its perceived intensity or strength. A hundred years earlier, Weber and Fechner developed a logarithmic law concerning the relationship between the strength of sensation and the size of the sensory stimulus.

Ziemniewska et al. investigated the effect of artificial (polyester) and natural (cellulose fibre) garments on the energy cost of a physical activity demonstrating that garment type influenced the subjects' performance. More hygroscopic fibres were considered more advantageous (Ziemniewska et al, 2010). However, opposite results have also obtained.

When discussing the influence of natural fabric garments on the human organism one should also mention similar research studies done on animals. Shafik - investigated the effect of different types of textile fabrics (polyester, cotton and wool) on dog hair growth (Shafik, 1993). The polyester-patch covered hair grew at a significantly lower rate and density than in the uncovered area. The author concluded that artificial fabrics generated electrostatic potentials which may have inhibited hair growth whereas cotton and wool did not induce such effects. The same researcher also studied the effect of different types of textiles - including wool - on canine reproduction (Shafik, 2008). He measured serum progesterone and electrostatic potentials detected on the skin. Bitches wearing polyester-containing pants showed diminished progesterone levels and ovulation disturbances. Thus, generation of highly disadvantageous electrostatic potentials by artificial fabrics has been confirmed as well as the nonexistence of such a negative impact in the case of wool.

### 2.3 Thermal and thermoregulatory significance of wool beddings for the improvement of sleep patterns

A large number of users of wool products maintain that wool textiles make a real difference in the comfort and quality of sleep. These effects can be accounted for by thermoregulatory mechanisms. As mentioned before, wool also helps maintain a fairly consistent **body temperature providing thermal comfort. The ability of textiles to ensure thermal comfort under some specific external conditions is referred to** as the Thermal Comfort Index. The parameter is affected by heat conduction, convection, evaporative heat loss and thermal absorption (Matusiak, 2010). Skenderi et al.'s investigations on fabrics made from natural fibre emphasize the capacity of fibrous materials to transfer water vapour and the significance thereof under various climatic conditions (Skanderi, et al, 2009).

Since wool properties counteract temperature changes, it can easily be concluded that wool beddings influence the thermoregulation of the human body during sleep. A number of changes occur during sleep in the thermoregulatory mechanisms of humans and animals. Core temperature rhythms and sleep propensity differ significantly across the day and night. According to Gilbert S. et al. thermoregulatory changes may provide a signal to the brain regions that regulate sleep (Gilbert, et al, 2004). These regions are anatomically and functionally connected with thermoregulatory mechanisms. Considerable relationships have been found between body temperature and sleep/wake rhythms. Patterns of heat exchange alter during sleep; metabolism decreases while heat loss increases resulting in a drop in body temperature (Szymusiak, 2009). Bach V et al. have found that **thermoregulatory** responses **depend** on the **sleep stage** and are affected by ambient temperature. Excessive cold and heat disturb sleep. Moderate pre-sleep body warming enhances SWS (Slow Wave Sleep) and improves sleep continuity. This might be of importance in depressive patients, whose sleep and thermoregulatory rhythms are disturbed (Bach, 2002). Sleeping in cool rooms improves the active stage of sleep, i.e., REM (Rapid Eye Movement), which is, however, related to a decrease in subjective sleep quality. Craig Heller believes that **sleep** and **control** of body **temperature** are intertwined. The comfort of sleep is strongly influenced by body and external temperatures; even slight changes in core temperature have an effect on sleep (Heller, 2005). The association between sleep and thermoregulation play an important role in several disease processes. Tsuzuki (Tsuzuki et al, 2004) investigated the effects of different temperature conditions and humidity on sleep. The subjects dressed in pyjamas and covered with a cotton blanket were assessed in 26°C 50% humidity and 32°C 80% humidity. Melatonin metabolite secretion and the duration of wakefulness were measured. The duration of wakefulness increased significantly at 32/80 while melatonin metabolite secretion in urine was lower at 32/80 than at 26/50, which was consistent with the decrease in sleep efficiency.

The significance of clothing insulation in outdoor sleeping infants during northern winter was emphasized by a team of researchers of the Institute of Health Sciences, University of Oulu, Finland. Inadequate thermal comfort resulting from deficient thermal insulation caused shorter sleep (Tourula et al, 2010).

Recognizing the unique characteristics of wool fabrics, their capacity to counteract temperature changes, insulation properties and an almost immediate impression of warmth in contact with the skin, beneficial effects of wool on thermoregulatory mechanisms and the quality of sleep can be considered certain.

The powerful impact of sleep on health has been well recognized. During deep sleep melatonin levels increase, a hormone which is an antioxidant and suppressant of tumor development through scavenging of free radicals. The incidence of somatic disease in patients suffering from insomnia increases twofold. Insomnia is also among the causes of the epidemics of diabetes, hypertension and lower immune function (Avidan & Zee, 2007).

### 2.4 Mechanical effects of wool on the skin

Mechanical effects of wool on the skin mostly depend on its properties including resiliency, elasticity, fiber thickness and density. To recap : *resiliency* is the property of a material which causes it to resist deformation and return to its original shape when wrinkled or distorted in any way. *Elasticity*, related to water-vapour absorption-desorption cycle, influences fibre flexural stiffness in contact with the skin (Beuth, 2004). The mechanical effect of wool on the skin also depends on fibre structure. Gualana, a new specialist technology used to manufacture wool fabrics, tumbling/brushing, chemical or gas (ozone) processing, and appropriate blend ratio in blends of cashmere with wool resulting in uniform fibre distribution in knitted fabrics, all these have clearly increased the utility value and beneficial effect of wool fabrics on the human body (Kun et al, 2010). At the same time these fabrics give the illusion of contact with the so called fleece.

A variety of nerve endings enables the skin to detect and differentiate the strength of mechanical stimuli. These include Merkel's touch corpuscles and free nerve endings found in the epidermis and papillary layer of the dermis, Meissner's corpuscles of the papillary layer of the dermis and Pacini's corpuscles scattered in subcutaneous tissues. Hair, rich in nerve endings, acts as a peripheral sensory organ. Even very gentle touch on the hair is transferred to mechanoreceptors in the skin (Bochenek & Reicher, 1989).

Fibre structure and arrangement in wool fabrics stimulate **superficial** sensory nerve endings with different strength. Compared to commonly used products (cardigans, socks, blankets), the amount of wool fibres per 1mm$^2$ results in far better stimulation of mechanoreceptors in the skin even if the stimulus is of mild or moderate strength. This type of stimulation produces an overly-calming and somniferous **effect; it also has a mild anaesthetic action.** The reflex need for touch on the body part that hurts (laying hands, gentle dubbing) provides evidence for the phenomenon. Stroking techniques of therapeutic massage cause a similar response (Field et al, 1996). Mechanical action of wool fibres on the skin also affects blood supply to this organ. Appropriate roughness of wool fabrics produces - during the use and due to non-irritating friction - some extra heat. This micromassage leads to neurohormone activation, which increases superficial blood flow and rises skin temperature. Thereby, thermal and insulation properties of wool fabrics become further enhanced. Using **thermography** and **thermistor** skin **temperature measurements**, Sefton et al. as well as Gieremek et al. (Sefton et al, 2010; Gieremek et al, 1991) confirmed the effect of therapeutic massage on skin temperature alterations directly associated with changes in peripheral blood flow in the treated areas as well as in adjacent not-massaged areas.

## 3. Thermal insulation and heat therapy in the treatment of some selected disease entities

Local use of thermal energy triggers several physiological reactions including increases in blood flow, muscle tension release, decrease of joint stiffness, oedema reduction, and, most importantly, pain relief. Impressions of temperature and pain are conveyed to higher centres through the same nerve tracts. General well-being associated with heat therapy is brought about by the release of endorphins and other neurotransmitters modifying the sensation of pain. The complex and interdisciplinary issues of the effect of heat on the skin became the domain of skin biothermomechanics. Thermomechanics also refers to mechanical tension (deformation) of skin collagen, also epidermal collagen. Thus, heat also modifies the sensation of pain via mechanical action (Xu et al, 2008).

The above mentioned effects of heat evoked through the use wool products occur to a greater or lesser degree in the prevention and treatment of several disease entities.

Interesting information concerning the prophylactic action of textiles in the management of decubitus ulcers has been provided by Irzmańska et al. (Irzmańska et al, 2010). The authors report that the factors contributing to the formation of bedsores comprise temperature, humidity and air flow. Antibacterial agents, type of bed padding and their effects on skin surface are also of importance. The influence of bed sheets mainly depends on fibre type, its structure and water absorption. A hypothermic and debilitated patient needs beddings made of low thermal conductivity fibres, which should also effectively transfer the excess of skin moisture to the ambient environment. Irzmińska (Irzmińska et al, 2010) have demonstrated significant differences in blood flow through skin regions particularly susceptible to decubitus ulcers formation.

While discussing the issue of thermal insulation it is also necessary to consider the patient's age. The ability to absorb and lose heat through the skin changes with age. Petrofski et al. investigated a group of subjects in the age range of 20 - 65 years to examine the effect of repeated local heat on skin blood flow and skin temperature over 3 sequential days. In the younger subjects, the blood flow response in the first 20 minutes of heat exposure was over 30% higher than that seen in the older subjects. However, the acclimatization response of the older participants was much slower (Petrofsky et al, 2010). Also, increased cutaneous blood flow helps reduce deep muscle tension (Roberts & Wenger, 1979).

Good effects of insulating use of wool underwear and beddings in patients with fibromyalgia were described by Kiyak (Kiyak, 2009). Almost all patients using wool bed liners and quilts reported significant alleviation of fibromyalgia symptoms. The effectiveness of heat therapy in fibromyalgia was emphasized by Löfgren M. and Norrbrick G. (Löfgren, & Norrbrick, 2009).

Maintaining thermal comfort is also of significant importance in Raynaud's disease. Neutralizing the influence of external temperature with warmers prevents the occurrence of vasoconstriction (Heller, 2006).

Thermal insulation and massage effect of wool fabrics is also important in chronic venous insufficiency. The condition causes blood pooling and congestion resulting in trophic changes, and, consequently, chronic dermatitis (Kelechi & Michel, 2007). Fornalczyk and Kuliński (Fornalczyk & Kuliński, 2008) point out the importance of physical therapy in the prevention and treatment of crural ulceration. The authors emphasize the significance of thermal effects of therapeutic ultrasound.

Following injuries to the extremities, Complex Regional Pain Syndrome often develops, also referred to as **Reflex Sympathetic Dystrophy** Syndrome. Blood vessels constrict due to increased sympathetic activity; thereby the **thermoregulatory** control of **skin** blood flow becomes impaired (Wasner et al, 2000). Maintaining thermal comfort of the dystrophic area helps reduce the suffering.

Vascular disturbances along with autoimmune processes also play a role in the development of neurodegenerative changes in patients with multiple sclerosis. Especially those over the age of 45 might benefit from appropriate thermal insulation to control microcirculation dysfunctions (D'haeseleer et al, 2011).

Peripheral circulation impairment, **manifested by decrease** in **skin temperature**, commonly develops as a complication of peripheral nerve injury. Physical therapy is usually complex; the rate of nerve fibre regeneration depends on providing the tissues of the affected segment with thermal comfort (Druschky, 1979). Standard therapy includes wearing warmers. Satoshi (Satoshi et al, 1991) investigated the effects of a chronic constriction injury to the sciatic nerve of rats accompanied by an abnormality of cutaneous temperature regulation. Pain resulting from peripheral circulation impairment was almost always associated with skin temperature changes. These results emphasize the need to warm the affected extremity. Hornyak et al. (Hornyak et al, 1990) have confirmed the relationship between regional denervation, sympathetic system and blood supply.

Damage to peripheral vessels accompanied by pain is also found in diabetic neuropathy and diabetic foot (PI-Chang, 2006). Providing gentle heat not only helps delay neurological changes but also promotes pain relief.

Pain is frequently associated with sympathetic system stimulation and almost always causes vasoconstriction and a resultant decrease in body temperature (Birklein et al, 1998). Long-term emotional stress may act as an additional factor activating the sympathetic nervous system; it often underlies psychosomatic syndromes including locomotor organ disorders. Dry or moist heat application provides therapeutic benefits (Fechir et al, 2009).

Heat therapy is a standard therapy in soft tissue contractures, which decrease joint mobility. Leung MS and Cheing GL demonstrated several beneficial effects of deep heating in the management of frozen shoulder (Leung & Cheing, 2008). The efficacy of such treatment may be increased by maintaining higher temperature of a given body part with the use of so called shoulder warmers.

Lespargot et al. opt for the use of local and general heat in the treatment of the upper and lower extremities of children suffering from cerebral palsy. Due to excessive muscle tension, using wool beddings might prove extremely beneficial in these children (Lespargot et al, 2000).

Heat, in the form of physical energy, has also been used in rheumatoid arthritis and arthroses. Ayling J. and Marks R. report multiple advantages of heat therapy in rhematoid arthritis especially as procedures preparing the patients to therapeutic exercises (Ayling & Marks, 2000).

Heat therapy procedures are most frequently applied in the treatment of degenerative spine disease. Thus, the importance of wool warmers should not be underestimated.

## 4. Wool products and their use in some selected disease entities

The use of wool products may be justified and their effect highly beneficial both through heat generation in contact with the skin as well as the prolongation of the thermal effect of therapeutic tissue overheating. Our own pilot studies aimed to determine the effect of natural wool seem to have confirmed the above mentioned benefits. The temperature of the wrist covered with a woolen pad after heat therapy decreased by approximately 1.5 to 2.0°C more slowly compared to the subjects who did not wear such pads. Prolonged tissue temperature elevation helps increase the efficiency of the therapeutic agent.

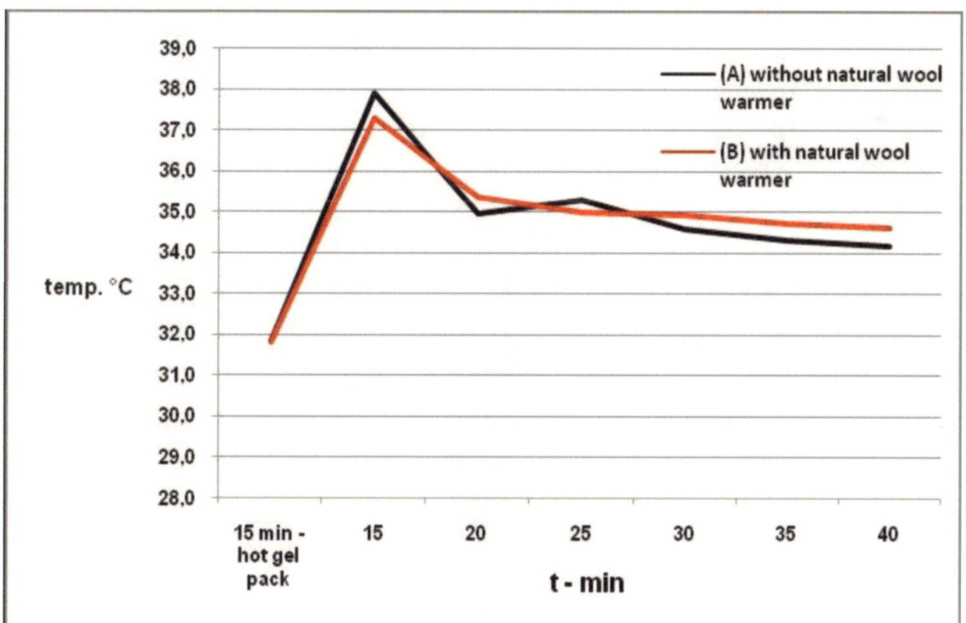


Fig. 2. Sample skin temperature of the carpal region before and after 15-minute hot gel pack with natural wool warmer (group A) compared to hot gel pack alone (group B); ambient temperature 23°C. Own material.

Due to the aforementioned health benefits (cutaneous blood flow improvement, increase of pain threshold and lowering muscle tension), wool products might play a role in preparing tissues for exercise as well as in strengthening the therapeutic effect. The influence of these fabrics on the efficiency and quality of sleep should be again emphasized since sleep has a powerful effect on the body's natural **regenerative processes** both in the case of somatic and psychic problems (depression).

Examples of wool products that may serve therapeutic purposes:

- *Body warmers* – quilts, bedspreads, lumbar belt, elbow, knee and shoulder pads, caps and gloves.

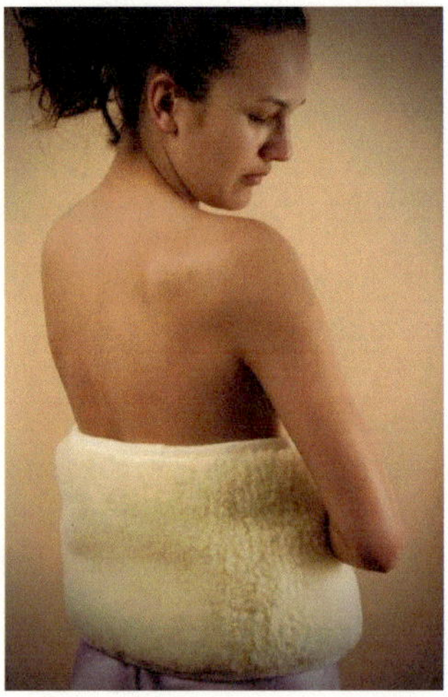

Fig. 3. Body warmers - lumbar belt.

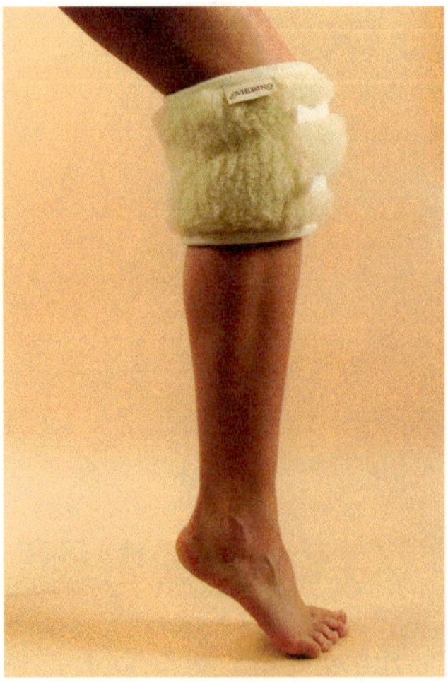

Fig. 4. Body warmers - knee pad.

- Physiotherapy flooring systems - mattresses and mats.
- Neck supports: regular and featured pillows, rolls, seats.

Fig. 5. Featured pillow.

The above products may have preventive or therapeutic, general or local uses (depending on product type) in the following conditions and disease entities:

- rheumatoid inflammatory disease - rheumatoid arthritis, ankylosing spondylitis;
- degenerative joint disease - coxarthrosis, gonarthrosis;
- degenerative disease of intervertebral joints - spondyloarthrosis;
- systemic connective tissue disorders (lupus erythematosus, systemic sclerosis);
- fibromyalgia;
- decubitus ulcers prevention;
- muscle, tendon, ligament, fascia and muscle-tendon attachment inflammatory conditions, eg., lateral epicondilitis (tennis elbow) medial epicondilitis (golfer's elbow);
- shoulder pain (impingement) syndrome, frozen shoulder;
- primary or secondary peripheral circulation disorders (skin, muscles);
- lower limb ischaemia;
- Sudeck's atrophy (reflex sympathetic dystrophy syndrome);
- diabetes-related pathology of the skin, muscles and the circulatory and nervous systems;
- peripheral nerves and brachial plexus paralysis;
- infantile cerebral palsy;
- wheelchair-bound patients with lower limb paralysis and trophic changes - seat or backrest padding;

- bone tissue disorders (osteoporosis, osteomalacia)
- depressive disorders.

Types of therapeutic interventions following which wool products should be used to prolong the thermal effect produced:

- partial paraffin bath;
- paraffin pack;
- warming gel packs;
- thermophore heating pack;
- Kenny's method of treating polio: hot packs, most heat wraps;
- electrotherapy - electric current is applied to decrease pain and muscle tension, e.g., diadynamic currents DF, CP, LP
- ultrasounds;
- shortwave diathermy;
- Sollux lamp infrared irradiation;
- classical massage;
- whirlpool and underwater massage.

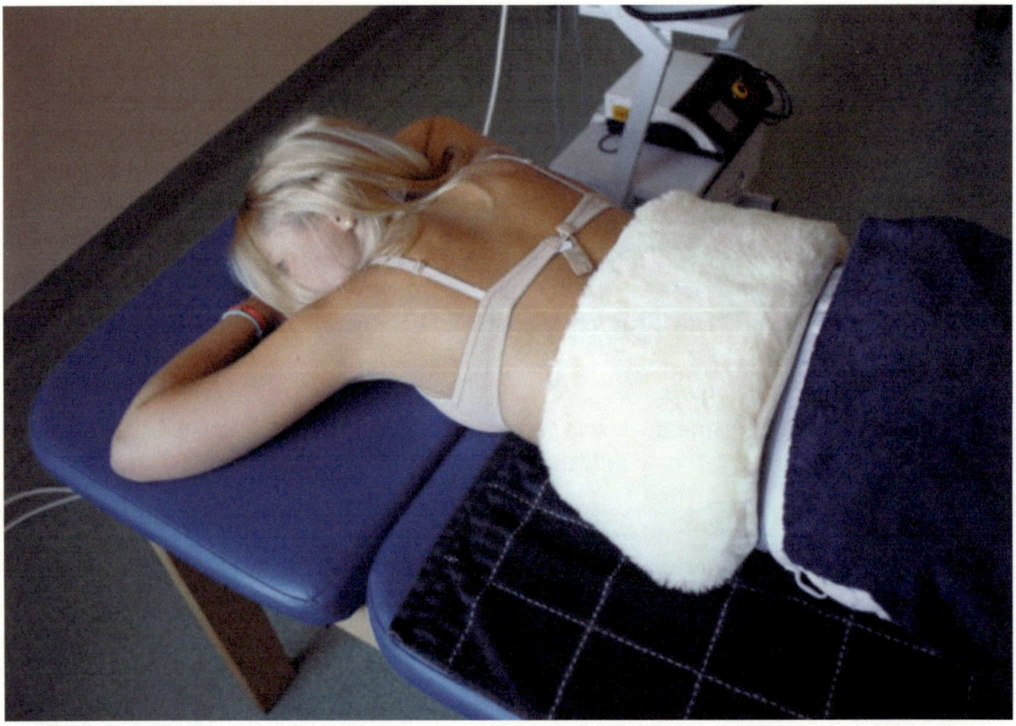

Fig. 6. Extension of the thermal effect using a lumbar belt after diadynamic current treatment.

Wool fabrics are most frequently used in the following functional disorders accompanying the above mentioned disease entities:

- increased or decreased muscle tension;
- paralysis and paresis;
- pain syndromes;
- peripheral vascular disease (especially of arterial and lymphatic vessels);
- sensorimotor disorders;
- morning stiffness pain.

Contraindications to wool products:

- all acute inflammatory conditions of soft tissues and locomotor organs;
- decubitus ulcers and other skin lesions;
- allergy to natural fabrics.

Other remarks:

Wool products used by infants, young children and individuals oversensitive to wool should be wrapped in cotton covers which prevent wool fibres from entering the eyes, and nasal / oral cavity.

## 5. Summary and conclusions

A considerable number of diseases are concurrent with primary or secondary peripheral, cutaneous and muscular circulation disorders. For example, primary circulatory disturbances result from peripheral vascular disease, which commonly affects the arteries supplying the leg and is mostly caused by atherosclerosis. Secondary changes occur in patients with limb paresis or paralysis following stroke, peripheral nervous system injury as well as in those suffering from muscle tension increase due to chronic pain (Moncur & Shields, 1987: Strass et al, 2002). Whether primary or secondary, the disorders result in reduced arterial and venous blood supply associated with lowered lymph flow dynamics.

All patients diagnosed with the above mentioned circulatory impairments (except those with accompanying acute inflammatory conditions) might benefit from the prevention of heat loss through the affected body part. It should be remembered that heat loss is directly proportional to the **temperature gradient between** the **skin** and surrounding environment (Straburzyński & Straburzyńska, 2000). Thus, the reduction of heat loss emerges as the basic objective of therapeutic interventions. The rate of peripheral nerve regeneration is positively affected by the temperature of surrounding tissues and metabolism, which directly depends on peripheral, and even capillary circulation (Straburzyński & Straburzyńska, 2000).

Thus, doctor's and physiotherapist's recommendations regarding the use of body warmers should be considered highly advisable from the medical point of view. Warmers enhance the capability of moisture absorption and ensure air exchange of the body part with its surroundings without excessive heat loss. Heat accumulates in the affected site and, through warming, entails its therapeutic effects. Increased sweating during exercise does not cause unpleasant cooling effects since **wool** fiber molecules **generate heat**.

Another important property of wool fibres is the **mechanical effect** in contact with the **skin**. Non-irritating roughness of wool fabrics induces beneficial mechanical stimulation of sensory receptors. Micromassage of reflex points and resulting **cutaneous vascular** responses stimulate microcirculation and β-endorphin secretion. Thereby, the sensation of pain becomes diminished (Bender et al, 2007; Grass, 1982; Strass et al, 2002; Wright & Slukak, 2001). It should be emphasized that the majority of locomotor system and visceral diseases, and especially chronic problems, cause certain areas of the skin (Head's zones) to develop microcirculation disturbances and hyper- or hypoesthesia through segmental spinal reflexes. These sensation disorders result in secondary ailments (alternative pathways of pathological stimulation). The above mentioned micromassage by wool fabrics facilitates the elimination of such unpleasant sensations. Roughness and thermal insulation also prove very advantageous during therapeutic exercise. It should also be remembered that different types of wool covers used on physiotherapy tables or floors additionally stimulate the skin, and thus perfectly enhance the effects of sensory integration therapy, and especially in children with neurological defects.

Therapeutic benefits of wool-covered orthopaedic appliances (e.g., neck supports) mainly depend on whether they are ergonomically designed. Adherence to biomechanical principles, i.e., appropriate size, shape and hardness of collars and belts used by individuals suffering from shoulder or spinal pain syndromes, ensures efficient sleep. These supports are also used during massage, physiotherapy and kinesitherapy interventions.

To sum up: the presented thermal, mechanical and ergonomic advantages of wool and specialist wool fabrics provide a large body of evidence to confirm that several properties of wool products facilitate the treatment and relief of numerous diseases and can thus be successfully used in medical care and **rehabilitation**.

Although specialist wool products alone may not be sufficient to achieve preventive and therapeutic objectives, they can certainly deliver excellent results as supplements to heat therapy, cryotherapy, electrotherapy and massage. The body's response to physical therapy lasts longer which results in shorter recovery time and lowers the costs of treatment.

## 6. References


[1] Avidan, AY. Zee, PC. (2007). *Sleep medicine handbook*. A. Wichniak, (Ed.), MediPage , ISBN 83-89769-50-3, Warsaw

[2] Ayling J, Marks, R. (2000). Efficacy of Paraffin Wax Baths for Rheumatoid Arthritic Hands. *Physiotherapy*, Vol. 86, No. 4, pp. 190-201, ISSN 0031-9406

[3] Bach, V. Telliez, F. & Libert, JP. (2002). The interaction between sleep and thermoregulation in adults and neonates. Sleep *Medicine Reviews*, Vol. 6, No. 6, December 2002, pp. 481-492, ISSN: 1087-0792

[4] Bender, T. Nagy, G. Barka, J. et all. (2007). The effect of physical therapy on beta – endorphin levels, *Europen Journal of Applied Physilogy*, Vol. 100, No. 4, pp. 371 – 82

[5] Beuth, B.(1968). *Textile materials technology*. Publisher of Light Industry and Food, Warsaw, Poland

[6] Birklein, F. Riedl, B. Neundörfer, B. & Handwerker, H. (1998). Sympathetic vasoconstrictor reflex pattern in patients with complex regional pain syndrome. *Pain,* Vol. 75, No. 1, pp. 93-100, ISSN: 0304-3959

[7] Bochenek, A. & Reicher M. (1989). Innervation of the skin. In: *Human Anatomy.* pp. 361-375, PZWL, ISBN 83-200-1230-9, Warsaw, Poland

[8] Chen, D.; Tan, L. Liu, H. Hu, J. LI, Y.& Tang F.(2010). Fabricating superhydrofilic wool fabrics. *Langmuir,* Vol. 26, No. 7, pp. 4675-4679, ISSN 0743-7463

[9] Ciesielska, I. Mokwiński, M. & Orłowska-Majdak, M. (2009). Influence of different kind of clothing material on selected cardiovascular, respiratory and psychomotor parameters during moderate physical exercise. *International Journal of Occupational Medicine Environmental Health,* Vol. 22, No. 3, pp. 215-226

[10] D'haeseleer, M. Cambron, M. Vanopdenbosch, L. De Keyser, J. (2011).Vascular aspects of multiple sclerosis. *Lancet Neurology,* Vol 10, pp. 657–66

[11] Druschky, KF.(1979). Conservative treatment of peripheral nerve lesion. *Fortschritte Der Medizin,* Vol. 97, pp. 1185-90

[12] Fechir, M. Schlereth, T. Kritzmann, S. Balon, S. Pfeifer, N. Geber, C. Breimhorst, M. Eberle, T. Gamer, M. & Birklein, F. (2009). Stress and thermoregulation: different sympathetic responses and different effects on experimental pain. *European Journal of Pain,* Vol. 13, No. 9, pp. 935-41, ISSN: 15322149

[13] Field, T. Robinson, G. Scafdi, F. Nawrocki, T. Goncalves, A. Pickens, J. Fox, N. Schanberg, S. Kuhn, C. (1996). Massage therapy reduce anxiety and enhances eeg pattern of alertness and match computations. *International Journal of Neuroscience,* Vol. 86, No. 3-4, pp.197 - 205

[14] Fornalczyk - Wachowska, E, Kuliński, W. (2008). Physical methods for prophylaxis and treatment of venous leg ulcers. *Acta Balneologica,* Vol. 50, No. 2, pp. 86-92

[15] Gieremek, K. Saulicz, E. & Nowotny, J. (1991). Local and reflexes changes in skin temperature under the influence of selected physical treatments. *Advances in Rehabilitation,* Vol. V, No. 3

[16] Gilbert, S. Cameron, J. Sally, A. & Drew, D. Thermoregulation as a sleep signaling system. *Sleep Medicine Reviews,* 2004, Vol. 8, No. 2, pp. 81-93, ISSN: 1087-0792

[17] Grass, D. (1982). Physical therapy and rheumatism of soft tissues. *Schweizerische Medizinische Wocheuschrift* , Vol. 112, No. 350, pp.1214 - 8

[18] Grycewicz, H.& Staniszkis, O. (1959). Wool. National Agricultural and Forestry Publisher, Warsaw, Poland

[19] Heller, HC. (2005). Temperature, Thermoregulation, and Sleep. In: *Principles and Practice of* Sleep *Medicine (Fourth Edition)*, Kryger, Meir H. Roth, Th. Dement, W C. pp.292-304, Saunders, ISBN 0721607977

[20] Heller, P.E. & Shear, N.H. (2006). Temperature-dependent skin disorders. Construction and Building Materials, Vol. 20, No. 6, July 2006, pp. 425-434, ISSN: 0950-0618

[21] Hornyak, ME. Naver, HK. Rydenhang, B. & Wallin, BG. (1990). Sympathetic activity influences the vascular axon reflex in the skin. *Acta Physiologica Scandinavica,* Vol. 139, No. 1-2, pp. 74-84

[22] Irzmańska, E. Lipp-Symonowicz, B. Kujawa, J. & Irzmański R. (2010). Textiles Preventing Skin Damage. *Fibres and Textiles in Estern Europe,* Vol. 18, No. 2 (79), pp. 84-90

[23] Kelechi, TJ, & Michel, Y. (2007). A descriptive study of skin temperature, tissue perfusion, and tissue oxygen in patients with chronic venous disease. *Biological research for nursing,* Vol. 9, No. 1, pp. 70-80, ISSN: 1099-8004

[24] Kiyak, EK. (2009). A new nonpharmacological method in fibromyalgia: the use of wool. *Journal of alternative and complementary medicine, Vol.* 15, No. 4, pp. 399-405

[25] Kun, Qian. Hongshun, Li. Haijian, Cao. Kejing, Yu. & Wei, Shen. (2010). Measuring the Blend Ratio of Wool/Cashmere Yarns Basedon Image Processing technology. *Fibres and Textiles in Eastern Europe,* Vol. 18, No. 4 (81), pp. 35-38

[26] Lespargot, A. Robert, M. & Khouri, N. (2000). Stretching the triceps surae muscle after 40 degrees C warming in patients with cerebral palsy. *Revue de chirurgie orthopédique et réparatrice de l'appareil moteur,* Vol. 86, No. 7, pp.712-7

[27] Leung, MS. & Cheing, GL. (2008). Effects of deep and superficial heating in the management of frozen shoulder. *Journal of Rehabilitation Medicine,* Vol. 40, No. 2, pp. 145-50

[28] Li Y. (2005). Perceptions of temperature, moisture and comfort in clothing during environmental transients. *Ergonomics,* Vol. 48, No. 3, pp. 234-48, ISSN 0014-0139

[29] Löfgren, M. & Norrbrink, C. (2009). Pain relief in women with fibromyalgia: a cross-over study of superficial warmth stimulation and transcutaneous electrical nerve stimulation. *Journal of Rehabilitation Medicine,* Vol. 41, No. 7, pp. 557-62

[30] Matusiak, M. (2010). Thermal Comfort Index as a Method of Assessing the Thermal Comfort of Textile Materials. *Fibres and Textiles In Easter Europe,* Vol. 79, No. 2, pp. 45 – 50

[31] Moncur, C. Shields, MM. (1987). Physiotherapy methods of reliving pain. *Baillieres Clinics Rheumatology, Vol.* 1, No 1, pp. 183-93

[32] Petrofsky, J. Bains, G. Prowse, M. Gunda, S. Berk, L. Raju, C. Ethiraju, G. Vanarasa, D. & Madani, P.(2009). Does skin moisture influence the blood flow response to local heat? A revoluation of the Pennes model. *Journal of Medicine Engineering and Technologies,* Vol. 33, No. 7, pp. 532-537, ISSN: 1464-522X

[33] Petrofsky, J. Goraksh, N. Alshammari, F. Mohanan, M. Soni, J. Trivedi, M. Lee, H. Hudlikar, AN. Yang, CH. Agilan, B. Pai, N. Chindam, T. Murugesan, V. Eun, Yim, J. & Katrak, V. (2010). The ability of the skin to absorb heat; The effect of repeated exposure and age. *Medical science monitor,* 2010 Dec 16;17(1):CR1-8, Available from www.ncbi.nlm.nih.gov

[34] Petrofsky, JS. Bains, G. Raju, C. Lohman, E. Prowse, M. Gunda, S. Madani, P. & Batt, J. (2009).The effect of the moisture content of a local heat source on the blood flow response of the skin. *Archives of Dermatological Research,* Vol. 301, No.8, pp.581-585. Epub 2009 May 5, ISSN: 0340-3696

[35] Pi-Chang, Sun. Hong-Da, Lin. Shyh-Hua, Eric Jao. Yan-Chiou, Ku. Rai-Chi, Chan. & Cheng-Kung Cheng. (2006). Relationship of skin temperature to sympathetic dysfunction in diabetic at-risk feet. *Diabetes Research and Clinical Practice.* Vol. 73, No.1, pp. 41-46, ISSN: 0168-8227

[36] Roberts, MF. & Wenger, CB. (1979). Control of skin circulation during exercise and heat stress. *Medicine and science in sport, Vol.* 11, No. 1, pp. 36-41

[37] Satoshi, W. Kajander K.C. & Bennett G.J.(1991). Abnormal skin temperature and abnormal sympathetic vasomotor innervations, in an experimental painful peripheral neuropathy. *Pain* Vol.46, No.3, September 1991, pp. 299-313, ISSN: 0304-3959

[38] Sefton, JM. Yarar, C. Berry, JW. & Pascoe, DD. (2010). Therapeutic massage of the neck and shoulders produces changes in peripheral blood flow when assessed with dynamic infrared thermography. *Journal of Alternative and Complementary Medicine,* Vol. 16, No. 7, pp. 723-732 ISSN: 1075-5535

[39] Shafik, A. (1993). Polyester but not cotton or wool textiles inhibit hair growth. *Dermatology,* Vol. 187, No. 4, pp. 239-242

[40] Shafik, A. (2008). An experimental study on the effect of different types of textiles on conception. *Journal of Obstetrics and Gynecology,* Vol. 28, No. 2, pp. 213 - 6

[41] Skanderi, Z. Salopek - Ćubrić, I. & Sdrjak, M. (2009). Water vapor permeability in a variety of climatic conditions. *Fibres and Textiles In Easter Europe,* Vol. 17, No. 2(73), pp. 72 - 75

[42] Straburzyńska - Lupa, A. & Straburzyński G. (2000). Thermotherapia. In: *Physiotherapy.* PZWL, ISBN 83-200-3372-1, Warsaw, Poland

[43] Strass - Blasche, G. , Ekmekcioglu, C. Vacariu, G. Melchart, H. Fialka-Moser, V. & Marktl, W. (2002). Contribution of individual spa therapies in the treatment of chronic pain; *The Clinical Journal of Pain,* Vol. 18, No. 5, pp. 302-9, ISSN 0749-8047

[44] Sudoł - Szopińska, I. & Chojnacka, A. (2007). Determination of conditions for indoor thermal comfort using the PMV and PPD indices. *Work Safety,* Vol.5, pp.19-23

[45] Szymusiak, R. Thermoregulation during Sleep and Sleep Deprivation. In: Encyclopedia of Neuroscience, 2009, pp: 971-975

[46] Tourula, M. Fukazawa, T. Isola, A. Hassi, J. Tochihara, Y. & Rintamäki, H. (2010). Evaluation of the thermal insulation of clothing of infants sleeping outdoors in Northern winter. *European Journal of Applied physiology,* Vol. 111, No. 4, pp. 633-640, ISSN: 1439-6319

[47] Tsuzuki, K. Okamoto-Mizuno, K. Mizuno, K. (2004). Effects of humid heat exposure on sleep, thermoregulation, melatonin, and microclimate. *Journal of Thermal Biology,* Vol. 29, No.1, January 2009, pp. 31-36, ISSN 0306-4565

[48] Wasner, G. Schattsschneider, J. Heck Mann, K. Maier, C. Baron, R. (200). Vascular abnormalities in reflex sympathetic dystrophy (CRPS I): mechanisms and diagnostics value. *Brain a journal of neurology,* Vol. 124, No. 3, pp. 587-99

[49] Wollina, U. Abdel-Naser, MB. & Verma, S. (2006). Skin physiology and textiles-consideration of basic interactions. *Current Problems in Dermatology.* Vol. 33. pp. 1-16, ISSN: 1421-5721

[50] Wright, A. & Slukak, A. (2001). Nonpharmacological treatment for muscoskeletal pain. *The Clinical Journal of Pain,* Vol. 17, No. 1, pp. 33-46, ISSN 0749-8047

[51] Xu, F. Lu, TJ. & Seffen, K.A. (2008). Biothermomechanics of skin tissues. *Journal of the Mechanics and Physics of Solids,* Vol. 56, No. 5, pp. 1852-1884, ISSN: 0022-5096

[52] Yiming, G. Ling, Ch. Xing, D.& Ning, P. (2010). Blood flow fluctuation underneath human forearm skin caused by local thermal stimuli of different fabrics. *Journal of Thermal Biology.* Vol. 35, No. 7, (October 2010), pp. 372-377, ISSN: 0306-4565

[53] Ziemniewska, M. Laurentowska, M. Bogacz, E. Kryscik, J. Domaszewska, K. & Ziemniewska, O. (2010). Influence of sportswear made from polyester and man-made cellulosic fibres on the energy cost of physical effort, *Fibres and Textiles in Eastern Europe,* Vol. 18, No. 3, pp. 94-99

[54] Żyliński, T. (1958). Learning about fiber. Publisher of Light Industry and Food, Warsaw, Poland

# Part 4

# Characteristics of Physical Therapy Interventions with Elderly Patients at Home and Within Clinical Settings

# 9

# Home Physiotherapy: The Relevance of Social Determinants of Health in the Development of Physiotherapy in the Home Environment

Berta Paz-Lourido
*University of the Balearic Islands, Spain*

## 1. Introduction

The current demographic, family and socio-sanitary trends require that many countries have to increase their effort for the delivery of health care at patient's home (World Health Organization [WHO], 2008). Among the services to be provided in the home is physiotherapy. This chapter will discuss some of the aspects that influence physiotherapy when performed in this particular domain, with special attention to the social determinants of health. The influence of social determinants of health has been given a significant emphasis in health promotion research in recent years, as well as in studies aimed at improving equity in health services.

In the first part of this chapter, the concepts of social determinants of health, equity in health, home care and home physiotherapy will be described. This will serve as the basis for the second part, a qualitative research project aimed at uncovering the perspectives of physiotherapists working in Primary Health Care about home physiotherapy. This is addressed to determine the varied influences of the physiotherapy intervention at home and its results. The relevance of the family-centred approach in home care requires a broader analysis of physiotherapy, highlighting the multi-dimensionality of home physiotherapy. This research cannot be generalised but elements may be transferred to similar contexts.

## 2. Social determinants, equity and health

It is widely acknowledged that heath is a multi-dimensional concept, and therefore transcends the mere absence of illness. The personal experience of health and wellbeing is influenced by several determinants, including the social determinants of health. The influence of social and environmental factors on health is not a recent concept, but the contribution of these social factors in achieving equity in health, requires a holistic approach to current health policies. The health care services must focus on the different health needs of the population in question. This implicates a great challenge for health professionals and policy makers to develop truly equitable health systems.

The consideration of health as a multi-dimensional phenomenon that does not depend exclusively or predominantly on the access to the specialised health services has brought important advances in the field of community health. Among them, the demystification of technology as a panacea to solve all illnesses, the consideration of disciplines beyond medicine and the recognition of the need for an interdisciplinary approach in health care. The model of health care has to focus on the health needs of people, understanding that they are influenced by many variables. This perspective highlights the changes required towards equity in health services, where a balance must be struck between the services offered and the ability to achieve them.

The social determinants of health are the conditions in which people are born, grow, live, work and age, including their relationship with the health system. The social determinants of health are mostly responsible for health inequities, the unfair and avoidable differences in health status seen within and between countries (Whitehead, 1988; Whitehead & Dahlgren, 2007). Within countries, the evidence shows that in general the lower an individual's socioeconomic position, the worse their health. There is thus a social gradient in health that runs from top to bottom of the socioeconomic spectrum. This is a global phenomenon, seen in low, middle and also higher income countries.

### 2.1 Social determinants of health

A consideration of health from a systemic perspective has implications for the organisation of health systems. The determination of the influence of contextual conditions such as housing, hygiene or working conditions have been the origins of public health, particularly during the industrial revolution.

In past decades there were attempts on behalf of the scientific community to demonstrate the multi-causal, ecological and systemic approaches to health. Nevertheless, health sciences have been clearly influenced by the theory of specific etiology of disease, which was in full agreement with the positivist paradigm of modern science (Anderson, 1998). Although this biomedical approach has been an educational and clinical source of intense criticism, its influences are routed in most of the academic curriculum of health professionals, where the biological factors of illness have greater relevance in their education than other cultural or social issues related to health.

However, recent concerns about equity in health are highlighting the diversity in health needs of the population. To achieve an equitable health system it is of utmost importance to identify the relationships between the health status and the influence of diverse determinants to it. At this point, beyond on the discussion about how to promote healthy lifestyles in the population, the current questions address how the social determinants of health impact in the adoption of healthy lifestyles. The approach to the social determinants of health means that social and economic status may influence the overall health status and the access to health care. Therefore, if the etiology of illness is focused on the cause of the illness, the social determinants of health are commonly described as "the cause of the cause", meaning the underlying social factors that contribute to health.

The recognition of the negative effects of the social determinants of health are being embraced by governments and are beginning to inform social and health policy. Some international events show the relevance of the social determinants of health on current discussions about what influences health. As an example, during the World Health Organisation Assembly held in 2004 the need to establish the Commission on Social

Determinants of Health (CSDH) was announced, which became operational in March 2005. This commission had the aim to generate recommendations that, based on the available evidence of the influence of social determinants in the health status of the population, could support interventions and policies to improve health and reduce health inequalities. The Pan American Health Organization (PAHO), has also turned its attention to this issue, led by countries such as Brazil and Chile, among others. During the annual meeting of the WHO/PAHO in September 2006, the agenda included discussion on the question of the determinants of health.

Some of the areas of discussion include the social gradient, stress, childhood living conditions, social exclusion, working conditions, unemployment, social support, addiction, access to food and transportation, among others (Lynch et al., 2000; Marmot & Wilkinson, 1999; Wilkinson & Marmot, 2003; Marmot, 2004; Whitehead & Dahlgren, 2007). Some of these concepts are explained below.

- Social Gradient: This concept refers to the different socioeconomic levels in a society. It can be explained with the image of a ladder. From the bottom to the top, each rung of the ladder indicates a socioeconomic level, where more disadvantaged groups are typically found on the lower rungs of the ladder. The existence of these social inequities in almost all countries poses the most serious change to improving the health of the population in general. The magnitude of the inequity experienced within a country will hold back other outcomes such as the advances in life expectancy.
- Social Exclusion: Social exclusion is the outcome of a loss of connection to the community in which one lives. Social exclusion is characterised by smaller social networks, fewer close relationships and reduced social support. Increased social interaction and participation in one's community perpetuates a sense of belonging and social connectedness. Both aspects have been linked to positive physical and psychological wellbeing, so their absence may have negative consequences.
- Living conditions: Living conditions may have a relevant impact on people's experience of health and wellbeing. The cost of privately renting suitable housing has a major impact on disadvantaged populations.
- Working conditions: Differences in exposure to unhealthy working conditions across the social spectrum is related to differences in the health status between socioeconomic groups.
- Access to Food: Access to nutritious food and clean drinking water is also vitally important for people's health and wellbeing, particularly in early life. A shortage of a variety of fresh healthy food is known to contribute to morbidity. However, processed and less healthy food outlets are often clustered within lower socioeconomic areas.
- Early life factors: The quality and of the environment in which a child lives from birth until early childhood determines their future development. Among other relationships, the importance of the brain development that takes place during early life may determine the school performance, considered to set the parameters of the outcomes as an adult.

Other determinants of health described in the literature are culture, education, the level of health literacy or access to health care (Canon, 2008). Nevertheless, the report from the Commission on Social Determinants of Health shows that social determinants of health are shaped by the distribution of power, money and resources (CSDH, 2008). That means that the key factor regarding social determinants of health is social justice at all levels. This highlights the need for social justice to be valued in parallel throughout the society to be integrated into policies and practices.

The identification of social determinants of health has been one of the pillars of the development of public health (Krieger, 2001), but the new global visibility on these determinants does resurface the issue of inequity in health and social justice, because many of the inequalities that decrease health are in fact avoidable (Whitehead & Dahlgren, 2007). However, a greater relevance of these concepts in research and the existence of an international committee for its study may not necessarily mean changing strategies in health policy. The level of evidence that we have today about the impact of social determinants in health is very important, but the outcomes for the people will depend on how this evidence is taken into account by politicians, managers, stakeholders and communities (Exworthy, 2008; Bambra et al., 2010).

### 2.2 Equity in health

Equity in health is multifaceted and represents the fair arrangements that allow equal geographic, economic and cultural access to available services for all in equal need of care (Whitehead, 1990). It is of utmost importance to describe the concept of equity to incorporate the differences in need for health services across different socioeconomic groups. This is because, given the social gradient in health status, the need for care tends to be greater among lower socioeconomic groups. Therefore, it is expected that this group makes greater use of public health services in comparison to more advantaged groups. When the reason for not using the services is poor geographic access or social or economic barriers, this would be considered unacceptable und unfair. This judgement regarding unfairness is based on universal human rights principles.

The inadequate access to essential health services is one of several determinants of social inequities in health (Whitehead & Dahlgren, 2007). The level of access to health services is related to the location and therefore physical availability of such service, but also to the economic and cultural access. The burden of payment for essential health services and drugs might reduce the possibility of access to the health system, in some cases exacerbated by other barriers such as language or cultural practices.

The level of health equity depends on the empowerment of individuals to change the unfair and steeply graded distribution of social resources to which everyone has equal claims and rights. Health inequities are avoidable inequalities in health between groups of people (Whitehead, 1988). These inequities arise from socially produced, modifiable, and unfair inequalities within and between societies (Whitehead, 1990; Marmot, 2004). Social and economic conditions and their effects on people's lives determine their risk of illness and the actions taken to prevent them becoming ill or treat illness when it occurs. The objective of equity in health is the elimination of systemic differences in health status between socioeconomic groups.

The end goal of equity in health care is to match services to the level of need, which may vary among the different groups (Whitehead & Dahlgren, 2007). In practice, the need to take action to reduce inequities and their root causes is becoming ever more present as a major public health issue. The equity in health systems should therefore be built upon equity principles:

- Public health services should not be driven by profit, and patients should never be exploited for profit.
- Services should be provided according to need, not ability to pay.
- The same high standard of care should be offered to everyone, without discrimination with respect to social, ethnic, gender or age profile.
- The values and equity objectives of a health system should be explicitly described, as well as the monitoring carried out to ensure these objectives are approached in the most efficient way possible.

Both equity in health and equity in health care are goals that clearly need a cross-sectoral approach. Health equity in all policies means that every aspect of government and the economy has the potential to affect health and health equity: finance, education, housing, employment, transport or health. While health may not be the main aim of policies in these sectors, they have strong bearing on health and health equity. Policy coherence is crucial; different government departments' policies must complement rather than contradict each other in relation to health equity.

### 2.3 Primary Health Care and Home Care

The conceptual model of Primary Health Care was stated in the International Conference of Alma Ata (WHO & Unicef, 1978) and gives great importance to the development of equitable health systems but also to the role of different sectors at national and local levels to tackle the diverse factors that contribute to illness and poverty. As stated at Alma Ata Conference (1978):

> "Primary health care is essential health care based on practical, scientifically sound and socially acceptable methods and technology made universally accessible to individuals and families in the community through their full participation and at a cost that the community and country can afford to maintain at every stage of their development in the spirit of self-reliance and self-determination."

The strategy of Primary Health Care highlights the need for a comprehensive health strategy not only providing health services, but addressing the underlying social, economic and political causes in poor health. From then until now, the model of Primary Health Care represents a more equitable system, sensitive to the needs of the population (Vüori, 1984; Starfield et al., 2005; Baum, 2007; De Maeseneer et al., 2005), but for this, its development in practice has to remain faithful to its founding principles. As stated in the Declaration of Alma Ata (1978), some of them are: (a) health inequity within a country is socially, politically and economically unacceptable and is therefore a concern both for government and wider society, (b) governments should adopt cross-sectoral strategies as part of national primary health care plans to address the health of all citizens, and (c) governments have a responsibility for the health of its citizens and this can only be achieved by the provision of adequate health and social measures.

Home care aims at satisfying people's health and social needs while in their home. Sociodemographic change and mobility trends affect home care needs. Life expectancy has risen sharply in the WHO European Region in the last few decades (WHO, 2008). The percentage of population over 65 years in Spain has increased from 10.58% in 1975 to 17.36% in 2010. Considering the data from the Organisation for Economic Cooperation and

Development (OECD), it is expected that by 2050 the percentage will be 35.7%, the highest in Europe. This means that one in three people in Spain will be over 65 years old. Though aging cannot be identified with illness, it is true that in a high percentage of the elderly the incidence and prevalence of diseases and their progressive functional limitation is directly affecting their life and relationships, increasing the demand of health services in their homes (Van Haastregt et al., 2000, WHO, 2008).

If empirical evidence shows a close relationship between age and disability, there is also a relationship to the level of dependency. This causes a reduction in participation as embodied in the dependence on help from others. Among the determinants of dependence are demographic factors such as aging, but also social factors such as the modification of the traditional family structure because of changes in the family dynamics such as more women working or increasing geographical distance between members of the same family. There are an increasing number of people living alone in their homes which also highlights the changing attitudes towards older members of society (Oliver, 1999; Sundström & Tortosa, 1999, WHO, 2008)

The recipients of home care are usually people who are not independently able to go to the health centre, so the intervention in homes may represent a good response to the challenge of aging and dependency of the population, although not all those that need home care receive it (Branick, et al.,2002; Kupper-Nybelen et al., 2006). Therefore, people treated at home are usually elderly and dependent, particularly vulnerable and sometimes invisible to the health care system. This may end in less priority on the provision of adequate services in level of health care, putting in risk the equity in health access.

People who stay at home may have different support systems. The closest is made of the family caregivers and informal support networks. Social services develop other activities, including the provision of home services or accommodation in nursing homes when required. The provision of formal care to people with similar dependency varies widely among countries, highlighting the dificulties to compare home care services internationally (Carpenter et al., 2004; Sørbye et al., 2009; Genet et al., 2011). Other public, private and non-profit agencies such as charities or associations very often serve the needs of patients at home by offering home services (Adelvert & Svetling, 1993; Hutten & Kerstra, 1996, WHO, 2008). Home physiotherapy can also be developed from different public and private agencies, but in the countries where a Primary Health Care system exists, it is usually considered part of the community approach in this primary level. Although specific interventions can be developed from specialist care, including hospitalisation in homes, continuity of care at home is one of the characteristics of Primary Health Care in Spain (Contel, 2000).

### 2.4 Home physiotherapy

Home physiotherapy in Primary Health Care includes a set of community-based activities performed at home, in order to detect, assess, support and give continuity to the health needs of individuals and their families, thus enhancing its autonomy. Therefore, in the home intervention, the physiotherapist's role relates to the the assessment, prevention and intervention, detecting compensatory mechanisms or ergonomic aspects to the patient, family and carers in their context (Von Koch et al., 1998). The goals of physiotherapy at home can focus on three areas: ensuring access to appropiate physiotherapy services, promoting the participation of the family and advising other professionals who perform home care.

Regarding users of home physiotherapy the elderly are the main - though not the only - beneficiaries, although changes in epidemiological patterns and the possibilities of technology transfer widen the availability of home physiotherapy (Elkan et al., 2001; Eriksson et al., 2010). Literature refers research studies and experiences in home physiotherapy envolving patients with diverse diagnosis. For some authors, people who have suffered a stroke or cerebral vascular accident (CVA) and have been discharged from hospital, should have access to patient or community services appropriate to their needs and may include home rehabilitation (Torsén et al., 2005; Turner-Stokes et al., 2005). The authors acknowledged that during home physiotheray other factors may influence the health outcomes, such as the physical and psycho-emotional aspects of the home (Hale et al., 2003). These elements have also been targeted in other studies, because home care also facilitates the development of greater involvement and control capability of the person (Von Koch et al., 1998; Widén et al., 2000). Due to that, the intervention of physiotherapy in the home is a valuable alternative in providing acess to health care services (Tornsén et al., 2005).

Home physiotherapy has been also studied from the scope of rehabilitation of people with cardiovascular and respiratory problems, where physiotherapy visits may decline the hospital admissions and reduce the length of hospital stay for these patients. But for some authors, the basic objectives to be achieved by a program of this kind go beyond the mere fact of reducing hospital admissions, focusing on quality of life, facilitating compliance and promoting a positive attitude towards independent life (Kirilloff et al., 1986; Bingöl et al., 2000; Jolly et al., 2003; Taylor et al., Smith, 2011).

For patients with osteoarticular pathology, given the rapidity which patients undergoing orthopaedic surgery return home, the intervention of physiotherapy at home after discharge from the hospital can determine the functional outcome for these patients (Deyle et al., 2005; Mangione et al., 2005). Discharging patients who have undergone recent hip or knee replacement from hospital without supervision may jeopardise the outcome of the prosthesis and thus their quality of life (Kuisma, 2002).

Most of the studies related to home physiotherapy focus on the effects of particular physiotherapy techniques to be applied to specific pathologies at the patient's home (Roddy et al., 2005; Nilsson et al., 2007). This kind of research has a great value to deliver evidence-based services of physiotherapy, although does not always consider how the social factors and the home environment intervene in the process (Paz-Lourido, 2008).

Another area of interest for research regarding home physiotherapy is the financial cost of this intervention in relation to a given profile of patient (Lloyd et al., 2010). However, the focus on the illness does not always show the real needs of people at home, which should be the starting point to describe the characteristics of the practices to be developed, the type of professionals required and therefore, the need for home physiotherapy.

## 3. The research study

Qualitative research is commonly used to gain insight into people's attitudes, behaviours, value systems, concerns, motivations, culture or lifestyles. This kind of research is widely employed in health research with the purpose of highlighting in -depth and contextualised data which can be used for policy formation.

The publication of qualitative research requires a description of the context in which research occurs in order to determine whether it can be transferable to other similar contexts. For this reason, this third part of the chapter will begin with this description and then details the research process and its results.

### 3.1 Context

This study was conducted in Majorca, the largest island in the Spanish archipelago known as the Balearic Islands, located in the western part of the Mediterranean Sea. The current structure of the Spanish health system was established during the transition to democracy during the 1970s and 1980s. The national health system is a decentralised organisation divided into 17 regions or autonomous communities, it is financed by taxes and with universal coverage. Health policy-making also lies at the regional level. The central government has the responsibility to promote coordination and cooperation in the health sector, as well as to ensure that the quality of all services is guaranteed. All autonomous communities have stipulated territorial subdivisions such as health areas and zones. Primary Health Care and specialised care services are organised in order to address the population's health needs (Durán et al., 2006)

The health care system is organised on primary and secondary levels, with an extensive network of Primary Health Care centres and hospitals throughout the region. The regulation for the transformation into the Primary Health Care system was developed in 1984, inspired by the Declaration of Alma Ata (1978). Nowadays, the Primary Health Care in Majorca is an integrated system composed by multidisciplinary teams, where general practitioners (GPs) have a gatekeeper role.

Primary Health Care includes care of individuals, families and the community at large through health promotion programmes, prevention, curative care and rehabilitation. The Primary Health Care reform in Spain was implemented in the country slower than expected, delayed by problems such as lack of resources and political priority, compared to other areas such as specialised care (Martín, 2003). Physiotherapists were integrated in Primary Health Care organisation in Spain in 1989, although the regulation of their role and functions in PHC was not clearly stated until 1991. Act 4/91, which regulates physiotherapy in Primary Health Care at the national level in Spain, considers the physiotherapist as a health professional with functions including direct intervention, prevention and education, research and management. The referral can come directly from GPs (and sometimes from specialists from hospitals). Attending to the Act 4/91, the physiotherapeutic interventions in Primary Health Care can be made in the health centre, in homes or in other contexts in the community. Nevertheless, GPs, nurses and social workers are considered to be the key professionals for home care and these services are provided all over the country (Martín, 2003).

Physiotherapy is considered to be an essential service to be provided in Primary Health Care and in specialised care in Spain. Nevertheless, the delivery of home physiotherapy as part of the PHC services differs among the regional communities. As a result, in some areas of the country home physiotherapy is not offered as a health service at the primary level of the health system. Despite laws regarding quality and cohesion of the Spanish Health System, the differences for Spaniards in access to health services depends on the autonomous community where they live, illustrating some of the negative effects of decentralisation and highlighting the need for coordination and cohesion among regions (Act 16/2003).

A total number of 22 physiotherapists work nowadays in Primary Health Care in Majorca[1]. They are located at physiotherapy units in some health centres to cover a population over 600.000 at this level of care. Since home physiotherapy service is not offered at this level of care, the health system provides free transportation of dependant people from their homes to the units located in any of the four public hospitals, which is specialised care and not Primary Health Care.

### 3.2 Study design

Qualitative studies include a range of research designs, attending to the theoretical perspectives, data collection instruments and methods of analysis. In the next paragraphs, the design of this study will be described.

#### 3.2.1 Aim

Taking into consideration a future implementation of home physiotherapy in the Balearic Islands, this study attempts to reflect how physiotherapists consider their role in this particular context, researching contextualised aspects that could inform policies for the provision of adequate home physiotherapy.

#### 3.2.2 Theoretical perspective

The theoretical perspective in this research is the social-critical paradigm. This paradigm has evolved since its inception in the Frankfurt School in the 1920s. The socio-critical paradigm was developed based on critical theory, which aims to transform the social structure and empower the disadvantaged.

The aim of qualitative research, under the influence of this paradigm, is to uncover the elements underlying social injustices in order to contribute in their transformation (Lincoln & Guba, 2000; Kincheloe & McLaren, 2005). Therefore, the researcher has to get involved in the processes of self-reflection and decision-making to inform policies, which are assumed in a responsible manner (Popkewitz, 1984). Critical consciousness is brought about not through intellectual effort alone, but through praxis, through the union of action and reflection (Freire, 1979).

This theoretical perspective has been selected for this research because it allows the understanding of the reality of physiotherapy in Primary Health Care as the praxis developed in this particular context, joining theory and practice. It is needed for the integration of knowledge, action and values to orientate the development of this public service.

---

[1] This number refers only to physiotherapists working in Primary Health Care. It doesn't include those working in specialised care and other private/public institutions.

### 3.2.3 Purposive sample

Participants in this study were 8 physiotherapists working in the Primary Health Care in Majorca. The purposive sample was selected taking into account the health centre where they worked and the personal and professional profile of the physiotherapists. A socio-demographic questionnaire was administered to the suggested participants, in order to select physiotherapists with different profiles regarding professional experience, level of expertise, age and location of the health centre (urban/rural).

Since home physiotherapy is not yet developed from Primary Health Care system in Majorca, the purposive sample included physiotherapists with previous experience in home care in the private sector.

### 3.2.4 Data collection

The main data collection tool was the semi-structured interview. The use of in-depth interviews was selected because they allow the participants' perspectives to be obtained and deepen the conceptualisations from which they are derived (Fontana & Frey, 2005).

The interviews were conducted until data saturation was achieved. This was the moment in the investigation in which participants' views are repeated, are redundant and no new information was provided. The interviews were recorded, transcribed and analysed. A research diary and a socio-demographic questionnaire were also used for data collection. These provided non verbal information and extra data regarding the participants profile, which was useful during the process of analysis.

### 3.2.5 Analysis

The data in qualitative research refers essentially to people, objects and situations. In this sense, the experiences are converted into words, typically compiled in texts (Huberman & Miles, 1994), whose transcription process also involves an analysis phase (Silverman, 2003). The first part of the analysis and interpretation of the information was produced in parallel with the collection of information during interviews, and later with a careful analysis of the registered data using discourse analysis. This kind of analysis highlights the social constructions that underlie what the interviewees say, taking into consideration that the speech is made up not only by ideas, ways of thinking and ways of speaking, but also practical, social and institutional relationships. The use of language is a specific method of social interaction and it is also analysed from multiple perspectives and schools of thought, both from the social sciences or the humanities and from psychology.

The discourse represents a set of linguistic practices that maintain and promote certain social relations. Some authors (McLure, 2003; Renkema, 2004) define it as critical discourse analysis, as this more clearly specifies that the analysis is not neutral or value-free. This may allow a critical analysis of the ideology that underlies it and its consequences in the context.

### 3.2.6 Triangulation

Qualitative research, like all research, has to be done with a precise methodological rigour. For Morse (1994) rigor in this type of research is given by various processes among which the adequacy of the selection of data to the theoretical framework of the study, obtaining data from multiple sources and methods, the addition of documentation to reconstruct the process by which the research team arrives at conclusions or the participation of other experts. This rigor has to be accompanied by an explanation of the reflective process to identify the strategies followed in the process and the motivations that lead to these (Janesick, 2000; Chesney, 2001; Malterud, 2001).

In this study, the triangulation of sources and methods was used. As to sources, a literature search was conducted using national and international literature. On the other hand, different data collection methods have been used ranging from the demographic questionnaire, individual interviews and research diary.

Another point of information for the triangulation has been the opinion of experts on both the geographical area studied, as well as other communities in the country and abroad. From the research process emerged discussion items that were presented at national and international forums incorporating elements thereby allowing for triangulation.

### 3.2.7 Ethical issues

This research has attempted to adjust the process to the ethical principles that avoid the risk that research can entail. In this sense, the participation of the group interviewed in the study was conducted after prior information of the objectives of the project, the institution that employs the researcher and the research process and expected benefits was given (Christians, 2005). Anonymity and confidentiality were assured at all stages in the process. Informed consent was obtained and other aspects regarding ethical issues in qualitative research were considered (Richards & Schwartz, 2002).

Therefore, although the knowledge gained from the study was expected to be used in the transformation of health services, the researcher explained that may not be a direct effect in terms of professional status or the organisation/implementation of any service. In this sense, the literature shows the existence of fragmentation between policy making bodies and research could potentially advice those who make decisions that affect the community (Rist, 2003).

Ethical issues have also been taken into account when presenting the results of the study, taking into consideration the confidentiality and anonymity of the interviewees. Ethical approval was received from the Committee on Bioethics of the University of the Balearic Islands.

### 3.2.8 Limitations of the study

This study refers to some health, social and educational aspects in reference to home physiotherapy, but many other issues may be considered in further research. The structure of health community services may be different in several countries, illustrating the different ways of developing the national health systems and therefore the professional practices developed. The structure of the care-giving system or the cultural concept of family among different areas should be also considered regarding the transferability of elements of this study to other contexts.

### 3.2.9 Results

Home physiotherapy was described in various ways by physiotherapists, since many factors from the home environment influence their perceptions. Some of those factors are the conditions in which families live, the availability of resources of formal or informal care, or coordination with other professionals that are also involved in home care.

> "Home physiotherapy is somewhat different. You never know what you'll find in a home. Here, in the health centre, you have everything organised in your way and you have somehow…control over everything. Here you have the equipment, other colleagues…you decide. But there, it is different. You arrive there and you have to adapt yourself to what you have there"(PT, 6)

The current crisis in informal care in many countries has major implications for the organisation of home physiotherapy. For many decades, it was traditionally accepted that the families had most of the responsibility in caring tasks, but this is not possible for many families any more. This was stated during the interviews with the physiotherapists.

> "If the family or a caregiver is there with the patient everything is much better. We could train the caregiver to keep what we have obtained during the session. But sometimes that is not possible. Sometimes the caregiver is more fragile than the patient, and sometimes there is nobody to help." (PT, 4)

Other issues related to physiotherapy were the perception that it is an intervention that requires a high level of physical exertion and the development of some techniques is more difficult in homes than in a clinical setting.

> "Yes it is true that in the field of physiotherapy, home work is hard, because you have to keep harmful positions in homes, which are often poorly adapted, but it is also true that a person who is dependent and it is not able to come independently to a primary care service ... or go to a hospital service, needs care at home".(PT, 3)

In general, the group interviewed mentioned specific aspects of home physiotherapy such as prevention of falls and prevention in people with chronic respiratory conditions, to prevent crises and relapse. The development of activities for assessing the actual and potential physical function to infer the level of autonomy and dependence were particularly noted, athough remained focused on the disease and disability.

> "I believe that every dependant patient in home needs and an initial assessment, and ... well ... then it should be decided what treatment may continue…and regarding physiotherapy, it should be analysed if it is necessary or whether the patient can benefit from that activity. But I consider home physiotherapy to be something more…educational." (PT, 7)

In this sense, the interviews with the various physiotherapists hint at two types of objectives for the intervention of physiotherapy. Some reflect the consideration that objectives are aimed at the recovery of the pathological processes that affect a person, and the other objectives are intended to prevent further complications from their dependency. In this sense, a distinction was verbalised between those in which interventions are expected to achieve observable results in the short term or those in which the focus is managing the conditions with the person is living with. With respect to the latter, their work was described as health education and advise.

There was a general perception that physiotherapy in Primary Health Care should cover those dependant people located in homes, but this kind of intervention seemed to have less prestige than the one developed in the health centre. The lack of human resources for developing home physiotherapy was the main reason explained, although some clearly expressed their dislike of home intervention. Their discourses on this topic gave visibility to the socio-economic determinants of health and the impact of inappropriate resources in public health services.

> "I am not sure if physiotherapy in Primary Health Care includes home visits. But nowadays, for us, is completely impossible to provide this service. We have a long waiting list and if I go to visit a home I am not treating six or seven people here. This is the situation." (PT, 8)

> "In my view home physiotherapy is a necessary service. We should offer it because it is part of Primary Care, and if we don't offer it, nowbody does it. Of course if you have money, you can pay a private service of physiotherapy in your home, but many people cannot afford it. If one day this service is established, I would have to go, I know, and I will go, but the truth is that I don't like it very much." (PT, 4)

The question of cost and inconvenience of transporting the dependents to physiotherapy services also appeared during the interviews. The costs of transportation of patients to the ambulatory services in hospitals were considered higher than the displacement of the physiotherapists to the homes. Transportation of patients for ambulatory physiotherapy was considered to have more disadvantages than advantages, which may eventually lead to patients giving up their treatment. Specific issues were highlighted such as the difficulty of overcoming architectural barriers to access to transport, the time constraints, delays, stress during transfer and the conditions of waiting in health care, especially for people with a delicate state of health.

> "Perhaps there are particular situations where being attended in the physiotherapy services in hospitals could benefit the patients, because maybe they are too lonely at home and it is good for them communicate with others. But often, it would be better for the patient and the physiotherapist to organise a home visit. We could also give advice to the patient, make recommendations to the family, care for the caregiver, and all these kind of things...But by now the system is like that. If patients are not independent enough, they are transported to get the physiotherapy" (PT, 1)

Communication with other professionals was seen as particularly relevant for working at home, but they noted that this collaboration is already poor in the health centre. They highlighted in particular that the communication between the physiotherapists, GPs and nurses is essential not only to provide the best care to the person and establish a continuity, but also to avoid conflicts. The role of physiotherapists in evaluating the patient's physical performance was seen as helpful for social workers when making decisions about the most appropriate social resources for individuals and their families. Nevertheless, most of the physiotherapists interviewed stated that other health care professionals had a fundamental lack of knowledge about the professional practice of physiotherapy, what was descibed as influencing to the lack of cooperation between different professional groups.

> "It seems that everybody knows what we do, but then you realise that nobody knows what physiotherapy is." (PT, 2)

Finally, other aspects such as research and teaching emerged in the interviews. Particularly some interviewees note that home care is a field that is provided in Primary Health Care and that it represents a scientific and professional development for the group of physiotherapists working in this level of care.

> "It is another area that can be developed, like any other area. If it is not developed, this professional field is lost" (PT,4)

The quotes reveal the perception that a professional area that is not exercised, missing or being taken over by other groups or health systems, compromises the future of the discipline and the professionals responsible for their development.

### 3.2.10 Discusion

One of the most important characteristics of home physiotherapy is that it is carried out in a different environment to clinical consultancy, as was highlighted in the results of this study. The home is considered in the Sundvall Declaration (1991) as a supportive environment for health, a community environment that gives people protection from threats to health, while allowing them to expand their skills and develop the autonomy to health (WHO, 1991). The home therefore represents the normal environment in which life has developed, and living in it, under appropriate conditions, is conceptually linked to the possibility of obtaining a better quality of life. Therefore, it must be understood not just as a physical space but also as a psychological and social concept, often linked with the warmth, security and protection from the pressures of work and public life (Bowlby et al., 1997). But on the other hand, when various health care professionals intervene in the home for a long period of time, changes occur in the meaning and habits that transcend the limitations of the home and which also involve the families (Angus et al., 2005; McGarry, 2010). A person with chronic illness in the family may affect the whole family functioning and at the same time, the relationship with the family may have an impact on the person with disability (Greenwood et al., 2009)

Not taking into account the characteristics inherent in the home in a broad sense are thus dismising the meanings that the home has for the person living there, and for the professional who has to intervene. In this study, the results highlight the perception of the home environment as an uncertain setting, where physiotherapists can have less control over the situation, pointing out changes in the power relations that may influence the therapy. Home physiotherapy requires them to be more creative when developing the appropiate physiotherapy treatments, but also to put in practice social skills to have an effective communication with the patients, families and other professionals.

Although the quality of life of those in the home can be influenced by factors such as loss of mobility, other aspects may be involved in the perception of quality of life. For example, the absence of family relationships, inability to perform activities of daily living and reduced social contacts and communication may also be associated with insufficient financial resources. For this reason, home physiotherapy intervention should not ignore these social determinants of health. The concept of home is not only steeped in the relations between its inhabitants, but is expanded to include the surrounding neighborhood with its network of social relations and services. In this way, family, neighbours, friends, associations or community groups, social and health services and schools can participate and influence in some way in the informal care that is performed in the home (Hanson & Pratt, 1988; Linström et al., 2002; Ollonqvist et al., 2007). This demostrates how relevant social networks are for the development of home care in general and home physiotherapy in particular.

It was a feeling among the interviewees in this research that it is very difficult to implement the service at people's homes mainly due to the lack of human resources and political priority. But they could see the benefits of home physiotherapy for those who could afford a private physiotherapy. This highlights the impact of social gradient in the access to health care and thus the issue of equity in health.

It is recognised that current demographic changes are already leading to greater pressure on health systems, since the rate of disability increases with age and has caused an increase in the number of dependents who need care at home (WHO, 2008). In this situation, it seems that many countries are committed to home care as a useful mechanism to address the health needs of users. However, at the same time, the number of studies on this topic have increased considerably in recent years, driven by improving the cost efficienty of specialised care developed in homes. This occurs at the expence of taking in consideration the larger systemic factors around the social determinats of heath and equity in Primary Health Care (Paz-Lourido, 2008).

Although several studies focus on assessing its benefits in comparison with other environments of intervention, one of the greatest difficulties in analysing the results of physiotherapy intervention in the home is in the low specificity of the studies about physiotherapy intervention (Ward et al., 2005). Without doubt, it must be recognised that health care in homes involves the interaction of multiple factors which are difficult to assess. It is challenging to depict such a multifaceted intervention, which influences many dimensions, not only physically but also psychologically and socially. As an example of this complexity, caregivers may be regarded as a resource for continuity of physiotherapeutic intervention, but on the other hand, they may be a group susceptible to a specific intervention due to the emotional and physical demands of the process of caring (Widén et al., 2000). Despite the complexities providing physiotherapy in the home it is deemed less prestigious that offering the same level of care within health care settings.

Several studies are making headway on the conceptual framework for physiotherapy practice at home, especially in terms of relational aspects between professionals and patients treated in the home environment (Von Koch et al., 1998; Wottrich et al., 2007). Other authors call attention to the value of structuring the intervention of physiotherapy on the needs and expectations of the person at home (Hale et al., 2003, Hale & Piggot, 2005). This suggests that the objectives of the intervention of physiotherapy at home should focus on the needs of the person therein situated, not only in observable physical findings or standardised protocol for intervention. This points to a future where research on physiotherapy intervention takes into consideration the cultural, emotional and social issues of home, beyond the focus on pathology and its costs.

### 3.2.11 Implications

Both the interdisciplinary and inter-sector approach have significant importance for professionals when working in the home environment, which means that the implementation or development of the service of home physiotherapy has political, organisational and educational implications. The general perspective of home intervention seems to be of high importance for working at patient's homes. This leads to the importance of inter-professional education, with influence in avoiding the stereotyped vision of physiotherapy that many social and health professionals still have, but also in a better integration of physiotherapy in Primary Health Care in the geographical context of this study.

The development of home physiotherapy requires not also technical expertise but also social skills in particular, to facilitate the communication and coordination between professionals of different health and social services. Taking into consideration the difficulties for the development of the needed inter-professional coordination, more focused efforts should be made to avoid the independent practices among Primary Health Care teams.

Creativity and social skills are also necessary to take advantage of this particular environment for the treatment and establish good communication with family, informal and formal caregivers. The home has emotional implications for patients and families that transcends the physical structure. That should be taken into consideration when deciding the need for health care.

Informal care has often shifted to formal care, and therefore it is possible to find new figures in the home arena, such as privately contracted caregivers or professionals from social services with different levels of responsibility in the continuity of care. This requires that the physiotherapist should understand the organisation of social services and the different professionals that intervene. Physiotherapists have an important role not only in direct intervention in the home, but also in the development of health programs for caregivers and other support networks. Nevertheless, the effects of home physiotherapy for patients may be influenced not just by the techniques developed by the physiotherapists, but also by the intervention of other health or social teams and the caring system as a whole.

The social factors of health such as living conditions, social gradient of social exclusion are related to issues such as health needs, equity in the access to health care or the role of the health system to satisfy health rights within society. The social conditions for the development of home physiotherapy and the professional role of the physiotherapists should be consciously integrated to develop the adequate services in the context.

## 4. Conclusion

In light of this chapter, the engagement in the home requires an understanding of the home as a particular environment, where the social determinants of health are more visible. Therefore, these aspects should be considered in reorienting a physiotherapy practice based on a more holistic approach to health as well as in the educational curriculum in physiotherapy. This new praxis can emerge from a more comprehensive theoretical perspective combined with supervised clinical practice in the home setting. Taking into consideration that many other health and social professionals may be required in home care and rehabilitation, well designed interprofessinal education is of critical importance for professionals and families. Finally, all these efforts should be designed in parallel to policies and strategies aimed to adapt the delivery of quality home physiotherapy balanced with the real needs of people, but also with cross sectoral strategies towards achieving equity in health.

## 5. References

Adalvert, E. & Svetling, I. (1993) *Balancing Pluralism: new welfare mixes in care for the elderly.* Viena: European Centre of Viena. Austria

Anderson, P. (1998).The importance of theories in health care. *British Medical Journal*, 317, 7164, pp.1007-1010. ISSN 14685833

Angus, J.; Kontos, P.; Dyck, I.; McKeever P. & Poland, B. (2005). The personal significance of home: habitus and the experience of receiving long-term home care. *Sociology of Health & Illness*, 27,2, pp. 161-187. ISSN 1467-9566

Bambra, C.; Gibson, M.; Sowden, A.; Wright, K.; Whitehealth, M. &Petticrew, M. 2010, Tacking the wider social determinants of health and health inequalities: evidence from systemic reviews. *Journal of Epidemiology & Community Health*, Vol. 64, 284-29, ISSN 1470-2738

Baum, F. (2007). Health for all now! Reviving the spirit of the Alma Ata in the twenty first century: An introduction to the Alma Ata Declaration. *Social Medicine*, Vol.2, No.4, pp. 34-41, ISSN 1557-7112

Bingöl, G.; Güneter, S.; Sarpel, T; Sur, S; Ufuk, D. & Yimaz, M. (2000). Efectos de un programa de rehabilitación diaria domiciliaria en niños asmáticos. *Allergologia et Immunopathologia*, Vol.28, No.1, pp.12-14. ISSN 0301-0546

Bowlby, S.; Gregory, S. & Mckie, L. (1997). "Doing home": Patriarchy, Caring and Space. *Women's Studies International Forum*, Vol.20, No.3, pp. 343-350. ISSN 0277-5395

Branick, L; Earnest, G; Long, L; Regnell, M; Warnecke, P. (2002). The home care therapist: the art of caregivig. *Caring*, Vol.21, No.2, pp.32-6.

Cannon, R. (2008). *The Social Determinants of Health*. SACOSS Information Paper (December 2008). South Australian Council of Social Service, ISBN 098-0-908293-53-7, Unley, Australia

Carpenter, I.; Gambassi, G.; Topinkova, E.; Schroll, M.; Finne-Soveri, H.; Henrard, J.C.; Garms-Homolova, V.; Jonsson, P.; Frijters, D; Ljunggren, G., Sørbye, L.W.; Wagner, C.; Onder, G.; Pedone, C. & Bernabei, R. (2004). Community care in Europe. The Aged in Home Care project (AdHOC). *Aging Clinical and Experimental Research*,Vol.16, No.4, pp. 259-69. ISSN 1720-8319.

Chesney, M. (2001). Dilemmas of self in the method. *Qualitative Health Research*, Vol.11, No.1, pp. 127-135. ISSN: 1552-7557

Christians, C.G. Ethics and politics in Qualitative Research. In: *The SAGE handbook of qualitative research*,Denzin N K & Lincoln YS, 139-164, Sage, ISBN 0-7619-2757-3, Thousand Oaks, CA, USA.

Contel, J.C. (2000). La atención a domicilio como modelo de atención compartida. *Atención Primaria*, 25, pp.526-527. ISSN 0212-6567

CSDH (2008). *Closing the gap in a generation: health equity through action on the social determinants of health. Final Report of the Commission on Social Determinants of Health*. World Health Organization. ISBN 978-92-4-156370-3, Geneva, Switzerland

De Maeseneer, J. et al. (2007). *Primary Health care as a strategy for achieving equitable care*. Health Systems knowledge Network. World Health Organization.

Denzin, N. K. & Lincoln, Y. S. (2003). Introduction: The discipline and practice of qualitative research. In: *Strategies of qualitative inquiry*, N. K. Denzin & Y.S. Lincoln (Eds.),1 – 45, Sage, ISBN 0-7619-2757-3, Thousand Oaks, CA, USA.

Deyle, G.D.; Allison, S.C.; Matekel, R.L.; Ryder, M.G.; Stang, J.M.; Gohdes, D.D.; Hutton, J.P.; Henderson, N.E. & Garber, M.B. (2005). Physical therapy treatment effectiveness for osteoarthritis of the knee: a randomized comparison of supervised clinical exercise and manual therapy procedures versus home exercise program. *Physical Therapy*, Vol.85, No.12, pp. 1301-17, ISSN 1538-6724

Durán, A.; Lara, J.L. & Van Waveren, M. (2006). Spain: Health system review. *Health systems in Transition*, Vol.8, No.4, pp.1-208. ISSN 1817-6127

Elkan, R.; Kendrick, D.; Dewey, M.; Hewitt, M.; Robinson, J.; Blair, M.; Williams, D. & Brummell, K. (2001). Effectiveness of home based support for older people: systematic review and meta-analysis. *British Medical Journal*, Vol.323, pp.1-9. ISSN1756 5833

Eriksson, L.; Lindström, B. & Ekenberg, L. (2010). Patients' experiences of telerehabilitation at home after shoulder joint replacement. *Journal of telemedicine and telecare*, 17, 1,pp.25-30. ISSN 1758-1109

Exworthy, M. (2008) Policy to tackle the social determinants of health: using conceptual models to understand the policy process. *Health Policy and Planning*, Vol.23, pp. 318-327. ISSN 1460-2237

Fontana, A. & Frey, J. (2005). The Interview: From Neutral Stance to Political Involvement. In: N. K. Denzin & Y.S. Lincoln (Eds.), *The Sage Book of Qualitative Research*, pp. 695-728, Sage, 978-07-6192-757-0, Thousand Oaks, USA

Freire, P. (1979). *Cultural action for freedom*. Cambridge, MA: Harvard Educational Review.

Genet, N.; Boerma, W.G.; Kringos, D.S.; Bouman, A.; Francke, A.L.; Fagerström, C.; Melchiorre, M.G.; Greco, C. & Devillé, W. (2011). Home care in Europe: a systematic literature review. *BMC Health Services Research*. Vol.30, No.11, pp.207-216. ISSN 1472-6963

Greenwood, N.; Mackenzie, A.; Wilson, N. & Cloud, G. (2009). Managing uncertainty in life after stroke: a qualitative study of the experiences of established and new informal carers in the fist 3 months after discharge. *International Journal of Nursing Studies*, Vol. 26, No. 8, pp. 1122-33. ISSN 1873-491X

Hale, L.; Piggot, D.; Bentley, M., Grawshaw, A. & Davis, H. (2003). Stroke rehabilitation-comparing hospital and home-based physiotherapy: the patient's perspective. *New Zealand Journal of Physiotherapy J*, Vol.31 No.2, pp. 84-92, ISSN 0303-7193

Hale, L.A. & Piggot, J. (2005). Exploring the content of physiotherapeutic home-based stroke rehabilitation in New Zealand. *Archives of physical medicine and rehabilitation*, Vol.86, No.10, pp. 1933-40, ISSN 0003-9993

Hanson, S. & Pratt, G. (1988). Reconceptualising the links between home and work in urban geography. *Economic Geography*, Vol.64, pp.299-318. ISSN 1944-8287

Huberman, A.M. & Miles, M.B.(1994). Data managment and Analisys methods. In: *Handbook of Qualitative Research*, N. K. Denzin & Y.S. Lincoln (Eds.), 428-444, Sage, Thousand Oaks, CA, USA

Hutten, J. & Kerkstra, A. (1996). *Home Care in Europe: A country-specific guide to its organization and financing*. Arena, ISBN 978-18-5742-336-5, Aldershot, UK

Janesick, V. (2003). The Chorreography of Qualitative Research Design. Minuets, Improvisations and Christallization. In: *Strategies of Qualitative Inquiry*, N. K. Denzin & Y.S. Lincoln, (Eds.), 46-79, Sage, Thousand Oaks, CA, USA

Jolly, K. et al. (2003) Home based versus hospital based cardiac rehabilitation after miocardial infarction or revascularisation: a randomissed controlled trial. *BMC Cardiovascular disorders*, 3,10, ISSN 1471-2261

Kincheloe, J.L. & McLaren, P. (2005). Rethinking Critical Theory and Qualitative Research. In: *The SAGE handbook of qualitative research*, N. K. Denzin & Y.S. Lincoln (Eds.), 303-342, Sage, ISBN 0-7619-2757-3, Thousand Oaks, CA, USA.

Kirilloff, L.H.; Carpenter, V.; Kerby, G.R.; Kigin, C. & Weimer, M.P. (1986). American Thoracic Society. Medical Section of the American Lung Association. Skills of the health team involved in out-of-hospital care for patients with COPD. *American Review of Respiratory Diseases*, 133, pp.948-949. ISSN 0003-0805

Krieger, N. (2001). Theories for social epidemiology in the 21st century: an ecosocial perspective. *International Journal of Epidemiology*, Vol. 30, pp.668-677, ISSN 0300-5771

Kuisma, R. (2002). A randomized, controlled comparison of home versus institutional rehabilitation of patients withy hip fracture. *Clinical Rehabilitation*, Vol.16, No.5, pp. 553-61, ISSN 1477-0873

Kupper-Nybelen, J.; Ihle, P.; Deetjen, W. & Schubert, I. (2006). Persons requiring long-term care: recommendations and utilization of rehabilitative therapies. *Zeitschrift für Gerontologie und Geriatrie*, Vol.36, No.2, pp. 100-108, ISSN 1435-1269

Lincoln, YS., Guba, EG., 2000. Paradigmatic Controversies, Contradictions and Emerging Confluences. In: *Handbook of Qualitative Research*, N. K. Denzin & Y.S. Lincoln (Eds.),163-188, Sage, Thousand Oaks, CA, USA

Lindstöm, M.; Merlo, M. & Östergreen, P.O.(2002). Individual and neighbourhood determinants of social participation and social capital in a public health perspective: a multilevel analysis of the city of Malmö, Sweden. *Social Science & Medicine*, Vol. 54, pp.1779-91 ISSN 02779536

Lloyd, J.M.; Martin, R.; Rajagopolan, S.; Zieneh, N. & Hartley, R. (2010). An innovative and cost-effective way of managing ankle fractures prior to surgery-home physiotherapy. *Annals of the Royal College of Surgeons of England*,92,7, pp. 615-18 ISSN 1478-7083

Lynch, J.W.; Smith, G.D.; Kaplan, G.A. & House, J.S. (2000). Income inequality and mortality: importance to health of individual income, psychosocial environment or material conditions. *British Medical Journal*, Vol. 320, 1200-4. ISSN 1756 5833

Malterud, K. (2001).Qualitative research: standards, challenges and guidelines. *The Lancet*, Vol. 358, pp.483-487. ISSN: 0140-6736

Mangione, K.K.; Crack, R.L.; Tomlinson, S.S. & Palombaro, K.M. (2005).Can elderly patients who have had a hip fracture perform moderate –to high- intensity exercise at home? *Physical Therapy*, 85, pp. 727-739, ISSN 1538-6724

Marmot, M. & Wilkinson, R. (1999). *Social Determinants of Health*, Oxford University Press, Oxford, UK

Marmot, M. (2004) Social causes of social inequalities in Health. In: *Public Health, Ethics and Equity*, Arnand S, Meter F, Sen A. pp. 15-21, Oxford University Press, Oxford

Martín, A. & Cano, J.F. *Atención Primaria. Conceptos, organización y práctica clínica*. Elsevier ISBN 84-8174-650-9, Barcelona, Spain

McGarry, J. (2010). Relationships between nurses and older people within the home: exploring the boundaries of care. *International journal of older people nursing,* Vol.5, No.4, pp.265-73. ISSN

McLure, M. (2003). *Discourse in Educational and Social Research,* Open University Press, Buckingham, UK

Mitchell, C.; Walker, J.; Walters, S.; Morgan, A.B; Binns, T. & Mathers, N. (2005). Costs and effectiveness of pre- and post-operative home physiotherapy for total knee replacement: randomized controlled trial. *Journal of Evaluation in Clinical Practice,*Vol.11,No.3, pp. 283-92. ISSN 1365-2753

Morse, J. (1994). Designing funded qualitative research In: *Handbook of Qualitative Research,* N. K. Denzin & Y.S. Lincoln, (Eds.), 220-235, Sage, Thousand Oaks, CA, USA

Nilsson, P.; Tom, E.; Baigi, A.; Marklund, B. & Mansson, J. (2007).A prospective pilot study of a multidisciplinary home training programme for lateral epicondylitis. *Musculoskeletal Care,* Vol.5, No.1, pp.36-50, ISSN 1557-0681

Oliver, M. (1999). The disability movement and the professions. *British Journal of Therapy and Rehabilitation,* Vol.6, No.8, pp. 377-379, ISSN 1354-8581

Ollonqvist, K.; Gronlund, R; Karppi, S.L.; Salmelainen, U.; Poikkeus, L. & Hinkka, K. (2007).A network-based rehabilitation model for frail elderly people: development and assessment of a new model. *Scandinavian Journal of Caring Sciences,* Vol.21, No., pp.253-61. ISSN: 1471-6712

Paz-Lourido, B. & Verger-Gelabert, S. (2008). La perspectiva comunitaria en la fisioterapia domiciliaria: una revisión. *Fisioterapia* , Vol.30, No.5, pp.231-237. ISSN: 0211-5638

Popkewitz, P. (1984). *Paradigm and Ideology in Educational Research.* The Falmer Press. ISBN: 978-1-59311-952-2, Philadelphia, USA

Renkema, J. (2004) *Introduction to discourse studies.* John Benjamin's Publishing, ISBN 9027232210, Amsterdam, Holland.

Richards, H.M. & Schwartz, L.J.( 2002). Ethics of qualitative research: are there special issues for health services research?, *Family Practice,*Vol.19, pp.135-139. ISSN 1460-2229

Rist, C.R. (2003). Influencing the policy process with qualitative research. In: *Collecting and interpreting qualitative materials,* N. K. Denzin & Y.S. Lincoln (Eds.), 619-644, Sage, ISBN 0-7619-2687-9,Thousand Oaks, CA, USA

Roddy, E.; Zhang, W. & Doherty, M. (2005). Home based exercise for osteoarthritis. *Annals of the Rheumatic Diseases,* Vol.64, pp.170-17, ISSN 1468-2060

Silverman, D. (2003). Analyzing Talk and Text. In: *Collecting and interpreting qualitative materials,* N. K. Denzin & Y.S. Lincoln (Eds.), 340-362, Sage, ISBN 0-7619-2687-9, Thousand Oaks, CA, USA.

Smith, K.M, McKelvie, R.S, Thorpe, K.E. &Arthur, A.M. (2011). Six-year follow-up of randomised controlled trial examining hospital versus home-based exercise training after coronary artery bypass graft surgery. *Heart,* Vol. 97, pp.1169-1174. ISSN 1468-201X

Sørbye, L.W., Garms-Homolová, V., Henrard, J.C., Jónsson, P.V., Fialová, D., Topinková, E. & Gambassi, G. (2009). Shaping home care in Europe: the contribution of the Aged in Home Care project. *Maturitas,* Vol. 62, No.3, pp.235-42. ISSN: 0378-5122

Starfield, B.; Shi, L. & Macinko, J. (2005) Contributions of Primary Care to Health Systems and Health, *The Milbank Quarterly*, Vol.83 No. 3, pp. 457-502, ISSN 1468-0009

Sundström, G. & Tortosa, M. (1999). Effects of rationing home-help services in Spain and Sweden, a comparative analysis. *Ageing and Society*, Vol.19, No.3, pp. 343-61. ISSN 1469-1779

Taylor, R.S.; Dalal, H.; Jolly, K.; Moxham,T. & Zawada, A. (2010). Home-based versus centre-based cardiac rehabilitation. *Cochrane Database of Systematic Reviews*, Issue 1. Art. No.: CD007130. DOI: 10.1002/14651858.CD007130.pub2.

Tornsén, A.M.; Holmqvist, L.W.; de Pedro-Cuesta, L. & von Koch, L. (2005). A randomised controlled trial of early supported discharge and continued rehabilitation at home after stroke: five-year follow-up of patient outcome. *Stroke*, Vol.36, No.2, pp. 297-303, ISSN 15244628

Turner-Stokes, L.; Nair, A.; Sedki, I.; Disler, P.B. & Wade, D.T. (2005). Multi-disciplinary rehabilitation for acquired brain injury in adults of working age. *Cochrane Database of Systematic Reviews*, Issue 3. Art. No.: CD004170. DOI: 10.1002/14651858.CD004170.pub2.

Van Haastregt, J.; Diederiks, J.; Van Rossum, E.; De Witte, L.P. & Crebolder, H. (2000). Effects of preventive home visits to elderly people living in the community: systematic review. *British Medical Journal*, Vol. 320, pp. 754-8. ISSN 1756 5833

Von Koch, L.; Wohlin Wottrich A. & Widen Holmqvist L. (1998). Rehabilitation in home versus hospital: the importance of context. *Disability and rehabilitation*, Vol. 20, No.10, pp. 367-372, ISSN 1464-5165

Vüori, H. (1984). Primary health care in Europe: problems and solutions. *Community Medicine*, Vol.6, 221-231.

Ward, D.; Drahota, A.; Gal, D.; Severs, M. & Dean, T.P. (2008).Care home versus hospital and own home environments for rehabilitation of older people. *Cochrane Database of Systematic Reviews*, Issue 4. Art. No.: CD003164. DOI: 10.1002/14651858.CD003164.pub2.

Whitehead, M. & Dahlgren, G. (2007). *Concepts and principles for tackling social inequities in health: Levelling up Part 1.* WHO Regional Office for Europe, Copenhagen, Denmark

Whitehead, M. (1988). *The health divide*. Inequalities in health. Penguin, London, UK

Whitehead, M. (1990). *The concepts and principles of equity and health.* WHO Regional Office for Europe, Copenhagen, Denmark

WHO (1978). *Declaration of Alma-Ata.* International Conference on Primary Heath Care. Alma Ata USSR.

WHO (1991). *Sundsvall declaration on supportive environments*. Geneva, Switzerland.

WHO (2004) *Commission on Social Determinants of Health: Imperatives and opportunities for change*. Geneva, Switzerland

WHO (2008). *Home Care in Europe.* World Health Organization. ISBN 978-92-890-4281-9 Copenhagen, Denmark

Widén, H.L.; von Koch, L. & de Pedro, J. (2000). Use of health care, impact on family caregivers and patient satisfaction of rehabilitation at home after stroke in southwest Stockholm. *Scandinavian Journal of Rehabilitation Medicine*, Vol.32, pp.173-179. ISSN: 0346-8720

Wilkinson, R. & Marmot, M. (2003). *Social determinants of health: The solid facts*, 2nd edition, World Health Organization, Copenhagen, Denmark

Wottrich, A.W.; von Koch, L. & Tham, K. (2007). The meaning of rehabilitation in the home environment after acute stroke from the perspective of a multiprofessional team. *Physical Therapy,*Vol. 87,No. 6, pp.778-88, ISSN 1538-6724

10

# Functional Challenges in the Elderly

A. C. Tavares and G. V. Guimarães
*Unidade Clínica de Insuficiência Cardíaca e Transplante – Instituto do Coração, Brazil*

## 1. Introduction

Aging reflects differences in geographic, culture, socio-economic, political and medical contexts, known as the increase in survival rates, with greater life expectancy. (Michel, 2004). Age itself is an independent morbidity and mortality risk factor for a long list of diseases, hospitalization, and length of hospitalization (Priebe, 2000).

The aging process combined with being sedentary enhances the loss of strength and muscular mass, decrease of balance, proprioception and mobility; all these symptoms combined predispose loss of independency and falls. Falls can evolve to fear of falling, to restriction of social activities and to impairment of daily activities performance. (Vreed, 2004)

Some diseases are frequently seen among the elderly. (Priebe, 2000) These include chronic heart failure (CHF), chronic obstructive pulmonary disease (COPD) arthritis and arthrosis, heart failure, (Okita, 1998)., hypertension, hypercholesterolaemia and diabetes mellitus (Ciolac and Guimaraes, 2002a).

The addition of those diseases to aging contributes to diminished functional development in elderly, (Okita, 1998) decreased life expectancy, greater risk of dementia (Richard, 2010), greater mortality risk and hospitalization length (Priebe, 2000).

## 2. Aging

Aging is a natural, intrinsic, detrimental, progressive and universal course of life which leads a subject to progressive functional aptitude loss (Papaléo, 2002) which corresponds to an independent morbidity and mortality risk factor for a long list of diseases, hospitalization, and length of hospitalization (Priebe, 2000).

Getting old represents physical, social, political and medical implications. Depending on the type and course of the developed morbidities, a subject can face physical suffering due to handicaps, which may increase government costs on social and health programs (Michel, 2004).

It's not totally clear how aging occurs. Flaw son automatic mistakes corrections, metabolic deviation, action of genes related to the protein synthesis and cells' programmed death (entitled apoptosis) implicate in structural and functional organs' alteration and they are responsible for their activity decline (Papaléo,2002; Alves, 2004).

## 3. Age related impairment in general

Changes in the elderly can be found in numerous interacting systems. Musculoskeletal, (Evans, 1993; Gill, 2004; Goldspink, 2005; Faulkner, 2007) cardiovascular,(Heath, 1981; Ciolac and Guimaraes, 2002b) pulmonary (Gill, 2004 ; Gardner, 2000) and neurological systems very often suffer impairment(Richard, 2010).

Decreases of 19 to 49% in muscle mass (sarcopenia (Ciolac and Guimaraes, 2002b) occur because of loss of myocytes (Goldspink, 2005) which leads to skeletal muscle atrophy. The gradual loss of 50% of muscle fibers happens in limb muscles (Faulkner, 2007).

There is also increase in bone porosity, cartilage degeneration, (Gardner,2000), flexibility and proprioception loss (Gill, 2004; Gardner, 2000).

Elderly frequently have loss in coordination, balance (Gardner,2000), a 2%-decrease in reflex velocity occurs each year (Gill, 2004) lower visual acuity, impaired central nervous system (CNS) function (Gill, 2004 ; Gardner, 2000).

Aging is associated with increased stiffness (reduced compliance) of large elastic arteries; impaired vascular endothelial function, including reductions in endothelium-dependent dilation (EDD), release of tissue-type plasminogen activator (fibrinolytic capacity) and endothelial progenitor cell number and function; increased intima-media wall thickness (IMT); and peripheral vasoconstriction (decreased basal leg blood flow); 40–50% differences in large elastic artery stiffness and compliance (Gatesa,2003; Seals, 2008).

Advancing adult age is associated with profound changes in body composition. Declining basal metabolic rate (BMR) at a rate of 3% each ten years happens (Evans,1993; Gardner, 2000; Gill, 2004) which may increase rates of cardiovascular and metabolic diseases (Fig. 1.) (Ciolac and Guimaraes, 2002a).

Of all the physiological changes that occur during the aging process, regarding quality of life and functional independence are declines in muscle strength and in aerobic capacity, (indexed as peak oxygen consumption-peak VO2) (Heath, 1981; Fleg, 1988; Katzel, 2001;Fleg,2005). As already known, peak VO2 is known to be dependent on age, gender, activity status and disease state (Morris, 1993) and to be a significant predictor of death (Myers, 1998).

The longitudinal rate of decline in peak VO2 with age is not linear but accelerates at higher age decades in both sexes. A longitudinal decline in peak VO2 was observed in each of the 6 age decades in both sexes; however, the rate of decline accelerated from 3% to 6% per 10 years in the 20s and 30s to 20% per 10 years in the 70s and beyond (Fleg,2005).

By age 75 years, over half of the functional capacity of the cardiovascular system, (peak VO2), has been lost (Tanaka, 1997; Hawkins, 2003).

Decline in peak VO2 in individuals can be explaned by the loss of muscle mass and changes in body weight and fat free mass (FFM), lifestyle habits, and development of subclinical and clinically apparent disease(Katzel, 2001) which occurs with advancing age (Fleg,1988; Fleg, 2005).

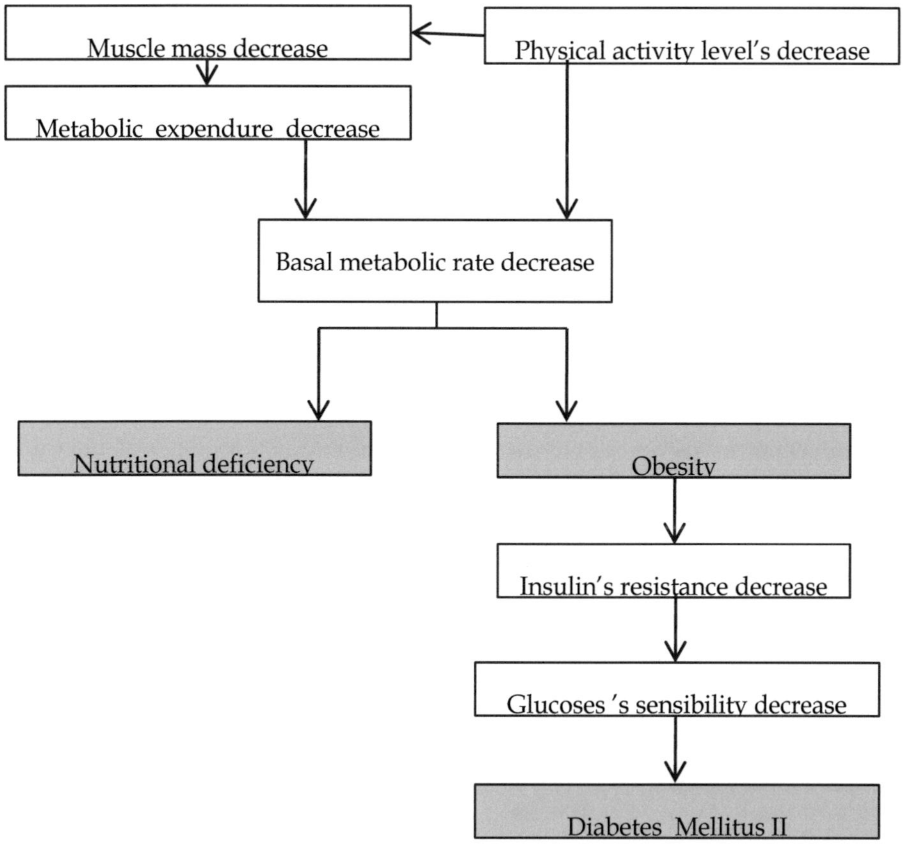


Fig. 1. Schematic representation of energetic metabolic reactions due to aging. Adapted from Ciolac and Guimaraes (2002a) with permission.

## 4. Age related functional decline

Functional ability decreases as people get old. Nonetheless, age is not the only contributing factor to function disability. Multiple factors influence functional decline in physiological systems, including aging, disuse and disease (Hawkins, 2003).

Thus, besides age (Morris, 1993), gender, weight, body surface area, thigh cross-sectional muscle area, peak $VO_2$ can also be correlated to functional capacity (Myers, 1998) (Heath, 1981).

However, among all these changes, the decrease in muscle mass and balance are the greater responsible for physical dependency (Avlund, 2003; Gill, 2004) because strength and balance are both necessary in order to develop any kind of duties with expected stability and agility(Brawley, 2003). As a result walking is the first affected daily activity, and its pace and speed are mostly influenced by lower limb strength (Nied, 2002, Pansa, 2003).

As a consequence of performance decline, elderly frequently experience falls (Blank, 2011). Falls can be common older person and can cause notable decreases of quality of life due to fear of falling, restricted mobility, loss of autonomy and bone fractures (Gill, 2004; Blank, 2011). Elderly people who have experienced bone fractures after falls have 45 to 60% of decline in their independency and around 80% of them become inactive due to psychosomatic aspects. (Marks, 2003)

## 5. Functional activity decline related inactivity

Inactivity itself (due to either fear of falling, loss in join mobility or decreased balance) is also a contributing factor for sarcopenia, muscle strength, balance, and quality of life decline (Avlund, 2003; Vreed, 2004) Indeed dependency during activities of daily living increases from three to five times as inactivity rises. (Frank, 2003)

## 6. Functional activity improvement

Recent trials suggest healthy older women are capable of exercising and increasing the exercise intensity by exercise training (ET) similar to young women (Fig. 2.) (Ciolac and Brech, 2010).

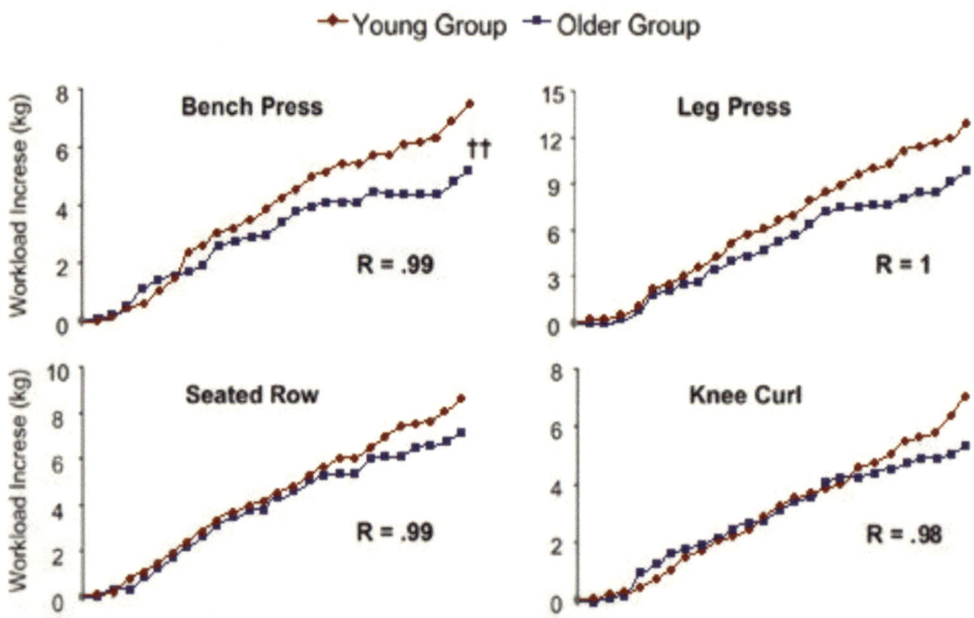


Fig. 2. Absolute workload increase curves for aerobic and resistance exercises. *Workload increase value in watts. †Tendency to be different from the younger group (p = 0.06). †† Significantly different from younger group (p < 0.01). Adapted from Ciolac and Brech (2010), with permission.

### 6.1 Selecting subjects

When selecting a population in which to introduce a physical activity program, the aims of the program need to be carefully considered. However, 65-year old people who have balance loss can be selected to a rehabilitation program (Vreede, 2004; Blank, 2011)

To evaluate the risk of falling older people need to have at least one of the following criteria: fall within the last 12 months; fear for falling; sitting to standing >10 sec; Timed-up-and-go-Test >10 sec (Blank, 2011)

However, the elderly in general would improve muscle strength and balance from exercise programs because it's known that they have progressive and declining loss of all organ functions over time. (Priebe, 2000; Papaléo, 2002).

### 6.2 Exercise in general

Strength training and weight-bearing exercises, functional power training, balance training, daily activity performance with motor coordination and proprioception, gait training with change of pace and direction while walking all contribute to functional activity improvement (Blank, 2011).

Strength training and weight-bearing exercises have perhaps greater adherence from the participants because they are more common to them. These exercises can fulfill their expectations and increase their self esteem (Vreede, 2004).

### 6.3 Muscle strength improvement

Limitation daily activities due to sarcopenia, loss in muscle strength can be partially or totally solved with the practice of strength training and weight-bearing exercise (Boulgarides, 2003; Seguin, 2003).

Weight-bearing exercise training is the most efficient intervention responsible for muscle mass and strength gain, besides it also guarantees increased functional reserve, cardiovascular conditioning, better sleep quality and lower fear of falling (Lord, 1995; Gardner, 2000; Borst 2004).

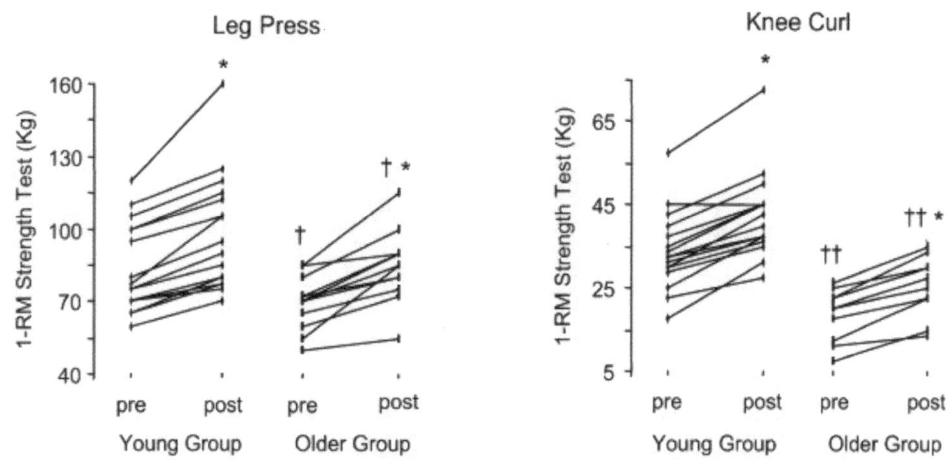


Fig. 3. Significantly difference between muscle strength (†p<0.05; ††p< 0.001).*Significantly different from before exercise training in the same group (p , 0.001). With permission of Ciolac and Greve (2011).

Generally strength training of antigravitational muscles is the most important measure to achieve better functional activity. Although these muscles are not considered essential for survival, they are extremely important for enabling people to carry out activities (Meuleman, 2000; Laughton, 2003). The most trained antigravitational muscles act on knee, hip and truck extension (Judge, 1994; Meuleman, 2000; Marks, 2003; Skelton, 2003;Vreede, 2004; Faulkner, 2007; Ciolac and Greve, 2011).

The 1-RMworkload, defined as the maximum weight that could be moved once through the full range of motion with proper form and without performing the Valsalva maneuver, is used in most populations, including the elderly (Ciolac and Greve, 2011).

Specificity is one of the fundaments of muscle strength improvement, however similarly gains can occur after resistance exercise, as seen in Fig. 3. (Ciolac and Garcez-Leme, 2010)

### 6.4 Balance improvement

Receptors in all musculoskeletal vestibular and visual ones systems are responsible for the body`s entire balance control (Lorrd, 1993;VanSwearingen,2001)Physical exercise may delay the synaptic function decline that is related to the aging process, thus increasing coordination and the performance of activities that require quick motor skills.(Lorrd, 1993;Fiatarone, 1994;Chiovatto, 2002;Ness, 2003).

Static and dynamic balance training can be improved by exercise and this leads to a decreased risk of falls as well as greater functional capacity. Figures 4 to 7 (Lorrd, 1993; Lord, 1995; Frank, 2003; Pansa, 2003).

Balance gain is effective when training different postural responses within functional activities. (Horak, 1997; Frank, 2003)but it is important to highlight that the key balancing point whenever in an upright position is in ankle mobility (Fig. 4.), thus, movement in this axis is essential (Chiovatto 2002; Brawley, 2003; Laughton, 2003; Marks, 2003).

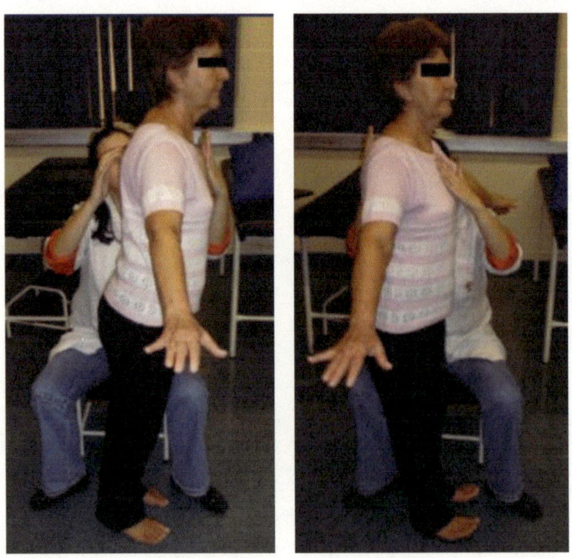

Fig. 4. Training balance through ankle mobility.

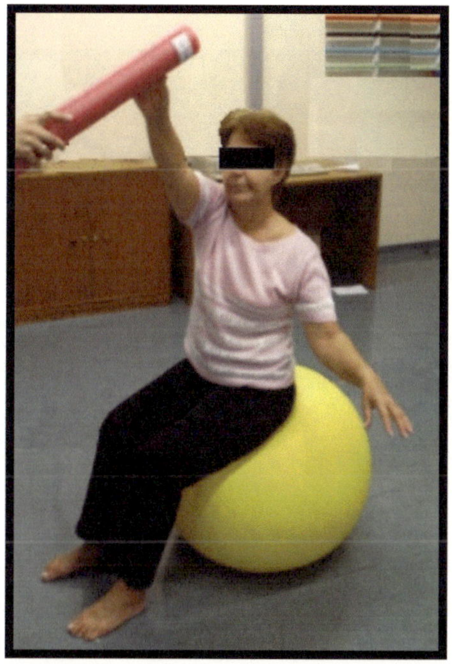

Fig. 5. Training balance on instable surface combined to upper limbs movements.

Fig. 6. Training balance on instable surface combined to upper limbs movements.

Whithin each session, proprioception gain is fundamental in order to obtain a better impact on postural control (Gauchard, 1999; Perrin, 1999). This can be achieved by the use of unstable surfaces (Fig. 4 to 8), which lower the ankle proprioceptive input and emphasizes the vestibular and visual system (Lorrd, 1993). Sensory awareness and balance can be even more directed whenever unipodal support is performed during the exercise program. (Judge, 1994; Ringsberg, 1999; Nied, 2002).

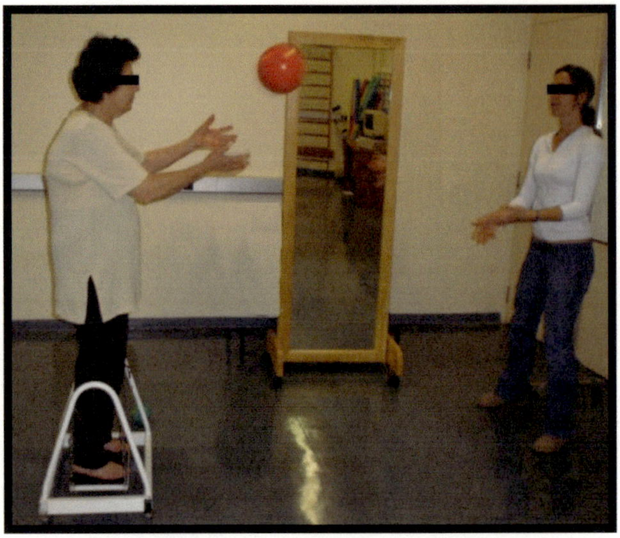

Fig. 7. Training balance on instable surface combined to upper limbs movements.

Fig. 8. Training balance and proprioception on instable surface.

Associating exercise with eye closed also contributes to a more gain in balance (Frank, 2003). Having the eyes closed forces the subject to use proprioceptive and vestibular sensors with priority, to correct posture and to adopt a more appropriate balance strategy (Gauchard, 1999; Perrin, 1999; Ringsberg, 1999; Frank, 2003).

### 6.5 Combined therapy

Getting old contributes to poor functional activity, it seems reasonable working with combined groups of exercise to improve elderly functional capacity (VanSwearingen, 2001; Ciolac and Guimaraes, 2002b; Papaléo, 2002; Alves, 2004).

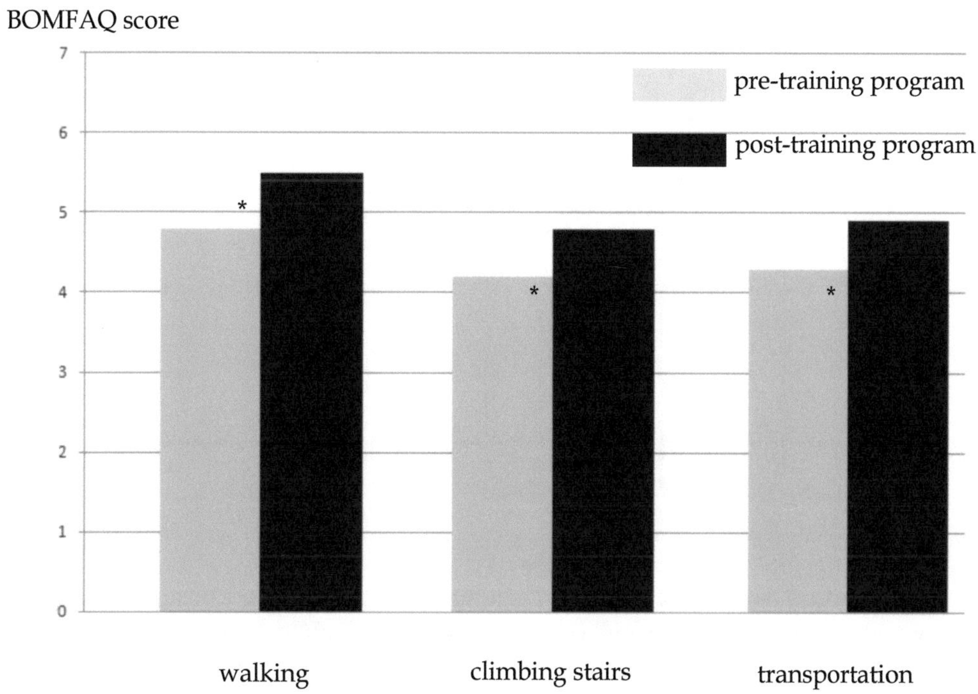


Fig. 9. Comparison of activities (walking, climbing stairs and getting off the transportation) after an exercise program. *Significantly different from before exercise training in the same group ($p < 0.05$). With permission of Tavares (2008).

Strength training, weight-bearing exercise and functional power training, plus balance training, daily activity performance with motor coordination and sensory awareness, gait trainingare all part of different elderly rehabilitation (Judge, 1994; Tavares, 2008; Ciolac and Brech, 2010; Blank, 2011).

Combined therapy with strength, balance and propriception training, plus weight-bearing exercise improves functional performance (Fig. 9) according to the Brazilian multidimensional Functional Assessment questionnaire (BOMFAQ) (Tavares, 2008).

Recent trials recommend a safe exercise training program (safe because no injuries or major muscle pain was observed) on cycle ergometry (CE, 65–75% of reserve heart rate), whole-body resistance exercise (RE, 60% of 1 repetition maximum [1RM]) and stretching during rehabilitation program. An increase of 5% is possible to be made in the exercise intensity, and one of 5–10% (1-5Kg) in the RE whenever adaptation occurs, without compromising this safety (Ciolac and Brech, 2010).

### 6.6 Hydrotherapy

Exercise in water may be a viable alternative for people whenever weight-bearing exercise are not be suitable for individuals with orthopaedic or musculoskeletal limitations, excess adiposity or other medical conditions (Meredith-Jones, 2011).

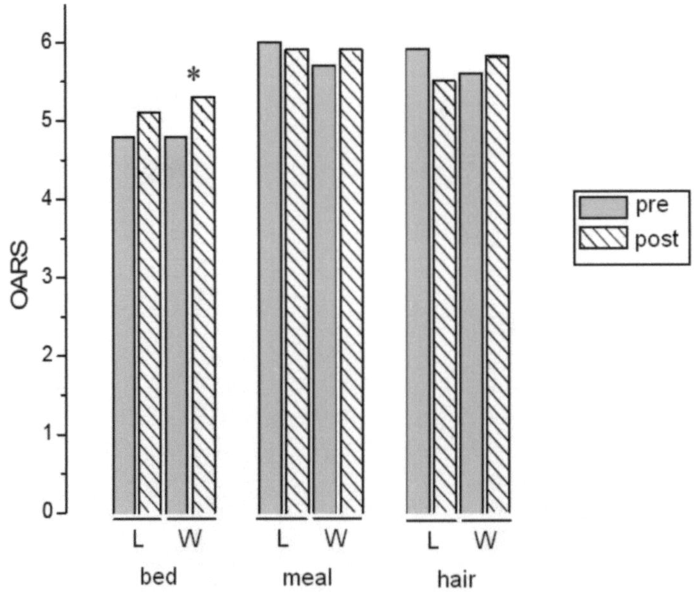


Fig. 10. Comparison of activities (lying and getting up the bed, having a meal; combing the hair) after an exercise programin land (L) and in water (W) *Significantly different from before exercise training in the same group (p <0.05). With permission of Tavares (2009).

Water's properties help participants to detect errors, correct them with more length of time, and experience a wider range of movement without an increase in the risk of injury due to a fall (Teffaha, 2011).

It is suggested that in water, people are allowed to have a larger band width of movement in exercise in which the participants could make mistakes, receive a feedback and, then, correct that error without an increase in their fear of injury through the use of these movement errors in water (Meredith-Jones, 2011; Teffaha, 2011).

The emphasis of an intervention in water has similarities to exercise in land. They all include warm-up, stretching, weight-bearing exercise on antigravitational muscles and balance training (Simmons, 1996; Vreed,2004; Teffaha, 2011).

A program based on exercises in the water improves the capacity of walking on the plane and of lying and getting out of bed after a 12-week, randomized, single-blind study in elderly as seen on Fig. 10 (Tavares, 2009).

Water exercise inducessignificant improvement in aerobic capacity, neuromuscular fitness and quality of life score (Simmons, 1996; Bocalini, 2010). Training needs to be constantly because a short length period of detraining (6 weeks) is responsible for returning to untrained levels the upper and lower body strength, agility, flexibility, and body balance (Bocalini, 2010).

## 7. Clinical improvement after rehabilitation

The elderly cluster that underwent clinical trials based on exercise found improved strength in more than 40% (Ciolac and Brech, 2010; Ciolac and Greve, 2011) improved neuromuscular fitness and quality of life score (Bocalini, 2010).

An activity program contribute to a positive cardiovascular risk reduction (Ortiz, 2010; Ciolac and Greve, 2011) because it decreases significantly systolic, diastolic blood pressure and pulse pressure (Ortiz, 2010; Pitsavo, 2011), it decreases significantly the left ventricular mass index (Pitsavo, 2011) and decreased atherogenic index and increased high-density lipoprotein cholesterol level (Ortiz, 2010). Besides, exercise shows significant improvement in aerobic capacity(Bocalini, 2010; Ciolac and Greve, 2011), and heart rate response (Ciolac and Greve, 2011)

## 8. Program specification for elderly, in general

Frequency trials have shown that functional capacity improvements often occur with 60 min training sessions (Perrin, 1999; Tavares, 2008; Tavares, 2009) with sessions ranging from twice a week (Perrin, 1999; Boulgarides, 2003; Gardner, 2000; Ciolac and Garcez-Leme, 2010) three times-per-week exercise-training programs. (Ciolac and Greve, 2011; Pitsavo, 2011This short variance of frequency is able to reveal beneficial effects of regular exercise training in the elderly (Pitsavo, 2011).

Some trials point out the muscle and balance improvement after both, 12-week exercise program (Bocalini, 2010; Boulgarides, 2003; Ciolac and Garcez-Leme, 2010) and a 16 one (Blank, 2011; Pitsavo, 2011).

Usually the duration of the activity program in elderly lasts up to 12 (Rydwik, 2005 Tavares, 2008; Tavares, 2009; Ortiz, 2010; Ciolac and Greve, 2011) to 16 months (Pitsavo, 2011) in order to reproduce clinical benefits. However exercise followed by 6 weeks without training induced significant improvement in aerobic capacity, neuromuscular fitness and quality of life score (Bocalini, 2010).

According to exercise intensity in elderly, most trials which have improved results have worked with large muscles and moderate to high intensity (Ciolac and Guimaraes, 2002b). A 60–80% of reserve heart ratewith an increase of 5% whenever adaptation occurs (Ciolac and Brech, 2010; Pitsavo, 2011) combined to a score between 11 and 14 on the perceived exertion scale can be used in a safe way whenever prescribing exercise (Guimarães, 2008).

## 9. Conclusion

Decline in functional capacity with age is expected due to the aging process itself combined with being sedentary. Careful measurement of quality of life and fear of falling needs to be included within the management program of elderly peoples` rehabilitation. Exercise addressing balance, body weight bearing, as well as motor coordination needs to be specifically addressed within elderly rehabilitation.

## 10. References

Alves RV, Mota J, Costa MCC, Alves JGB. (2004) Physical fitness and elderly health effects of hydrogymnastics. Rer Bras Med Esporte.10: 38-43.

Avlund K, Vass M,Hendriksen C. (2003) Onset of mobility disability among community-dwelling old men and women. The role of tiredness in daily activities. Age and Ageing. 32: 579-84.

Blank WA, Freiberger E, Siegrist M, Landendoerfer P, Linde K, Schuster T, et al. (2011) An interdisciplinary intervention to prevent falls in community-dwelling elderly persons: protocol of a cluster-randomized trial [PreFalls]. BMC Geriatrics.11:7.

Bocalini DS, Serra AJ, Rica RL, Santos L. (2010) Repercussions of training and detraining by water based exercise on functional fitness and quality of life: a short-term follow-up in healthy older women Clinics. 65(12):1305-9.

Borst SE. (2004) Interventions for sarcopenia and muscle weakness in older people. Age and ageing. 33: 548-55.

Boulgarides LK, McGinty S, Willett J, Barnes C. (2003) Use of clinical and impairment - based testes to predict falls by community - dwelling older adults. PhysTher, 83: 328-39.

Brawley LR, RejeskiJ, King A. (2003) Promover atividade física para idosos. Os desafios para mudanças comportamentais. American Journal of Preventive Medicine. 25: 172-83.

Chiovatto, J. (2002) Reabilitação em geriatria. In: Papaléo, M. Gerontologia. A velhice e o envelhecimento em visão globalizada. SP: Ed. Atheneu. 324-47.

Ciolac EG. Guimaraes GV. (2002) Síndrome metabólica: abordagem do educador físico.RevSocCardiol do Estado de São Paulo.6(supl A):15-26.

Ciolac EG. Guimaraes GV. (2002) Importância do Exercício físico no idoso. RevSocCardiol do Estado de São Paulo.6(supl A):15-26.

Ciolac E, Brech G, Greve JMD. (2010) Age does not affect exercise intensity progression among women. J Strength Cond Res.24(11):3023-31.

Ciolac EG, Garcez-Leme LE, Greve JMD.(2010) Resistance Exercise Intensity Progression in Older Men. Int J Sports Med. 31: 433 - 8.

Ciolac E, Greve JMA. (2011) Exercise-induced improvements in cardiorespiratory fitness and heart rate response to exercise are impaired in overweight/obese postmenopausal women. Clinics. 66(4):1-7.

Evans S, Campbell WW.(1993) Sarcopenia and Age-Related Changes in Body Composition and Functional Capacity. J. Nutr. 123: 465-8.

Faulkner JA, Larkin LM, Claflin DR, Brooks SV. (2007) Age-related changes in the structure and function of skeletal muscles. Clinical and Experimental Pharmacology and Physiology. 34: 1091–6.

Frank JS, Aftab E, Patla AE. (2003) Balance and mobility challenges in older adults. Implications for preserving community mobility. American Journal of Preventive Medicine. 25: 157-63.

Fiatarone MA, O'Neill E, Ryan ND. (1994) Exercise training and nutritional supplement for physical frailty in very elderly people. New Engl J Med.330(25):1669-75.

Fleg JL, Lakatta EG.(1988) Role of muscle loss in the age-associated reduction in $VO_2$max. J. Appl. Physiol. 65: 1147–51.

Fleg JL, Morrell CH, Bos AG, Brant LJ, Talbot LA, Wright JG, et al. (2005) Accelerated Longitudinal Decline of Aerobic Capacity in Healthy Older Adults. Circulation 112:674-82.

Gatesa PE, Tanakaa H, Gravesa J, Seals DR. (2003) Left ventricular structure and diastolic function with human ageing Relation to habitual exercise and arterial stiffness. European Heart Journal.24: 2213–20.

Gardner MM, Robertson MC, Campbell AJ. (2000) Exercise in preventing falls related injuries in older people: a review of randomized controlled trials. Br F Sports Med. 34:7-16.

Guimarães GV, Carvalho VO, Bocchi EA. (2008) Reproducibility of the self-controlled six minute walking test in heart failure patients. Clinics. 63 (2): 201-6.

Goldspink DF. (2005) Ageing and activity: their effects on the functional reserve capacities of the heart and vascular smooth and skeletal muscles. Ergonomics. 48(11)1334 - 51.

Hawkins AS, Wiswell RA. (2003) Rate and Mechanism of maximal oxygen consumption decline with aging. Implication for exercise training. Sports Med. 33(12):877-88.

Heath GW, Hagberg JM, Ehsani AA .(1981) A physiological comparison of young and older endurance athletes. J Appl Physiol.51:634–40.

Katzel LI, Sorkin JC, Fleg JL.(2001) A Comparison of Longitudinal Changes in Aerobic Fitness in Older Endurance Athletes and Sedentary Men. J Am Geriatr Soc. 49:1657–64.

Laughton C, Salvin M,Katdare K, Nolan L, Bean J, Kerrigan DC,et al. (2003) Aging muscle activity, and balance control: physiologic changes associated with balance impairment. Gait and Posture.1-8.

Lord SR, Ward JA, William P,Strudwick M. (1995)The effect of a 12-month exercise trial on balance, strength, and falls in older women: a randomized controlled trial. J Am Geriatr Soc. 43:1198-206.

Lorrd SR, Caplan G, Ward J. (1993) Balance, reaction time, and muscle strength in exercising and nonexercising older women: a pilot study. Arch Phys Med Rehabil. 74:837-9.

Marks R, Allegrante J, MacKenzie CR, Lane J. (2003) Hip fracture among the elderly: cases, consequences and control. Ageing.2: 57-93.

Meredith-Jones K, Waters D, Legge M, Jones L. (2011) Upright water-based exercise to improve cardiovascular and metabolic health: a qualitative review. Complement TherMed.19(2):93-103.

Meuleman J, Brechue M, Kubilis P, Lowental D. (2000) Exercise Training in the Debilitated Aged: Strength and Functional Outcomes. Arch Phys Rehabil.81:312-8.

Michel JP, Robine JM.(2004) A "new" general theory of population ageing. The Geneva Papers on Risk and Insurance.29: 667-75.

Morris CK, Myers J, Froelicher VF, Kawaguchi T, Ueshima K, Hideg A. (1993) Nomogram based on metabolic equivalents and age for assessing aerobic exercise capacity in men. J Am Coll Cardiol.22:175-82.

Myers J. Gullestad L. Vagelos R. Do D. Bellin D. Ross H, et al. (1998) Clinical, Hemodynamic, and Cardiopulmonary Exercise Test Determinants of Survival in Patients Referred for Evaluation of Heart Failure. Ann Intern Med.129:286-93.

Nied R, Franklin B. (2002) Promotion and prescribing exercise for the elderly. Am Fam Physician, 65: 419-26.

Okita K, Yonezawa K, Nishijima H, Hanada A, Ohtsubo M, Kohya T, et al. (1998) Skeletal Muscle Metabolism Limits Exercise Capacity in Patients With Chronic Heart Failure. Circulation.98;1886-91.

Ortiz LG, Grandes G, Pérez AS, Montoya I, Valiente JAI, Rodríguez JIR, et al. (2010) Effect on Cardiovascular Risk of an Intervention by Family Physicians to Promote Physical Exercise Among Sedentary Individuals. RevEsp Cardiol.63(11):1244-52.

Pansa F, Gardil T,Chiarello B,Driusso P. (2003) Treino de equilíbrio em mulheres idosas. Rev. Fisioter. UNICID.2(2):89-99.

Papaléo M, Ponte JR. (2002) Envelhecimento: Desafio na Transição do Século. In: Papaléo, M. Gerontologia: A velhice e o envelhecimento em visão globalizada. SP: Ed. Atheneu, 3-12.

Perrin P, Gauchard G, Perrot C, Jeandel C. (1999) Effects of physical and sporting activities on balance control in elderly people. Br F Sports Med;33:121-6.

Pitsavo C, Chrysohoou C, Koutroumbi M, Aggeli C, Kourlaba G, Panagiotakos D, et al. (2011)The Impact of Moderate Aerobic Physical Training on Left Ventricular Mass, Exercise Capacity and Blood Pressure Response During Treadmill Testing in Borderline and Mildly Hypertensive MalesHellenic J Cardiol.52: 6-14.

Priebe HJ. (2000) The aged cardiovascular risk patient.Br J Anaesth. 85:763-78.

Richard E, Ligthart SA, Charante EPM, Gool WA. (2010) Vascular risk factors and dementia - towards prevention strategies The Netherland Journal of Medicine.68(10):284-90.

Rydwik E, Kerstin F, Akner G. (2005) Physical training in institutionalized elderly people with multiple diagnoses - a controlled pilot study. Archives f Gerontology and Geriatrics, 40: 29-44.

Seals DR, De Souza CA, Donato AJ, Tanaka H. (2008) Habitual exercise and arterial aging. J Appl Physiol.105: 1323–32.

Seguin R, Nelson ME. (2003) Os benefícios do treinamento de força em adultos. American Journal of Preventive Medicine, 25: 141-9.

Simmons V, Hansen PD. (1996) Effectiveness of water exercises on postural mobility in the well elderly: an experimental study on balance enhancement. J Gerontol.51:M223-7.

Skelton DA, BeyerN. (2003) Exercise and injury prevention in older people. Scand J MedSci Sports, 13: 77- 85.

Tanaka H, De Souza CA, Jones PP, Stevenson ET, Davy KP, Seals DR. (1997) Greater rate of decline in maximal aerobic capacity with age in physically active vs. sedentary healthy women. J. Appl. Physiol. 83(6): 1947–53.

Tavares AC, Sacchelli T. (2008)Avaliação da atividade funcional em idosos submetidos à cinesioterapia em solo. Rev Neurocienc.17(1):19-23.

Tavares AC, Sacchelli T. (2009) Comparação de cinesioterapia em solo e em água em idosos. Rev Neurocienc.17(3):213-219.

Teffaha D, Mourot L, Vernochet P, Ounissi F, Regnard J, Monpère C, et al. (2011) Relevance of water gymnastics in rehabilitation programs in patients with chronic heart failure or coronary artery disease with normal left ventricular function. J Card Fail.17(8):676-83.

VanSwearingen JM, Brach J. (2001) Making geriatric assessment work. Selecting useful measures. Phys Ther. 81:1233-52.

Vreede P, Samson M, Meeteren N,Bom JV, Duursma S, Verhaar H. (2004) Functional tasks exercise versus resistance exercise to improve daily function in older women: a feasibility study. Arch Phys Med Rehabil. 85: 1952-61.

# 11

# Evidence-Based Management in the Rehabilitation of Osteoporotic Patients with Fragility Fractures

Umberto Tarantino, Irene Cerocchi, Federico Maria Liuni, Pietro Pistillo, Cecilia Rao and Monica Celi
*University of Rome "Tor Vergata", Italy*

## 1. Introduction

Osteoporosis is a common skeletal disease in older populations, leading to more than a million fractures annually in the United States (Rockville, 2004) and Europe (O'Neil, 1996). Epidemiologic studies show that increased fracture risk in adults begins around the age of 45 years, particularly in women, progressing notably more with each decade of aging (Siris et al., 2010). About half of adult women and one third of adult men will sustain a fracture (Moayyeri,2008).

Non vertebral fractures represent 75% of osteoporotic fractures seen in clinical practice (Tosteson et al.,2005). The incidence of non-vertebral fractures, especially of the hip, increases rapidly with age (Piscitelli et al.,2010). The progressive aging of the population, observed during recent years, inevitably led to an increase in age-related diseases such as osteoporosis (OP). Patients with fragility fractures need to be prioritized for health and social security concerns. Fractures in older adults are often referred to low trauma or fragility fractures because they tend to occur as the result of a fall from a standing height (or lower) or from a minimally traumatic event that would not necessarily have resulted in a fracture in a younger person, or in the same individual at a younger age. They have also been referred to as osteoporotic fractures, since epidemiologic evidence has shown that low bone mineral density (BMD) is associated with an increased population based risk of fractures (Siris et al.,2006). The risk of fractures seems to be determined by a balance between bone strength and propensity for falls, which in term are determined by the frailty of the patient (Bergman,2007). Falls are the most significant risk factor for fractures, with 90% of hip fractures occurring as a direct result of a fall (Jarvinen,2008). About half of adult women and one third of adult men will sustain a fracture ( Moayyeri,2008). Lifetime risk after the age of 50 years for sustaining hip, distal radius, or proximal humerus fractures is 25% in men and 55% in women (Ahmed,2009).

Fragility fractures may have important consequences such as hospitalization with long periods of immobility, the need to undergo surgery, increased risk of disability and partial or complete loss of autonomy during ordinary activities of daily life and related economic burden.

After a femoral fragility fracture, only one out of three older people who survive returns to his previous level of independence, 50% require long-term help with routine activities and cannot walk unaided, 25% require full-time nursing home care and 20% die within one year

(Heinrich et al.,2010). Vertebral fractures are the most common osteoporotic fractures, after hip fractures, affecting 25% of the elderly female population. The acute pain after a new vertebral fracture traditionally is managed with rest and analgesic therapy (Rossini et al.,2010). The presence of vertebral fractures is associated with a decrease in the quality of life and with a higher risk of further vertebral and non -vertebral fractures. Once the first vertebral fracture has occurred there is a higher risk (equal to 20%) of further vertebral fractures in the first year ("vertebral fracture cascade") (Briggs et al.,2007). Clinical studies, in the last years, identified a significant increase in the incidence of ankle fractures in the elderly population, in particular in postmenopausal women and this led the authors to include these among the fragility fractures classification (Eric,2007).Regardless of the type of fragility fracture, the majority of women do not receive any treatment during the year following a trauma. Failure to implement or late implementation of preventive measures, as well as poor treatment compliance, leads to the deterioration of the health economic outcome. In the past decade, there has been an increasing interest in research on exercise for older people recovering from fragility fractures; however, the extent of this literature remains unclear. In the elderly osteoporotic patient, it is necessary to add to these objectives more specific ones: prevention of new fractures, increase of bone mass, maintenance of improved bone quality, improvement of balance to avoid falls, maintenance of effectiveness during the years of treatment. Bone response to mechanical stimuli must also be taken into account. This is different in osteoporotic individuals compared to healthy ones, as well as is the effect of rehabilitation exercises to improve balance and muscle strength. Patients with fragility fractures should be offered a coordinated multidisciplinary rehabilitation program with the specific aim of regaining sufficient function to return to their pre-fracture living arrangements. Many functional benefits are associated with exercise and physical activity participation by older adults, including improved cognitive, cardiovascular functioning, strength and balance, as well as reduced risk for falls and bone density loss (Nelson et al.,2007). However, there is little information on the safety, efficacy, or effectiveness of exercise prescription in older adults after a fracture.

In this chapter we report results of recent studies in the international literature regarding the complex and multifactorial rehabilitation process to be carried out in the osteoporotic elderly patient with a femoral or non-femoral fragility fracture. We wanted to emphasize how the post-operative management in these patients requires a different approach from that of a young patient, with an integration of different aspects such as multidisciplinary approach, specific exercises, educational and social measures and appropriate drug treatment support.

## 2. Physical activity and rehabilitation in the elderly

Fractures in the elderly—particularly fractures of the appendicular skeleton—result from two processes: a loss of skeletal integrity and an increased risk of falls (Nordstrom et al., 2011).

However, little attention has been given to the targeting of extra-skeletal factors to prevent fractures in selected individuals. In the management of patients at increased risk of fracture because of osteoporosis or extra-skeletal risk factors, measures of musculoskeletal rehabilitation should be considered as a prelude to, or even in conjunction with, pharmacotherapy to optimize musculoskeletal health, improve quality of life, and reduce the risk of fracture and fracture recurrence (Pfeifer et al., 2004).

Rehabilitation is a goal-oriented and time-limited process that focuses on making a functionally impaired person reach an optimal mental, physical and social functional level. The aim is on one hand to restore the functional level of people who sustained a fracture as a consequence of falling; on the other hand, when falling is not combined with fractures, the goal is to avoid the latter by educating the high risk groups.

With aging, a decrease in reaction time is expected, and therefore the ability to respond rapidly and effectively is reduced in older people compared to younger adults. Studies of reaction time in stepping have typically observed a delay in step initiation and execution timing in older people. Co-ordination time has also been linked to upper extremity fracture risk, as elderly people often delay in breaking the fall by outstretching their hand (Dionyssiotis et al., 2008).

Before starting a tailored program for any individual patient, correct medical history and clinical examination must be performed. The following information should be collected: weight/height, menarche/menopause, nutrition, pharmacological therapy (past and present), level of activity, previous fractures, history of falls, risk factors for secondary osteoporosis. The following are considered high risks for secondary osteoporosis: liver and severe chronic kidney pathology, steroid drugs (>7,5 mg for more than 6 months), malabsorption (e.g. Crohn's disease), rheumatoid arthritis, systemic inflammatory syndromes, hyperthyroidism, primary hyperparathyroidism, antiepileptic drugs.

The rehabilitation protocol must be global and specific. The objectives of the general therapeutic exercise are: prevention of fractures in all skeletal sites, increase of bone mass in all districts, maintenance or improvement of bone quality, improvement of balance with reduction of falls, maintenance of effectiveness during the years of treatment (Mangone, 2010). Important potentially modifiable risk factors for falling in community-dwelling older adults are: mental status, psychotropic drugs, multiple drugs, environmental hazards, vision, lower extremity impairments, balance, gait status; for institution-dwelling older adults: mental status, depression, urinary incontinence, hypotension, hearing, balance, gait, lower extremity impairments, low activity level (exercise less than once a week), psychotropic drugs, cardiac drugs, analgesics and use of a mechanical restraint. Non-modifiable risk factors (i.e., hemiplegia, blindness) also exist (Moreland et al., 2003). Interventions to prevent falls may be planned to reduce a single intrinsic or extrinsic risk factor of falling or be broadly focused to reduce multiple risk factors simultaneously.

Single evidence-based interventions include exercise, reassessment of medications and environmental modification. Graded reductions in the risk of hip fracture were found in women who performed moderate-to-vigorous activities for at least 2 h/week or who reported more hours of heavy chores per week. In contrast, the more hours a woman spends sitting per day, the higher the risk is of hip fracture: women who sat for at least 9 h/day had a 43% higher risk than those who sat for 6 h/day. Although the dose–response relation with physical activity was established for hip fractures, it was less apparent for wrist and vertebral fractures.

However, many functional benefits are associated with exercise and physical activity participation by older adults, including improved cognitive, cardiovascular, strength, and balance functioning, as well as reduced risk for falls and bone density loss.

However, there is little information on the safety, efficacy, or effectiveness of exercise prescription in older adults after a fracture (Freehan et al., 2011).

The intervention of "exercise prescription" is broadly defined as physical activity, exercise or active rehabilitation prescribed by a physician, physical therapist or occupational therapist, or other allied health professional. The concept of "exercise prescription" was broadly defined to fit with the concept of exercise prescription as defined by the ACSM position statement (1999) as "the process whereby a person's recommended regimen of physical activity is designed in a systematic and individual manner." Physical activity was defined as bodily movement produced by contraction of skeletal muscles that leads to increased energy expenditure, whereas exercise was defined as planned, structured and repetitive movements focused on improving or maintaining physical fitness (Andreoli et al., 2001).

Using the framework of the International Classification of Functioning, Disability and Health (ICF) (WHO, 2001), therapeutic exercise was defined as exercise prescribed by a medical (physician), rehabilitation (physical therapist or occupational therapist), or other allied health professional to address an individual's structural or functional impairments (e.g., in range of motion, flexibility, strength, balance), limitations in activity (e.g., activities of daily living, dexterity, walking speed/distance, walking up or down stairs), or restrictions in participation in life situations (e.g., work, sport, other life roles). Older people who had recurrent falls should be offered long-term exercise and balance training. Tai Chi is a promising type of balance exercise, although it requires further evaluation before it can be recommended as the preferred method for balance training. Tai Chi, which consists of slow, rhythmic movements emphasizing on the trunk rotation, weight shifting, co-ordination, and a gradual narrowing of the lower extremities' position, is thought to be an excellent choice of exercise for the elderly. There is experimental evidence from both cross-sectional and longitudinal studies that Tai Chi exercise has beneficial effects on balance control and that postural stability is improved more by Tai Chi than by other types of exercise. However, Tai Chi has not been shown to reduce falls in frailer older people, so cannot be recommended for fall prevention to a group who had hip fractures and are likely to be frail. Those with a history of Colles' fractures and with only mild deficits of strength and power are more likely to benefit. It is the slow, smooth, 3-D nature of Tai Chi with its transitions of stance that challenge balance that will help reduce a person's risk of falls; if the person is too frail to lift one foot off the ground and move it forward, then the Tai Chi must be adapted so much that perhaps more static balance work is more appropriate to start with.

A review by Gilespie L.P. et al. (2001) described the results of 13 randomized controlled trials of physical exercise or physical therapy to prevent falls in elderly people. One of the studies reported that participants exposed to Tai Chi intervention had a lower rate of falling than controls (risk ratio, 0.51; 95% CI, 0.36–0.73). Tai Chi exercises were performed on a weekly basis for 15 weeks with supervision by an instructor. In addition, subjects were requested to try Tai Chi exercising for 15 minutes. Pooled data from three studies with a total of 566 community-dwelling 80 years old women using the same individually tailored program of progressive muscle strengthening, balance retraining, and a walking plan, indicated that this intervention reduced the number of individuals sustaining a fall over a 1-year period (pooled relative risk [RR], 0.80; 95% CI, 0.66–0.98).

A trial published by Jensen et al. reported that a multidisciplinary program of both general and resident-specific tailored strategies reduced falls and fall-related injuries in persons 65 years of age living in residential-care facilities. The strategies comprised educating staff, modifying the environment, implementing individual exercise programs, supplying and repairing aids, reviewing drug regimens, providing free hip protectors, and having problem-solving conferences after falls (Jensen et al., 2002). However, as aging is related to reduced physical functioning, exercise prescription for fall prevention, except balance and strength training, should surely include exercises to increase the functional capabilities in all elderly people (Judex and Rubin, 2010). The suggested solutions are low intensity balance exercises (tandem walking and standing on one's foot) combined with co-ordination exercises. Individuals who are frail, severely kyphotic or suffer from pain or poor balance, may benefit from water exercise (hydrotherapy). The elderly are also advised to undergo strengthening exercises of the quadriceps, hip abductors/extensors, back extensors and the arm muscles. Back strength is also significantly lower in persons with osteoporosis than in healthy ones. As reported by Sinaki et al. (2002), strengthening of the paraspinal muscles can reduce the risk of vertebral fractures. They found that progressive, resistive back strengthening reduced the risk for vertebral fractures in women 58–75 years old. As for fractured patients, although there are trends concerning studies for faster recovery in mobility, functioning and pain reduction (without risk of increased complications) when using early post-surgical motion and/or functional use and/or weight bearing, there is still insufficient or conflicting evidence from randomized trials to be able to recommend any one early mobilization or exercise intervention over another.

### 2.1 Rehabilitation after hip fractures

Over 98% of hip fractures are associated with falls. A pro-active approach to prevent falls should receive at least as much attention as drug therapy for osteoporosis, in hip fracture patients. However this area of care is often neglected. The concept of frailty has received increasing attention in recent years as neither BMD nor clinical risk factors (i.e. age and weight) can fully quantify the risk of osteoporotic fractures in the elderly. Frailty is a state of poor well being, related to muscle weakness and sarcopenia, poor endurance, a low level of physical activity, easy exhaustion and a slow gait (Fried et al., 2001). Pre-fracture cognitive impairment, functional dependency, and co-morbidities will negatively influence an older adult's ability to regain pre-fracture functioning. Slow functional recovery and increased risk of progressive functional decline within the first year after fracture are most notable after hip fracture, with marked decline in physical, mental, and emotional functioning at 3 months post-fracture and over the next few months. One year after hip fracture, 50–55% of people will have a residual walking disability, with many never returning to pre-fracture or age-matched control functional status. In addition, 15–30% of community-dwelling people will be living in institutional care 1 year after the hip fracture. The potential physical, emotional, and social consequences of any fracture in an older adult are different and wide ranging. Many fractures are managed surgically, with older adults more vulnerable to complications arising from surgery and hospitalization. Successful operative treatment of hip fractures allows optimization of post-injury mobility and functional recovery. Rehabilitation after surgical stabilization of a hip fracture is crucial in order to restore pre-fracture function and decrease the risk of long-term institutionalization, which can be 25% during the first year post-fracture.

Evidence-based clinical practice guidelines suggesting possible treatments and rehabilitation pathways for hip fracture patients, agree that it would be best if they underwent multidisciplinary rehabilitation (Scottish Intercollegiate Guidelines Network, 2002; Chilov et al., 2003). Multidisciplinary rehabilitation can be defined as the combined and coordinated use of medical, social, educational and vocational measures for training or retraining the individual to the highest possible level of function. Physical activity and exercise form, part of the post-hip fracture rehabilitation in the elderly, also serve to increase muscle mass and strength, improve body function, reduce risk of falls, and contribute to a better quality of life. Immobilization accelerates bone loss and should be avoided as much as possible. Nonetheless the minimal level of physical activity and exercise required to prevent bone loss remains a matter for debate (Moayyeri, 2008).

In hip fracture patients with reduced mobility and poor balance, careful evaluation is required before exercise is prescribed: without adequate balance training the subject may be at higher risk of falls and hence fractures. In post-hip fracture subjects with poor mobility, poor motivation, and easy fatigability, whole-body vibration could be an alternative to conventional exercise. Whole-body vibration can induce strengthening of the muscles as well as improvement of the BMD in postmenopausal women. According to a meta-analysis by Cameron I. et al. (2000), there is no conclusive evidence that coordinated multidisciplinary inpatient rehabilitation is more effective than conventional hospital care (no rehabilitation professionals involved) for older patients with hip fracture. Because many community-living older persons who fracture a hip eventually return home, much of post-fracture rehabilitation occurs at home; thus, little is known about effective ambulatory strategies for the rehabilitation of geriatric patients after hip operation.

In a randomized controlled trial by Tinetti et al. (1999), a home-based systematic multicomponent rehabilitation strategy was no more effective for promoting recovery than usual home-based rehabilitation (Cameron et al., 2000). The Australian guidelines suggest a coordinated rehabilitation program that starts just after admission and provides opportunities for early supported discharge, as long as the patient's mobilization is established (Chilov et al., 2003). Frail patients should follow an inpatient program and are instructed to continue rehabilitation for some time after their discharge. All participants should be nutritionally assessed, so that they receive the recommended protein and energy supplementation. It should be noted that early assisted ambulation should begin 48 hours post-operatively. The majority of studies focus primarily on mobilization introduced within the first 3–4 weeks following fracture, which is consistent with the adverse functional (structural impairment and activity limitation) consequences associated with limb immobilization. The primary clinical concern with early post-fracture mobilization is the possibility of impairing the quality and rate of fracture healing, potentially causing delayed union, non-union or malunion. Hip fracture patients should immediately start breathing exercises so that pulmonary secretions are drained, thus reducing the risk of atelectasies and other complications deriving from the pulmonary system. "Pump like" energetic exercises (ankle pumps) and dorsal/plantar flexion of the foot, knee joint flexion, exercises for the hip and thigh, abduction exercises for the gluteal muscles and exercises for the quadriceps are important. Exercises of the upper extremities and trunk must also be part of the rehabilitation program, so that the patient can move in bed, stand up from a chair and later on be able to mobilize himself by using crutches or a stick. Abdominal and dorsal muscles should also be exercised isometrically and then energetically, in order to minimize the risk of low back pain during weight-bearing exercises. There are some limitations on the range of motion (ROM) after surgery, depending on the surgical procedure (hip fractures stabilized with internal fixation do not require ROM precautions). As soon as the patient who undergoes a prosthetic replacement regains his vigilance, he is instructed to avoid: a) hip flexion greater than 70-90°, b) external rotation of the leg, c) adduction of the leg past midline. Toward this end, the patient must be instructed: a) not to bend forward from the waist more than 90°, b) not to lift the knee on the side of the surgery higher than the hip, and c) not to cross the legs, neither at the knee, nor the ankle. These precautions should be maintained for approximately 12 weeks. Because dislocations occur usually within the first 30 days more clinical studies are needed to determine the optimum length of time to maintain hip precautions (Kayali et al., 2006). By the third day after surgery, the patient should start training from a sitting position. During transfer from bed to chair, the hip must be abducted. Weight-bearing should start later on, from the 6$^{th}$ to the 10$^{th}$ day when the patient is capable of standing on his feet by himself. Most patients are more likely to start using a walking frame and then progressively move to using crutches.

In the partial weight-bearing stage of rehabilitation, the operated hip is allowed to bear only a load of 20-50% of body weight. Partial weight-bearing should be preserved for 6-12 weeks. Gradually, the patients will be allowed to start walking on crutches for 4-6 weeks. Complete weight-bearing depends on the surgical procedure. Usually, complete weight-bearing is scheduled after a period of 6 weeks following total hip arthroplasty and after 3 months following open reduction and internal fixation (Rucco, 2003).

Physicians should be ready to treat life-threatening medical complications such as cardiopulmonary, deep venous thrombosis and ischemic episodes, but also variable complications such as hip pain, uneven limb length, heterotopic ossification, pressure sores and neurological complications that are likely to occur during hip fracture rehabilitation. Patients are urged to keep on training even after they are discharged and their period of rehabilitation is over. After their strength is regained, they should follow individually tailored and targeted training for dynamic balance, strength, endurance, flexibility, gait and functional skills, training to improve 'righting' or 'correcting' skills to avoid a fall, backward-chaining and functional floor exercises (Skelton et al., 2005).

Hip fracture patients should also have Occupational Therapy (OT) training for skills adaptation and a home visit to get individualized support to improve the ability to perform activities of daily living and to speed up both mental and social recovery. Most studies conducted on training after hip fracture conclude that combined training with task-specific and functionally based exercises may be a sensible way of retraining leg strength, balance and gait ability in elderly people after a hip fracture. The training thus may include a variety of gait exercises, step exercises, stair climbing, and rising from and sitting down on a chair. Most studies of the effectiveness of rehabilitation on functional recovery after hip fracture have involved acute or subacute rehabilitation facilities, targeting patients on wards and in post-discharge settings. In a randomized intervention study by Hauer et al. (2002), a 3-month progressive resistance and functional training program increased strength and functional performance during rehabilitation after hip fracture. Binder et al. (2004) found that six months of extended outpatient rehabilitation including progressive resistance training improves physical functioning, quality of life and reduces disability compared with low intensity home exercise in community dwelling men and women over 65 years. A randomized controlled trial of 49 patients aimed at evaluating the health benefits of an individualized nutrition support program and/or a progressive resistance lower limb training program for older adults admitted to the hospital following a fall-related lower limb fracture. Subjects were randomly allocated to receive a 12-week resistance training intervention using latex-free resistive elastic bands supervised by a physiotherapist. The progressive resistance exercises involved training of the hip and knee extensors, hip abductors, ankle plantar- and dorsi-flexors. The trial provided evidence that this form of resistance training is well suited to an older, frail lower limb fracture group. Exercise adherence remained high in hospital, in residential care and community settings but it did decline slightly without regular supervision. Progression within the exercise program was steady, with most participants reaching very similar maximal band levels for injured and non-injured sides. This training modality appears to be a suitable alternative for this patient group where traditional methods are limited due to pain, weakness and limited mobility post-hip fracture. It is also an inexpensive option, patients can be educated whilst in hospital and monitored infrequently resulting in reduced travel costs (therapist to home or patient to facility) and resistive bands cost very little. The overall adherence rate was 95%. However it is important to acknowledge that adherence in the study was highest during the supervised period (99.7%; weeks 1-6) compared with the unsupervised period (91.3%; weeks 7-12). This could possibly reflect a decrease in motivation that accompanied a reduction in physiotherapist contact from three times per week initially to only once per week in the final six weeks of the intervention. This reduced supervision may also have had an adverse effect on the social component of regular interaction with the physiotherapists that the program facilitated, which can be an important motivating factor in elderly women. The high adherence rates achieved are also likely a result of the training program being tailored to suit each individual, based on initial strength and pain level of both the injured and non-injured limb (Miller et al., 2008).

Di Monaco et al. (2011) tested the effect of a multidisciplinary program for fall prevention in 95 women with a hip fracture. They received a multidisciplinary intervention during inpatient rehabilitation, which consisted of one to three hours a day for five days a week of physical exercise to improve strength and balance, recommendations and training on the use of assistive devices, training in activities of daily living conducted by physical and occupational therapists. Additionally, 45 women also received a home visit by an occupational therapist at a median of 20 days after discharge. The absolute risk of falling in the population study was significantly lower than that previously reported in hip fracture survivors. Nonetheless, uncorrected environmental and behavioral risk factors and poor adherence to targeted recommendations for fall prevention significantly predicted the risk of falling during a six-month follow-up in community-dwelling women who sustained a fall-related hip fracture. Thus, improving adherence to the recommendations is a major goal to prevent falls.

### 2.2 Rehabilitation after vertebral fractures

Vertebral fractures have a high impact on the quality of life and their occurrence is related to digestive and respiratory morbidities, anxiety, depression and death. Thereby, the incidence of a vertebral fracture should be followed by a limited period of bed rest, to avoid the hazards of deconditioning, accelerated bone loss, deep venous thrombosis, pneumonia, pressure sores, disorientation and depression. A rehabilitation program is necessary and helps prevent deformity by strengthening anti-gravity muscles and promoting postural retraining. Breathing exercises promote thoracic expansion and improve the heavily degraded pulmonary function found in patients with spinal osteoporotic fractures (Schlaich et al., 1998). Instruction on the proper way of lifting things, as well as how to appropriately use a walker or a cane, could be beneficial and thus is strongly recommended (Bonner et al.,

2003). Patients with fractures should perform low-intensity exercise and gentle strengthening programs (e.g., Tai Chi and hydrotherapy) and are strongly recommended to avoid high impact exercise or movements, to avoid suffering new vertebral fractures. Forward bending of the spine or flexion exercises, especially in combination with twisting, should be avoided.

According to Bassey, this includes several old favourite exercises which are now considered outdated, namely straight-leg toe touches and sit ups (or crunches) for strengthening the abdominal muscles. Sinaki and Mikkelsen reported that the latter are associated with a dramatically increased rate of vertebral fracture in osteoporotic women (89% compared to 16% of those who did extension exercises). As the acute fracture pain subsides, a walking program can begin with gentle strengthening exercises focusing on spinal extensor muscles. A carefully supervised rehabilitation program should be started after 3 to 4 months, to strengthen the spinal extensor and abdominal muscles more aggressively (Sinaki, 1995). Back strengthening exercises can reduce thoracic hyperkyphosis, vertebral fracture, loss of height, and pain of the anterior rib cage, which are the most disfiguring consequences of osteoporosis. Improvement of back strength reduces the kyphotic posture that can occur with osteoporosis and aging. Development of hyperkyphotic posture does not only predispose the patient to postural back pain but can also increase the risk of falls. In subjects with hyperkyphosis, compensatory use of hip strategies rather than ankle strategies occurs during incidents of momentary challenges of balance (Sinaki and Lynn, 2002).In a controlled trial, use of a proprioceptive dynamic posture training (PDPT) program improved balance in osteoporotic subjects with kyphosis. Chronic pain may be caused by vertebral fractures or may be a result of postural deformities, such as hyperkyphotic or scoliotic changes in the spine, with inappropriate stretching of ligaments. Strong back muscles are significantly correlated with a decreased risk of vertebral fractures and kyphosis.)

In patients with severe kyphosis, pressure of the lower part of the rib cage over the pelvic rim causes considerable flank pain and tenderness and compromises breathing. With healthy posture, there is sufficient space between the lower ribs and the iliac crest, so no contact occurs, even on lateral bending of the trunk. In severe osteoporosis with compression fractures, substantial dorsal kyphosis, and loss of height, iliocostal contact occurs. Therefore, helping the patient to decrease kyphotic posturing through recruitment of back extensors for provision of better dynamic-static posturing can reduce pain, increase mobility, reduce depression, and improve the patient's quality of life.

Sacral insufficiency fractures necessitate milder physical therapy, reduction of weight bearing with use of gait aids, and orthoses. Even though there is a lack of specific studies comparing various types of orthoses, it is widely accepted that all spinal orthoses, whether made of cloth, metal, or plastic, or whether rigid or flexible, use a three-point pressure system. Traditionally, spinal orthoses have been used in the management of thoracolumbar injuries treated with or without surgical stabilization. The vast majority of orthoses, however, are used in persons with back pain and spinal deformities.

In the United States alone, 250,000 corsets are prescribed each year. Kaplan et al. (1996) found that rigid bracing is not necessary for managing postural osteoporotic back pain, and indeed, a weighted kypho-orthosis was more effective for patient compliance and pain relief (Kaplan, 1996). Moreover, the use of rigid thoracolumbar braces in osteoporosis is limited by factors such as short stature (147 cm), atrophy of trunk muscles, hiatal or inguinal hernia, moderate to severe obesity, scoliosis caused by osteoporosis and compression fractures, and restricted respiration, leading to low compliance.

Lantz and Schultz (1986) described increased electrical activity of back muscles when a lumbosacral orthosis is worn, supporting the concept of the so-called `biofeedback` as an underlying principle of efficacy. Stronger back muscles may decrease the angle of kyphosis and thus improve body height. This result may be associated with better posture and a correction of the center of gravity, which then results in less body sway. Given that body sway is a well-documented risk factor for falls and fall-related fractures, this change of the center of gravity may be accompanied by a lower rate of falls and nonvertebral fractures. Given the widespread use of orthoses in various diseases, there is an urgent need for controlled clinical trials to further elucidate functions and applications of these technical devices.

### 2.3 Rehabilitation after Colles' fracture

Colles' fracture is the most common fracture in women over 40. The reason for prescribing physiotherapy after a fracture of the distal radius is that it serves primary mobilization which is the most important principle of fracture management.. Complications of Colles' fracture may be: rigidity, algodystrophy, or functional limitation. Treatment planning considers the following aspects: type of fracture, stability, presence of possible associated lesions, and type of patient.

Physical therapy after a Colles' fracture consists of muscle strengthening, range of motion exercises, wound healing and scar adhesion. In the rehabilitation pathway consideration is given to the type of surgical intervention. Depending on the type of intervention, we can have different times of limb immobilization: the conservative treatment lasts 5 weeks,osteosynthesis with a plate lasts 2 weeks,external fixing: neutralization for the first 4 weeks, active movements for the successive 3 weeks. The rehabilitation treatment starts during the period of immobilization whatever the treatment (invasive or otherwise). Early reduction of edema is of primary importance in determining hand function. Elevation of the hand above the heart's level and active range of motion exercises are instructed to facilitate the pumping action of hand muscles to decrease swelling. A 15 minute hand-wrap with paraffin should be followed by exercise of equal duration. The hand should be kept in both cold and hot water in order to augment venous return. The articulations not involved in the process must be mobilized actively and passively in order to maintain the integrity of the joint capsule and to prevent tendon thickening and adhesions. The rehabilitation protocol states that at the end of the immobilization period, it is necessary to proceed to active, passive and counter resistance exercises (with or without tutors) under the guidance of the rehabilitation therapist. Exercises must be progressive, but must be initiated early, on the first or second day. Exercise programs consist of transverse scar massage, passive range of motion and progressive resistive exercises, massage and active range of motion exercises, which focus on strengthening both extrinsic and intrinsic muscle groups of the hand.

Global mobilization of the wrist includes: flexion and extension, abduction (ulnarization) and adduction (radialization), longitudinal rotation, prono-supination of the forearm, and mobilization against resistance for regaining of strength. Sessions should be daily for 30 minutes for 2 weeks. Specific exercises aim at producing tension of the periarticular and articular elements. They involve flexion with ulnar inclination (subject standing with forearm in supination, the dorsum of the hand facing the table top and forearm vertically pressing on the table), and extension associated with radial inclination (subject standing, palm of hand on the table and forearm in vertical position, with arm and forearm pronated). Exercises at home are an essential part of the rehabilitation treatment, to be performed at least 4 to 6 times a day for 2 to 3 weeks. Specific exercises are Fingers 'wall walking', bilateral paper ripping, circular 'dusting', simple 'blackboard writing' and drawing tasks, various opposition and pinching are among the most recommended exercises. These activities are graded according to resistance, type of motion and grasp strength. Splinting helps develop the range of motion. Physical therapy should be followed by occupational therapy for 3 weeks (Christensen et al., 2001). Thus in dealing with the rehabilitation treatment of Colles' fracture it is fundamental not to limit ourselves to the specific post-fracture protocol but also to establish a rehabilitation pathway which foresees the global care of the patient with fractures due to fragility fractures.

| **Clinical practice guidelines for rehabilitation** | **Objectives of the procedure** |
|---|---|
| Mobilization and early assisted ambulation | • Avoids immobilization and bone loss<br>• Improves bone healing quality and rate |
| Breathing exercises | • Drainage of pulmonary secretions<br>• Reduction of pulmonary complications |
| Multidisciplinary rehabilitation | • Increases muscle mass and strength<br>• Improves body function<br>• Early supported discharge<br>• Evaluation of medical complications |
| Adequate balance training | • Reduces risk of falls<br>• Reduces risk of a second fracture |
| Physical activity | • Reduces risk of falls<br>• Increases muscle mass and strength<br>• Increases BMD |
| Nutritional statement | • Adequate protein and energy supplementation |
| Educational and social measures | • Better quality of life<br>• Improvement in the ability to perform daily activities<br>• Improves risk awareness<br>• Speeds up mental and social recovery |

Table 1. Summary of guidelines and aims of rehabilitation

## 3. Conclusion

Current osteoporosis practice guidelines recommend that any adult 40 or older who sustains any fracture should be investigated and treated for low BMD, as well as educated about lifestyle and nutritional factors, including exercise and physical activity participation, as a way of potentially modifying their subsequent risk of fractures. In the past decades, there has been increasing interest in research on exercise for older people recovering from fragility fractures; however, the extent of this literature remains unclear. This poses a challenge for determining whether or not there is sufficient evidence to develop practice recommendations, whether or not a systematic review is feasible to conduct and, if there is a gap in knowledge, if that gap exists. Exercise is a complex intervention, and a major challenge to exercise prescription is the identification of the optimal exercise program elements or strategies that contribute to favorable treatment outcomes. Researchers investigating the effects of exercise on BMD have reported increases in BMD in pre-menopausal women participating in low-intensity regular exercise and following the introduction of an exercise regimen. Weight-bearing activities, such as walking or running, have a greater effect than non-weight-bearing activities, such as cycling and swimming, whereas a reduction in mechanical loading, that is, bed rest or space flight, leads to bone loss (Andreoli et al. 2011).

Equally important is the identification of exercise program elements or strategies that do not contribute to more favorable outcomes or are not cost-effective. Although there were trends across all these studies towards more favorable outcomes after specialized, multidisciplinary care programs and more favorable ambulatory and functional outcomes following early mobilization with weight bearing/ambulation, aerobic, balance, and strengthening exercises, there was still insufficient or conflicting evidence from randomized trials to be able to recommend any one mobilization strategy or exercise intervention over another after hip fracture.

There have been few studies that have evaluated resistance training in frail older adults, but in those that have been reported, it appears that high intensity training is more successful than low intensity training in achieving gains in strength. Nevertheless translation of these findings into a clinical setting has been difficult. The provision of therapeutic exercise programs in hospitals to post-operative patients has a poor evidence base and the approach varies from therapist to therapist. There is uncertainty about whether patients recovering from an operation are able to adhere to a resistance strengthening program which commences within a week of fracture when they are often recovering from an operation and dealing with pain and discomfort. There are, however, consistent trends in the literature suggesting: 1) Improved functional outcomes with acute, inpatient, multidisciplinary rehabilitation care programs following hip fracture. 2) Better ambulatory and functional outcomes with early mobilization, weight bearing/ambulation, aerobic, balance, and strengthening exercise interventions following hip fracture. 3) Faster recovery in mobility, functioning and pain without risk of increased complications with early post-surgical motion and/or weight bearing following ankle fractures, and early motion and/or functional use following non-displaced proximal humerus fractures. 4) That supervised or formal therapy may be no more effective in regaining long term mobility and function than unsupervised or home exercise programs following a period of immobilization for distal radius fractures. Prevention of falls and refractures may be even more effective when multiple risk factors are taken into account simultaneously. Most multifactorial fall prevention programmes have been successful in reducing the incidence of falls and risk factors of falling, especially when prevention has been individually tailored and targeted to populations at high risk of falling.

Multifactorial interventions should include: a) among community-dwelling older persons: (i.e., those living in their own homes) gait training and advice on the appropriate use of assistive devices, review and modification of medication, especially psychotropic medication, exercise programs with balance training as one of the components, treatment of postural hypotension, modification of environmental hazards and treatment of cardiovascular disorders, b) among older persons in long-term care and assisted living settings: staff education programs, gait training and advice on the appropriate use of assistive devices and review and modification of medications, especially psychotropic medications.

Patients with fractures should perform low-intensity exercise and gentle strengthening programs (e.g., Tai Chi and hydrotherapy) and are strongly recommended to avoid high impact exercise or movements, to avoid suffering new vertebral fractures.

Key areas for further research related to exercise prescription following hip fracture included identifying which exercise interventions and/or components of a care program are likely to have long-term, cost-effective and beneficial impacts on functioning and health-related quality of life.

## 4. References

Ahmed LA, Schirmer H, Bjornerem A, Emaus N, Jorgensen L, Stormer J, Joakimsen RM (2009) The gender- and age-specific 10-year and lifetime absolute fracture risk in Tromso, Norway. Eur J Epidemiol 24:441–448

Andreoli, A.; Monteleone, M.; Van Loan, M.; Promenzio, L.; Tarantino, U.; De Lorenzo, A.; (2001) Effects of different sports on bone density and muscle mass in highly trained athletes. *Med Sci Sports Exerc,* Vol.33, No.4, pp. 507-11

Andreoli, A.; Celi, M.; Volpe, S.L.; Sorge, R.; Tarantino, U. (2011) Long-term effect of exercise on bone mineral density and body composition in post-menopausal ex-elite athletes: a retrospective study. *Eur J Clin Nutr, advance online publication, 15 June 2011*

American College of Sports Medicine (1999). Position stand: the recommended quantity and quality of exercise for developing and maintaining cardiorespiratory and muscular fitness, and flexibility in healthy adults. Med Sci Sports Exerc. Vol. 30. pp. (975–979).

Bergman H, Ferrucci L, Guralnik J, Hogan DB, Hummel S, Karunananthan S, Wolfson C. (2007) Frailty: an emerging research and clinical paradigm—issues and controversies. J Gerontol A Biol Sci Med Sci 62:731–737

Bonner F.J. Jr, Sinaki M., Grabois M., Shipp K.M., Lane J.M., Lindsay R., Gold D.T., Cosman F., Bouxsein M.L., Weinstein J.N., Gallagher R.M., Melton L.J. III, Salcido R.S. & Gordon S.L. (2003). Health professional's guide to rehabilitation of the patient with osteoporosis. Osteoporos Int. Vol. 14. Suppl.2. pp (S1-22).

Briggs AM, Greig AM, Wark JD. (2007).The vertebral fracture cascade in osteoporosis: a review of aetiopathogenesis. Osteporos Int, 18:575-584.

Cameron I., Crotty M., Currie C., Finnegan T., Gillespie L., Gillespie W., Handoll H., Kurrle S., Madhok R., Murray G., Quinn K. & Torgerson D. (2000). Geriatric rehabilitation following fractures in older people: A systematic review. Health Technol Assess. Vol. 4. pp. (1–111).

Chilov M., Cameron I.D. & March L.M. (2003). Evidence-based guidelines for fixing broken hips: An update. Med J Australia. Vol. 179. pp. ( 489-92).

Christensen O.M., Kunov A., Hansen F.F., Christiansen T.C. & Krasheninnikoff M. (2001). Occupational therapy and Colles' fractures. Int Orthop. Vol. 25. pp (43-45).

Di Monaco M., Vallero F., De Toma E., Castiglioni C., Gardin L., Giordano S. and Tappero L. (2011). Adherence to recommendations for fall prevention significantly affects the risk of falling after hip fracture: post-hoc analyses of a quasi-randomized controlled trial. Eur J Phys Rehabil Med. Vol. 47. pp (1-2).

Dionyssiotis Y.,. Dontas I.A, Economopoulos D. & Lyritis G.P. (2008). Rehabilitation after falls and fractures. J Musculoskelet Neuronal Interact. Vol. 8, No. 3, pp. (244-250).

Feehan L. M., Beck C.A. Harris S.R., MacIntyre D. L. & Li L. C. (2011). Exercise prescription after fragility fracture in older adults: a scoping review. Osteoporos Int. Vol. 22. pp. (1289–1322).

Fried L.P., Tangen C.M., Walston J., Neuman A.B., Hirsch C., Gott Diener J., Seeman T., Tracy R., Kop W.J., Burke G. & McBurnie M.A.; Cardiovascular health study collaborative research group (2001). Frailty in older adults: evidence for a phenotype. J Gerontol A Biol Sci Med Sci. Vol. 56. pp. (M146–M156).

Hauer K., Specht N., Schuler M., Bartsch P. & Oster P. (2002). Intensive physical training in geriatric patients after severe falls and hip surgery. Age Ageing. Vol. 31. pp (49–57).

Heinrich S., RappK.,Rissman U.,Beker C.,Konig H. & Heinrich H. (2010). Cost of falls in old age: a systematic review. Osteoporosis Int.; Vol. 21. Pp (891-902)

General Rockville, U.S. Department of Health and Human Services (2004) Bone, Health and Osteoporosis: A Report of the Surgeon General. U.S. Department of Health and Human Services, Office of the Surgeon

Gillespie L.D., Gillespie W.J., Robertson M.C., Lamb S.E., Cumming R.G. & Rowe B.H. (2001). Interventions for preventing falls in elderly people. Cochrane Database Syst Rev 3:CD000340.

Järvinen TL, Sievänen H, Khan KM, Heinonen A, Kannus P. (2008) Shifting the focus in fracture prevention from osteoporosis to falls.BMJ. Jan 19;336(7636):124-6.

Jensen J., Lundin-Olsson L., Nyberg L. & Gustafson Y. (2002). Fall and injury prevention in older people living in residential care facilities: A cluster randomized trial. Ann Intern Med. Vol. 136. pp. (733–741).

Judex, S.; Rubin, C.T. (2010) Is bone formation induced by high-frequency mechanical signals modulated by muscle activity? J Musculoskelet Neuronal Interact, Vol.10, No.1, pp. 3-11

Kaplan R.S., Sinaki M. & Hameister M. (1996). Effect of back supports on back strength in patients with osteoporosis: A pilot study. Mayo Clin Proc. Vol. 71. pp (235–241).

Kayali C., Agus H., Ozluk S. & Sanli C. (2006). Treatment for unstable intertrochanteric fractures in elderly patients: internal fixation versus cone hemiarthroplasty. J Orthop Surg (Hong Kong). Vol. 14. pp. (240-244).

Mangone G., Postiglione M. & Pasquetti G. (2010). Rehabilitation in peripheral non femoral fractures: a review. Clinical Cases in Mineral and Bone Metabolism. Vol. 7, No. 1, pp. (48-50).

Miller M.D., Foley A., Gunn S.M. & Crotty M. (2008). Progression and adherence to an individually prescribed and supervised resistance training intervention in older adults recovering in hospital from lower limb fragility fracture. Patient Preference and Adherence . Vol. 2. pp (107–113).

Moayyeri A. (2008). The association between physical activity and osteoporotic fractures: a review of the evidence and implications for future research. Ann Epidemiol. Vol. 18, No. 11. pp. (827–835).

Moreland J., Richardson J., Chan D.H., O'Neill J., Bellissimo A., Grum R.M. & Shanks L. Evidence-based guidelines for the secondary prevention of falls in older adults. (2003). Gerontology. Vol. 49, pp. (93-116).

Nelson ME, Rejeski WJ, Blair SN, Duncan PW, Judge JO, King AC, Macera CA, Castaneda-Sceppa C (2007) Physical activity and public health in older adults. Recommendation from the American College of Sports Medicine and the American Heart Association. Circulation 116:1094–

Nordström, P.; Eklund, F.; Björnstig, U.; Nordström, A.; Lorentzon, R.; Sievänen, H.; Gustafson, Y.; (2011) Do Both Areal BMD and Injurious Falls Explain the Higher Incidence of Fractures in Women than in Men? Calcif Tissue Int, Vol.89, No.3, pp. 203-10

O'Neil TW, Felsenberg D, Varlow J, Cooper C, Kanis JA, Silman AJ (1996). The prevalence of vertebral deformity in european men and women: the european vertebral osteoporosis study. J Bone Miner Res 11: 1010-1018.

Ozcivici, E.; Luu, Y.K.; Adler, B.; Qin, Y.X.; Rubin, J.; Judex, S.; Rubin, C.T. (2010) Mechanical signals as anabolic agents in bone. Nat Rev Rheumatol, Vol.6, No.1, pp. 50-9

Piscitelli P.,Chitano G., Greco M., Benvenuto M., Sbenaglia E., Colì G., Migliore A., Granata M., Iolascon G., Gimigliano R., Baggiani A., Distante A., Tarantino U., Rizzuti C., Brandi M.L (2010). Pharmaco-economic issues in the treatment of severe osteoporosis. Clinical Cases in Mineral and Bone Metabolism; 7(1): 61-64

Pfeifer, M., Sinaki M., Geusens P., Boonen S., Preisinger E. & Minne H.W. for the ASBMR Working Group on Musculoskeletal Rehabilitation. (2004). Musculoskeletal Rehabilitation in Osteoporosis: A Review. J Bone Miner Res. Vol. 19, No. 8, (Published online on May 10, 2004), pp. (1208 -1214).

Rossini M., Viapiana O., Gatti D., De Terlizzi F.,Adami S. (2010) Capacitively Coupled Electric Field for Pain Relief in Patients with Vertebral Fractures and Chronic Pain. Clin Orthop Relat Res 468:735–740

Rucco V., Visentini A. & Pellegrini E. (2003). The rehabilitation project in hip arthroplasty patients. Eur Med Phys. Vol. 39. pp (45-57).

Schlaich C., Minne H.W., Bruckner T., Wagner G., Gebest H.J., Grunze M., Ziegler R. & Leidig-Bruckner G. (1998). Reduced pulmonary function in patients with spinal osteoporotic fractures. Osteoporos Int. Vol. 8. pp (261-267).

Scottish Intercollegiate Guidelines Network (2002). Prevention and Management of Hip Fracture in Older People. A National Clinical Guideline. Scottish Intercollegiate Guidelines Network, Edinburgh. Guideline 52. http://www.show.scot.nhs.uk/sign/guidelines/published/index.html

Sinaki M. Musculoskeletal rehabilitation. In: Riggs B.L., Melton L.J. (Eds.) (1995). Osteoporosis: Etiology, Diagnosis, and Management, 2nd ed. Philadelphia: Lippincott-Raven.

Sinaki M. & Lynn S.G. (2002). Reducing the risk of falls through proprioceptive dynamic posture training in osteoporotic women with kyphotic posturing: A randomized pilot study. Am J Phys Med Rehabil. Vol. 81. pp (241–246).

Sinaki M., Itoi E., Wahner H.W., Wollan P., Gelzcer R., Mullan B.P., Collins D.A. & Hodgson S.F. (2002). Stronger back muscles reduce the incidence of vertebral fractures: A prospective 10 year follow-up of postmenopausal women. Bone. Vol. 30. pp (836–841).

Siris ES, Brenneman SK, Barrett-Connor E, Miller PD, Sajjan S,Berger ML, Chen YT (2006) The effect of age and bone mineral density on the absolute, excess, and relative risk of fracture in postmenopausal women aged 50–99: results from the National Osteoporosis Risk Assessment (NORA). Osteoporos Int 17:565–574

Skelton D.A., Dinan S., Campbell M. & Rutherford O. (2005). Tailored group exercise (Falls Management Exercise - FaME) reduces falls in community-dwelling older frequent fallers (an RCT). Age Ageing. Vol. 6. pp (636-9).

Tinetti M.E., Baker D.I., Gottschalk M., Williams C.S., Pollack D., Garrett P., Gill T.M., Marottoli R.A. & Acampora D. (1999). Home-based multicomponent rehabilitation program for older persons after hip fracture: A randomized trial. Arch Phys Med Rehabil. Vol. 80. pp. (916–922).

World Health Organization (2001). International classification of functioning, disability and health. WHO, Geneva, Switzerland.

# Part 5

# Effects of Psychological Factors, Body Awareness and Relaxation Techniques in Physical Therapy Interventions

# 12

# Focus on Psychological Factors and Body Awareness in Multimodal Musculoskeletal Pain Rehabilitation

G. Gard
*Luleå University of Technology, Sweden*

## 1. Introduction

Musculoskeletal pain disorders are substantial health problems in many countries and consume a large proportion of the health services. These disorders are the main causes of occupational disability and sick leave in Sweden (Bergström et. al., 2007, 2009; Vingård et. al., 2004; Dionne et. al., 2007). A varied sickness panorama can be seen within primary health care in Sweden, with a high prevalence of stress-related pain disorders. These disorders have increased in the Swedish population in recent years and are common causes of sickness absence and disability pensions, with a higher prevalence in women than in men. The sick leave rate and stress-related disorders have increased most among women working within healthcare, teaching, social work, schools, child care and care of the elderly. In addition to sick leave, the level of sickness presence is also high in these sectors (Vingård et. al., 2004). This may be explained by organizational changes and downsizing and/or high physical and psychosocial workload and the problems must be viewed from a systems perspective. The development of stress-related disorders can be described as a process over years with symptoms of fatigue and musculoskeletal tensions and pain that develops into psychosomatic and more severe musculoskeletal pain diagnoses for example persistent pain, chronic fatigue syndrome and emotional exhaustion. The diagnosis chronic fatigue syndrome implies that the aetiology can be long-term stress emanating from working life or life situation (SBU, 2010).

## 2. Multimodal pain rehabilitation according to cognitive behavior therapy principles

Multimodal pain rehabilitation according to cognitive behaviour therapy (CBT)-principles has been shown to reduce pain and stress levels and increase return to work (Van der Klink et al., 2001; Foster et. al., 2009). An analyse of 48 intervention studies showed reduced pain and stress levels by cognitive behaviour interventions, multimodal programs and organisational interventions. The individually tailored interventions were the most effective in reducing pain and stress levels (Van der Klink et al., 2001).

Today psychological factors are important for the understanding and treatment of persistent pain disorders (Linton, 2000; Goossens et. al., 2005; Linton et. al., 2005). Persistent pain is a specific condition and should be treated in that order with respect to previous diagnosis (SBU, 2010). A holistic perspective on the problem is needed for successful return to work, including salutogenic factors, psychological factors, regular physical activity and bodily treatments (SBU 2010; Flensborg-Madsen et. al., 2006). Persistent pain is defined as persistent or recurring pain with an endurance of at least three to six months (SBU, 2010). To live with persistent and long-lasting pain require support and coaching. The role of the multimodal team is to meet, see and guide the client towards increased health and work ability.

Research has shown that the use of CBT- principles in persistent pain rehabilitation can improve social and physical function and reduce pain significantly (SBU rapport 177: 2006). To focus on supporting and coaching clients own healthy coping strategies has been shown to be a significant factor to reduce clients´ pain (Foster et. al., 2009; Dionne et. al., 2007). A multimodal pain rehabilitation intervention according to CBT-principles focuses on client participation in the rehabilitation process by clear goal formulations (goals/subgoals) and a stepwise process with tailored progression according to clients abilities. To formulate realistic and concrete subgoals, to use adequate dosage and give regular feedback is important for positive rehabilitation results. To receive positive feedback when goals are reached increases clients´ self-efficacy, motivation for change and goal attainment. In multimodal pain rehabilitation the clients´ formulate their goals together with their therapist which has been shown to be important for positive rehabilitation results (Foster et. al., 2009; Denison et. al., 2004; Dionne et. al., 2007).

## 3. The importance of psychological factors within multimodal pain rehabilitation

Within effective multimodal rehabilitation it is important to consider the importance of salutogenic and psychological factors for a successful return to work (Denison et. al., 2004; Gard & Larsson; 2003a, b; Grahn, 1999). Salutogenic factors means factors of importance for experiencing coherence in life. Salutogenic factors have been identified as important resources for health. The salutogenic factors have been defined as the experience of meaningfulness, comprehensiveness and ability to cope with events in life. A high sense of coherence is associated with good health and capacity to withstand mental stress (Flensborg-Madsen et. al., 2006).

To consider psychological factors in rehabilitation means to identify and consider:

1. clients´ responsibility, opportunity to influence and participate in rehabilitation,
2. clients´ opportunities for pain control,
3. motivational factors, self-efficacy and empowerment factors in the rehabilitation process. (Linton, 2000; Jensen et. al., 2007; Foster et. al., 2009; Gard & Larsson., 2003a, b, Larsson & Gard., 2003; Arnesson & Ekberg, 2005).

High client participation increases clients´ control, activity, health and well-being (SBU, 2006, 2010). Clients´ pain control can be improved by physical and joyful activities which reduces the experience of pain. Joyful activities can reduce pain by changing the mental focus towards joyful activities. Multimodal physical training programmes have been shown to teach clients´ how to cope with and reduce persistent pain. There is evidence that physical activity can improve muscle strength endurance, coordination, stability and reduce persistent pain (SBU, 2006). By increased control of the situation for the patient, it is possible to participate in the rehabilitation to a higher extent and learn more (Theorell et. al., 2005; Arnesson & Ekberg, 2005; Foster et. al., 2009;Dionne et. al., 2007).

To consider motivational factors in rehabilitation is important for good results (Gard & Larsson 2003a; Grahn, 1999). Motivation can be defined as everything that drives and sustains human behaviour and motivation for change. Motivation is influenced by a combination of personal and social factors, such as having individually formulated goals, expectations for the future and self-efficacy (Gard & Larsson 2003a; Grahn, 1999).

Self-efficacy is also an important psychological factor in rehabilitation. By increasing an individual's self-efficacy, the likelihood of higher motivation for a behavioural change increases. Self-efficacy is related to goal commitment. Several studies have been published on the effectiveness of self-efficacy enhancing interventions (Arnesson & Ekberg, 2005). Self-efficacy and coping with anxiety and fear can be focused in pain rehabilitation by use of CBT-principles with positive results (Foster et. al., 2009)). Perceived self-efficacy to perform physical tasks, meet role expectations, obtain support and maintain job security has been shown to be important for health and workability. The process through which people gain greater control over decisions and actions affecting their health is frequently associated with Bandura's concept of self-efficacy, i.e. one's confidence in performing a particular behaviour and in overcoming barriers to that behaviour (Denison et. al., 2004; Grahn, 1999, Foster et. al., 2009).

A focus on clients´ personal resources such as experience of control, self-efficacy, trust and confidence have found to be important for maintaining and improving health and work ability (Denison et. al., 2004; Jensen et. al., 2007; Linton et. al., 2005). What a person wishes, is clearly connected with views on own possibilities and own competence and what one 'can manage'. For successful work rehabilitation, perceived self-efficacy to perform physical tasks, meet role expectations, obtain support and maintain job security has to be focused on (Denison et. al., 2004; Grahn, 1999, Foster et. al., 2009; Lindström & Eriksson, 2005). Thus, personal resources such as one's ability to assess and understand the situation, to find a meaning in moving in a health promoting direction and having the capacity to do so, seem to function as 'brokers' moderating how health is affected by stressful situations (Lindström & Eriksson, 2005). The demand-control-social support model also indicates that these relations are very important for good health. Social support such as emotional support by family and close relatives, social support at work in terms of employers and colleagues are also important for return to work. The most critical predictors of the outcome of rehabilitation have been found to be the individual's own expectations about vocational return, perceived health, and participation in the treatment situation (Dworkin et. al., 2005; Grahn et al. 1999; Theorell et.al.., 2005; Leijon et. al., 2006).

## 4. Information and Communication Technology (ICT) can be a tool to promote health in a multimodal context

"E-Health can be defined as strategies to improve health and well-being, work ability and quality of life by the use of information and communication technology and different electronic devices. It is a tool to ensure information, participation, choice and empowerment to people within the area of health and health care "(www.ehtel.org). Research using E-health solutions have shown its` possibilities to improve quality, effectiveness and safety of care as well as to facilitate active client participation in health and disease management (IST.2008), "E-Health can be defined as strategies to improve health and well-being, work ability and quality of life by the use of electronic devices and equipment and information and communication technology. It is a tool to ensure information, participation, choice and empowerment to people within the area of health and health care "(www.ehtel.org). Research using E-health solutions may improve quality, effectiveness and safety of care, facilitate active client participation and open up for new opportunities in health and disease management (IST, 2008).

Researchers have shown that health and well-being can be improved by interventions at the appropriate organisational level with the use of E-health tools (Foster et. al., 2009; Leville et. al., 2009, Lorig et. al., 2001). Practice development can be performed through the use of E-health applications in an on-going practice development process. Important steps in such a practice development process are to establish a network and agree on goals, priorities and strategies (Tolson & Kelly, 2006). Today, we have an on-going project at Luleå University of Technology with the purpose of evaluating a tailored web-based support programme to complement multimodal rehabilitation in primary health care for patients with long-term non-specific neck/shoulder/back pain. The specific objectives are to study the:

1. Effects of the program on self-perceived workability and days of sick-leave among the clients
2. Effects on costs and resource utilization in the primary health care system
3. Effects on pain and function in the musculoskeletal system, self-efficacy and coping with pain, general health and well-being among the clients and
4. Satisfaction and perceived usability, regarding both staff and clients, with the web-based support system.

As clients require increased knowledge and information as well as increased participation in decisions concerning their own rehabilitation, a web-based program may imply a faster and earlier access to professional rehabilitation which will increase the effectiveness of multimodal rehabilitation. It is important to get as early access to primary health care as possible to reduce waiting lists and long term sick leaves and persistent pain conditions.

## 5. The importance of body awareness

To work with body awareness through body experiences has been shown to be a good way to preserve health and can lead to freedom from muscular stiffness, easier breathing and increased mental awareness (Lundvik-Gyllensten et. al., 2010). Body awareness is focused in different traditions within physiotherapy, for example within psychotherapeutic bodily traditions inspired by Reich (1949) and Lowen (1975) and within body awareness therapies (Mattsson, 1998; Roxendahl, 1985). The aim of body awareness therapies are to integrate the body in the total experience of the self (Mattsson, 1998; Roxendahl, 1985). Within the therapy the body is seen as a functionally integrated entity, where different parts are dependent on one another. Psychosomatic symptoms can be seen as an expression of imbalance in the body. To focus on increasing the balance in the body is important. To find one's stability in relation to graviity can imply deeper breathing and deeper contact with emotions (Lundvik Gyllensten et. al., 2010; Rosberg, 2000).

Basic Body Awareness Therapy, BBAT, has developed in Scandinavia and aims to restore body awareness and movement coordination, integrating physical, physiological, psycho-socio-cultural and existential aspects. The effectiveness of BBAT has been studied in different contexts, individual as well as group contexts. BBAT can reduce pain and increase health-related quality of life. The meaning of body awareness has also been studied from a patient perspective (Lundvik Gyllensten et. al., 2010). The meaning of having body awareness can be expressed as being embodied, to be in contact with and being within one`s body, feeling alive. This awareness influenced the relation to others and one´s participation in society. The meaning of living in ones' body was to become more aware of the body and to experience oneself from the inside as a means to know one's needs. The key was the fact that bodily experiences always are in the present moment. The experience of the body, the balance and stability of the physical self were basic experiences that were connected to the conception of well-being and control. To understand one's emotions and needs through the awareness of the body were seen as the base for self-confidence, trust in oneself and the ability to take care of oneself and ones needs physically and mentally. Living in relation to others was a need, a need which included recognition of the embodied self and bodily contact. Improved body awareness seemed to lead to being more satisfied and at peace with oneself, participating more actively in life. Problems with body awareness seemed to lead to a feeling of not being alive, missing something important in life. Working with the body in physiotherapy practice should include an awareness of that; the awareness of the body is inseparable from one's identity and has a direct connection to the experience, of the embodied identity (Lundvik Gyllensten et. al., 2010).

The meaning of body awareness was constructed by the informants as becoming more in contact with the body in order to strengthen the identity. To experience oneself, through being aware of the body from the inside was a means to be oneself in a deeper way. (Lundvik Gyllensten et. al., 2010).

*"Body awareness to me is not to see myself from the outside, but to be inside of myself. I am the receptor that is catching signals from the inside of me as well as from the outside. Breathing has to do with my vitality and tells me that there is life in me. My breathing is part of what I experience from inside of me. For me body awareness is a centred awareness that includes my body, is in the body…"*(Lundvik Gyllensten et. al., 2010).

## 6. The importance of body trust

Body trust can be defined as the trust in relation to one's own body, a new concept in physiotherapy (Hedlund & Gard, 2000). Body trust is together with body awareness a resource towards increased health (Mattsson, 1998; Rosberg, 2000). Increased body trust can lead to an increased mastery of the life situation. Trust is generally defined as the state where one individual is emotionally stable to other people's behaviour. Trust takes time to establish, because frequent interactions between trustor and trustee is needed (Blois, 1999). Body trust and body awareness can be seen as two dimensions of the body ego, a body dimension, body awareness and a psychological dimension, body trust, both influencing the level of perceived health (figure 1).

Having body trust has been shown to be related to feelings of joy and security, to be proud of the body and to feelings of stability and strength. Individuals with body trust use the body as a tool to realise their own goals, to satisfy their needs in life and to communicate with others (Hedlund & Gard, 2000). Body trust is also a prerequisite to reduce muscle strain and muscle tension. It is only when you trust your body that you can actually relax and reduce muscle tension.

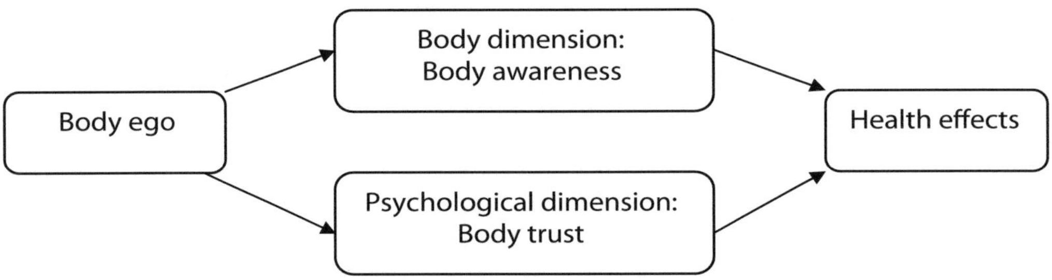


Fig. 1. The body trust health model. The body ego with two dimensions, a body dimension, body awareness and a psychological dimension, body trust, both is influencing the level of perceived health.

The concept of body ego is well-established and defined as a human being`s total perception of identity. It can be observed in psychotherapeutic assessment. A strong body ego has been shown to increase reality orientation and adaptation and the understanding of oneself and one's emotions (Roxendahl, 1985). The coordinating function of body ego is mental awareness, which regulates the psychomotor and perceptual aspects of movements. Mental awareness can be trained to increase opportunities of self-experiences in all areas of life (Mattsson, 1998). Roxendahl developed the concept of body ego in relation to physiotherapy. Quality of movement is related to body function and to the experience of the body as a part of the identity. A person with a well-functioning body ego is rooted in reality and can express herself through body movements (Lundvik Gyllensten et.al, 2010). From a biomedical aspect this means to find one's centre in relation to graviity. From a psychological perspective this means emotional security, trust and existential roots in life. Having a strong body ego means an opportunity to have a deeper understanding of oneself, one's emotions, one's reality and existence as a human being.

A questionnaire study was conducted among 68 physiotherapy students at Lund University describing students´ perceptions of the concept "Body trust". The results showed that all students perceived that the concept was a valuable concept. It was described as "to be satisfied with and in harmony with one's own body", "trust in how the body functions", "increased body awareness", good self-confidence and to be able to take care of the body (Table 1). In the questionnaire study the question "What are the consequences of increased body trust" was also asked. The consequences of increased body trust were described as increased self-confidence, increased harmony and joy in life, increased security in social situations and increased mastery in life situations. These descriptions of the concept "body trust" support the notion that this concept can be seen as parts of the psychological aspects of the body ego. It was associated with self -confidence, satisfaction with the body, harmony and mastery in life.

| Groups of answers | Freq. |
|---|---|
| To be satisfied with and in harmony with one's own body | 27 |
| Trust in how the body functions | 21 |
| Increased body awareness | 19 |
| Good self confidence | 26 |
| To take care of the body | 13 |
| To accept the body | 12 |
| To listen to the body and to the breathing | 10 |
| To relax | 9 |
| To use the body as a tool and to use body language | 8 |
| Increased mastery in life situations | 8 |

Table 1. The characteristics of the concept "Body trust" (N=68).It was possible to give more than one answer.

The concept of body trust can also be seen as a prerequisite for developing emotional intelligence, the cognitive skills required to effectively monitor emotions. I am of the opinion that a basic level of body trust is necessary for self-awareness, for managing emotions so that they are appropriate, and for motivating oneself in rehabilitation. A basic level of body trust is needed to identify emotions in others and to handle relationships. Self-awareness can be defined as knowing one's internal states, preferences, resources and intuitions. It can be described as

1. emotional self-awareness: recognizing one's emotions and their effects and
2. accurate self-assessment: knowing one's strengths and limits.

In physiotherapy, basic training for increased body trust and self-awareness can to a higher extent be included in body awareness therapies. Trust in your body can also be seen as a prerequisite for trusting other people. Body awareness and Body trust can be increased by focusing on the body in therapy (Gard, 2005). Focusing is a body-centred method that can be used to listen to feelings by becoming aware of body sensations that carry meaning. Trust can be increased through listening to a patient's life story, body language and meta communication.

The concept of body trust can be used to more fully integrate the body in the total experience of the self. Today body awareness therapies include not only body awareness, but also expressions of emotions and personality. The concept of "body trust" can be used to focus these emotional, psychological and social dimensions in body awareness therapy and in movement therapy. Being relaxed or tense is just as much an emotional as a physical condition. Reducing tension is not only mobilising willpower and ability. It is also to recognize suppressed emotions and to be willing to give up control. The psychological understanding of muscular tension is the regulation of emotions. Muscular tension allows emotions to be controlled· Through body experiences patients' can train to give up control, without anxiety or fear. This can be combined with mental and cognitive training with the aim of increasing self-confidence and a more positive attitude towards the own body. Here the concept of body trust can be valuable. As physiotherapists we are role models for patients in that we have a positive and accepting attitude towards our own body. When working with the body ego in treatment, the identity can be strengthened and can be focused on both body experiences and increased body trust and confidence.

## 7. The importance of emotional awareness in multimodal rehabilitation

Today emotional intelligence has become increasingly important to discuss. It can be defined as the cognitive skills required to effectively monitor and self-regulate emotions (Goleman, 1995).

Emotional intelligence contain abilities in five domains:

1. knowing one's emotions, self-awareness - recognizing a feeling as it happens;
2. managing emotions so that they are appropriate, which is an ability that builds on self-awareness;

To fulfil these five criteria and be emotionally intelligent is a prerequisite for being a professional physiotherapist. To have the capacity to help patients in situations of emotional crises require the capacity to deal with emotions in patients in a professional way (Gard & Gyllensten, 2004)

Sweden is today, together with many other countries, a multicultural country. Therefore, there has been a research interest to invent factors of importance for a professional relationship and good interaction between physiotherapists and patients in different health care contexts. Efforts have been made to consider cultural and religious factors in the interaction situation and to accept cultural diversity in patient treatment. The research has shown that it is important to be sensitive to the patients´ religious beliefs, norms and values (Gagnon et. al., 2004; Gard, 2003; Norregaard, 2001).

In most countries there exist specific pain clinics where immigrants with traumatic experiences can be treated, for example torture victims. The aim of the physiotherapy treatment for persons who have undergone torture is to relieve or reduce pain, correct musculoskeletal dysfunctions, teach the clients to cope with pain and regain body awareness. A good interaction and communication with each refugee is needed to optimize the treatment. Interview studies of physiotherapists treating torture victims have shown that the following factors are important for a good interaction with torture victims;

1. Characteristics of the physiotherapist such as being open and listening,
2. A capacity to handle negative emotions,
3. Professional and therapeutic support services,
4. Ability to tailor the interaction to meet patients´ needs and
5. Ability to shape environment factors.

These factors have to be considered to improve the interaction between PT's and persons who have undergone torture (Gard, 2009). The physiotherapists opportunity to shape environment factors, structure and time, were particularly important for tortured patients. Frames, structure and enough time to build confidence and trust have to be in focus in their treatment. An increased self- control and coping ability in daily life are needed by many refugees and may improve their mental health (Bates & Rankin-Hill, 1994). Also language factors were important for a good interaction, for example to have a good translator service and to develop a personal relation to the translator. In this context, where all patients speak a foreign language, it is reasonable to believe that getting correct information about the torture history and to develop a tailored treatment for each individual is dependent upon good translation and communication. Other factors, such as the ability to understand symbolic meanings, metaphors and humour and interpret it correct are also important to consider (Bäärnhielm, 2000). To give increased time for conversations about factors that increase a person's "sense of coherence", factors that can increase a person's feelings of comprehensibility, manageability and meaningfulness in life may also increase a person's confidence and trust in a rehabilitation situation as they have been shown to be important "generalized resistance resources" preserving good mental health ( Cederblad et. al., 1995).

Factors of importance for good interaction between physiotherapists and patients in a primary health care context have been identified, which is of interest for multimodal rehabilitation. The results showed that 1) prerequisites for good interaction and 2) interaction factors were important for positive treatment results (figure 2). Interaction factors are important for establishing contact, ways of contact, frames and the therapeutic process ( Lundvik Gyllensten et. al., 2000). These factors are all relevant to consider in multimodal rehabilitation. Physiotherapists´ working with patients with a multicultural background need also to consider how they as physiotherapists can communicate with and activate the resources of the patient in the best way"? Expertise in cross-cultural communication is highly needed among physiotherapists working in multimodal rehabilitation.

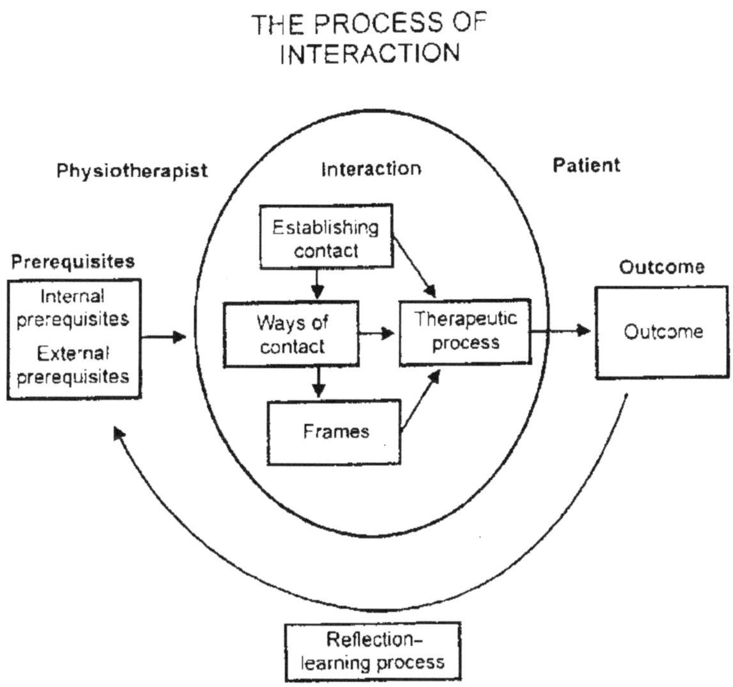


Fig. 2. Factors of importance for good interaction between physiotherapist and patient from physiotherapists´perspective ( Lundvik Gyllensten et. al., 2000).

## 8. Conclusion

It is important not to forget psychological factors and body awareness in multimodal rehabilitation. A health promotion perspective is useful, focusing psychological health promoting factors in the patients´ life and work situation as well as body awareness promoting factors in the rehabilitation situation.

## 9. References


Arnesson H, Ekberg K. Evaluation of empowerment processes in a workplace health promotion intervention based on learning in Sweden. *Health Promotion Internationa.l* 2005; 20 (4): 351-359.

Bates M, Tankin-Hill L 1994 Control, culture and pain. Soc.Sci.Med 39, 5:629-645

Bergstrom G, Bergstrom C, Hagberg J, Bodin L, Jensen I. A 7-year follow-up of multidisciplinary rehabilitation among chronic neck and back pain patients. Is sick leave outcome dependent on psychologically derived patient groups? *Eur J Pain.* 2009.

Bergstrom G, Bodin L, Bertilsson H, Jensen IB. Risk factors for new episodes of sick leave due to neck or back pain in a working population. A prospective study with an 18-month and a three-year follow-up. *Occup Environ Med.* 2007;64:279-87.

Bäärnhielm S 2000. Making sense of suffering. Illness meaning among somatizing Swedish women in contact with local health care services. Nordic Journal of Psychiatry 54:6:423-430.

Blois K. Trust in business to business relationships: An evaluation of it's status. Journal of Management Studies, march, 1999

Cederblad M, Dahlin L, Hagnell O, Hansson K 1995. Coping with life span crisis in a group at risk of mental and behavioural disorders: from the Lundby study. Acta Psychiatrica Scandinavia 91:322-330.

Coons SJ, Rao S, Keininger DL, Hays RD.A comparative review of generic quality of life instruments *Pharmacoeconomics* 2000;17:13-35

Cherniss C. Bringing emotional intelligence to the workplace. Journal of Work Environment and Nordic Ergonomics, 2000: 1: 33-46.

Denison E, Asenlof P, Lindberg P. Self-efficacy, fear avoidance, and pain intensity as predictors of disability in subacute and chronic musculoskeletal pain patients in primary health care. *Pain* 2004;111:245-52.

Dionne CE, Bourbonnais R, Frémont P, Rossignol M, Stock, SR, Nouwen A, Larocque I, Demers E. Determitants of "return to work in good health" among workers with back pain who consult primary care settings: a 2-year prospective study. *Eur Spine J* 2007; 16: 641-655.

Dworkin RH, Turk DC, Farrar JT, et al. Core outcome measures for chronic pain clinical trials: IMMPACT recommendations. *Pain.* 2005;113:9-119.

Feise R Jm Menke JM. Functional Rating Index. A new valid and reliable instrument to measure the magnitude of clinical change in spinal conditions. *Spine* 2001:26:78-87

Flensborg-Madsen T, Ventegodt S, Merrick J. Sense of coherence and physical health. The emotional sense of coherence (SOC-E) was found to be the best-known predictor of physical health. *ScientificWorld Journal.* 2006 Jun 22;6:2147-57.

Foster G, Taylor S, Eldridge S, Ramsay J, Griffits C. Self-management education programmes by lay leaders for people with chronic conditions .An interventions review. *Cochrane database of systematic reviews,* 2009; issue 3

Gard G. Factors important for good interaction in physiotherapy treatment of persons who have undergone torture. A qualitative study. Physiotherapy Theory and Practice 2007; 23 (1):1-9.

Gard G. Body Awareness therapy for patients with fibromyalgia and chronic pain. Disability and Rehabilitation 2005;27:725-728

Gard G & Gyllensten A. Are emotions important for good interactions in treatment situations? Physiotherapy Theory and Practice 2004;20:107-119

Gard, G. and Larsson, A. Focus on motivation in the work rehabilitation planning process. *Journal of Occupational Rehabilitation, 2003a;*13 (3).

Gard, G. and Larsson, A. How can the rehabilitation planning process at the workplace be improved? *Journal of Occupational Rehabilitation*, 2003b;13(3)

Gard G, Lundvik-Gyllensten A., Salford E, Ekdahl C 2000 Physical therapists emotional expressions in interviews about factors important for interaction with patients. Physiotherapy, 86, 5, 229-240.

Goleman D. Emotional intelligence, New York: Bantam, 1995

Goossens ME, Vlaeyen JW, Hidding A, et al. Treatment expectancy affects the outcome of cognitive-behavioral interventions in chronic pain. *Clin J Pain* 2005;21:18-26.

Grahn B. (1999). Quality of life, motivation and costs in multidisciplinary occupational rehabilitation. Thesis. Lund University, Lund, Sweden.

Hedlund L, Gard G. Body trust. Nordic Physiotherapy 2000; 4: 1-8.

Hockings RL, McAuley JH, Maher CG. A systematic review of the predictive ability of the Orebro Musculoskeletal Pain Questionnaire. *Spine*, 2008, Jul 1;33(15):E494-500.

Lundvik Gyllensten A, Gard G, Salford E, Ekdahl C. Interaction between patient and physiotherapist.A qualitative study reflecting the physiotherapist's perspective. Physiotherapy Research International 1999; 4: 2: 89-107.

Lundvik Gyllensten A, Skär L, Miller M, Gard G. Embodied identity - A deeper understanding of body awareness. Physiotherapy Theory and Practice 2010;27(7):439-446

IST *ftp://ftp.cordis.lu/pub/fp7/ict/docs/ict-wp-2007-08_en.pdf* 2006

Jensen MP, Keefe FJ, Lefebre JC, Romano JM, Turner JA.(2003) One and two-item measures of pain beliefs and coping strategies.Pain 104:453-469

Jensen MP, Turner JA, Romano JM. Changes after multidisciplinary pain treatment in patient pain beliefs and coping are associated with concurrent changes in patient functioning. *Pain*. 2007;131:38–47.

Jensen IB, Linton SJ. Coping strategies questionnaire (CSQ): Reliability of the Swedish version of the CSQ. *Cognitive Behaviour Therapy* 1993; https://webmail.lu.se/owa/redir.aspx?C=e5f5af41d1a144669f5e673a5412c0db&URL=http%3a%2f%2fwww.informaworld.com%2fsmpp%2ftitle%7edb%3dall%7econtent%3dt713926011%7etab%3dissueslist%7ebranches%3d2222:139 - 145.

Larsson, A., & Gard, G. How can the rehabilitation planning process at the workplace be improved? A qualitative study from employers' perspective. *J Occup Rehabil 2003;* 13 (3): 169-181.

Leveille SG, Huang A, Tsai SB, Allen M, Weingart SN, Iezzoni LI. Health coaching via an internet portal for primary care patients with chronic conditions: a randomized controlled trial. *Med Care*. 2009 Jan;47(1):41-7.

Lorig KR, Ritter P, Stewart AL, Sobel DS, Brown BW Jr, Bandura A, Gonzalez VM, Laurent DD, Holman HR. Chronic disease self-management program: 2-year health status and health care utilization outcomes. *Med Care*. 2001 Nov;39(11):1217-23.

Linton SJ, Hallden K. Can we screen for problematic back pain? A screening questionnaire for predicting outcome in acute and subacute back pain. *Clin J Pain*. 1998;14:209-215.

Linton SJ, Boersma K, Jansson M, et al. The effects of cognitive-behavioral and physical therapy preventive interventions on pain-related sick leave: a randomized controlled trial. *Clin J Pain* 2005;21:109-19.

Linton SJ. A review of psychological risk factors in back and neck pain. *Spine* 2000;25:1148-1156.

Linton S J, Boersma K. Early Identification of Patients at Risk of Developing a Persistent Back Problem: The Predictive Validity of The Örebro Musculoskeletal Pain Questionnaire. *Clinical J Pain* 2003;19:80-86.

Lindström B, Eriksson M. Salutogenesis. *J Epidemiol Community Health* 2005;59:440–442.

Lowen A. *Bioenergetics. The revolutionary therapy that uses the language of the body to heal the problems of the mind.* New York: Coward, McCann and Geoghegan Inc, 1975.

Lundvik Gyllenstan A, Gard G, Salford E, Ekdahl C. Interaction between patient and physiotherapist.A qualitative study reflecting the physiotherapist's perspective. Physiotherapy Research International 1999; 4: 2: 89-107.

Rabin R, de Charro F. EQ-5D - a measure of health status from the EuroQol group. *Ann Med.* 2001;33:337-43

Reich W. Character analysis.3rd enlarged ed. New York: Farrar, Straus and Giroux, 1949.

Mattsson M. Body awareness. Applications in physiotherapy. Medical dissertation Umeå University 1998

Rosberg, S. Body, being and meaning in a physiotherapeutic perspective. Ph.D. thesis, University of Gothenborg 2000.

Roxendahl, G. Body Awareness Therapy and the Body Awareness Scale, treatment and evaluation in psychiatric physiotherapy. University of Gothenburg, Sweden, 1985.

Theorell T, Westerlund H, Alfredsson L, Oxenstierna G. (2005). Coping with critical life events and lack of control—the exertion of control. *Psychoneuroendocrinology.* 2005;30, 1027–1032.

SBU. Rehabilitering vid långvarig smärta. En systematisk litteraturöversikt. Stockholm:. Statens beredning för medicinsk utvärdering (SBU); 2010. *SBU-rapport* nr 198. ISBN 978-91-85413-34-8. (Rehabilitation for long-lasting pain conditions. A systematic literature review. The Swedish Council on Technology Assessment in Health Care, SBU).

The Swedish Council on Technology Assessment in Health Care, *SBU-report* 177: 2006

Sullivan M, Karlsson J, Ware JE. The Swedish SF-36 health Survey.I. Evaluation of data quality, scaling assumptions, reliability and construct validity across general populations in Sweden. *Soc Sci Med.* 1995; 41: 1349-58.

Shaw W, Endresen R, Linton S, Huang Y, Pransky G.3:rd place PREMUS best paper competition:development of the return to work self-efficacy questionnaire - psychometric properties and predictive validity.Scan J Work Environ Health 2011, 37(2):109-119

Toumi K, Illmarinen J, Jahkola A, Katajarinne L, Tulkki A: WAI -- A method for work ability assessment. Swedish version Helsinki, *Finland National Institute for Working Life*; 1998.

Tolson D, Kelly T. Constructing a new approach to developing evidence-based practice with nurses and older people. *Worldviews on Evidence –based nursing* 2006; 3 (2) 62-72

Van der Klink JJ, Blonk RW, Schene AH, van Dijk FJ. The benefits of interventions for work-related stress. *Am J Public Health.* 2001 Feb;91(2):270-6.

Varekamp I, Verbeek JHAM, van Dijk FJH. How can we help employees with chronic diseases to stay at work? A review of interventions aimed at job retention and based on an empowerment perspective. *Int Arch Occup Environ Health;* 2006; 80: 87-97.

Vingård E, Alexandersson K, Norlund A. Chapter 10. Sickness presence. *Scand J Public Health.* 2004;32(5): 216-221.

13

# Relaxation Techniques and States – Applications to Physical Therapy

Leslaw Kulmatycki
*Health Promotion Department, University School of Physical Education in Wroclaw, Poland*

## 1. Introduction

The chapter below is dedicated to relaxation techniques. It is an attempt to explain their main theoretical concepts and techniques and to describe some benefits of their application. The chapter is addressed primarily to professional people who work with patients, but may also interest those non-professionals who want to improve their general health and well-being. The material is divided into nine sub-chapters, each describing a different aspect of the technique and practice of relaxation. First, these techniques are described as part of the application of holistic concepts to health and illness, and of relaxation techniques grounded in such concepts. Subsequently, relaxation is defined and analyzed as both a process and a state of awareness. Within the chapter, there is a description of the dimensions, levels, and the two main ways of attaining the deepened state of relaxation. The final section describes five sets of relaxation techniques and examples of their implementation in therapeutic settings.

## 2. The unity of body, mind and spirit

Hippocrates was convinced that illness stems from natural causes which can and should be studied. However, he qualified this by stating that it required the broadest study of a patient's entire environment and should encompass many factors. One of his works included in Corpus Hippocraticum, is titled "On Airs, Waters and Places." It is a treaty about human ecology explaining that optimal health and well-being depends on many environmental factors, and above all, on wise management of one's life. Relaxation and meditation create the possibility of heightened insight into one's life and one's broader environment. Contemporary man, overwhelmed by modern technology, should be encouraged to rely more on his inner resources, and those of natural environmental resources available to him. The development of practical methods of relaxation is a manifestation of this search for balance. Their purpose is to provide the feeling of greater control over the flow and type of reactions to undesirable stimuli and to magnify the effect of those which can help us achieve a healthy development and well-being.

There exist known cases where mental disturbances impact upon physical health causing "physio-pathology." The body may revert to this language, expressing as physical symptoms what is at root a mental disturbance. Some forms of convulsions, paralysis, blindness or muteness, are examples of these conversions known as "dissociative disturbances" - illnesses caused by some type of psychological distress. The medical literature uses the term "somatization" to describe this general process. An example of the above are organ neuroses, including a wide range of heart ailments, respiratory problems, and some disorders of the digestive system (Pelletier 1977; Astin et. al. 2003; Harrington 2008). Often the psychological factor is the only cause of ailments. It is known that the patient's unresolved, negative feelings caused by problems or conflicts are transformed into physical symptoms, while there are no changes in the body organs that could explain their origin. However, the psychological origin is discernible.

The conventional notion of health and disease was rebutted in the work of another holistic thinker, Aaron Antonovsky. Antonovsky was a strong proponent of well-person care, believing that doctors and therapists should focus on "strengthening the health pole" rather than on treating disease. Yet Antonovsky did not emphasize balance; in fact, he described the normal condition of the human body as a mess characterized by entropy and disturbances in homeostasis. This condition of abnormal 'normality' arises from the dynamic of life itself, which involves activity, motion, and variation as integral parts. Within this theoretical framework it is also an error to posit health and disease as static and opposite conditions. Health and disease are matters of degree, and largely depend on the individual himself. In order to approach the pole of health, one must be sensitive to the signs of good and ill health, and one should adopt practices which will strengthen immunity and promote well-being. In this view, the person should be able to understand the most important stressors in life, and realize these have the power to undermine one's vitality (Antonovsky, 1993).

The psychosomatic nature of health and illness furnishes the rationale for a therapeutic approach directed at the mind. The close association of body and mind requires the patient to be an active participant in the healing process. At a minimum, he should be aware of the fundamentals of a healthy lifestyle. In addition, each person should have the opportunity to learn techniques and methods to gain more control over his health and well-being. Until the very end of life, even the sick retain an element of health and some ability to improve their condition. When illness is present, a patient may compensate for it, or do so with the support of friendly people. Blaxter has concluded that throughout history, the notion of disease has varied from one period to the next, that views of health and illness are largely social constructs (Blaxter, 2004). Different ages have been preoccupied with different diseases, and the same disease has been viewed quite differently depending on the century or even the decade (Herzlich, 1985; Herzlich & Pierret, 1987).

## 3. History of western relaxation techniques

Relaxation techniques and methods have always been rooted in the prevailing philosophical and religious systems of a given society. Australian Aborigines, Native Americans, Africans, or Siberian shamans, Christian mystics, saints, Indian, Chinese, Tibetan Buddhists, and Taoists have each developed their own relaxation methods. The condition of deep relaxation was not considered to be sacred as such, but was viewed as a prelude to transcendental experience. It seems as though the rationalist West is the only culture in history that considers spirituality to exist outside the realm of everyday life. Only in our culture does the phrase "human nature" appear to be an oxymoron. It is as if Western man has a longing to experience a deeper subjectivity, but does not want to give up the habits that impede spirituality. The material world, alluring and diverting, is difficult to abandon.

The yoga of India provides the earliest, comprehensive account of relaxation and its broader connection to mind-body states. The inspiration of many contemporary techniques is found here. In the West, the precursors of relaxation technique worked in psychosomatic medicine or in the specialties like integrative psychology. They were able to see the relationship between the scientific and spiritual mind sets, denying that the two were in contradiction profiles the scientists who affirmed that psychology could affect the body and its functioning (Harrington 2008). Their work demonstrated the practical implications of this idea, such as the discovery of the placebo effect, the benefits of positive thinking, and the use of oriental techniques in therapeutic environments. Other more sensational phenomena such as exorcism, mental suggestion, and trance states were also included as indicative of the mind-body connection. In Europe after Franz Anton Messmer, a Viennese physician, had claimed to establish the effect of so-called "animal magnetism" on bodily fluids, the field gained a more empirical dimension with the discovery of hypnosis by James Braid, a physician from Manchester. This was the epoch of Sigmund Freud and Carl Jung, the great twentieth century investigators of the human psyche. Specialists in hypnosis such as the Italian psychiatrist (and Jung disciple) Roberto Assagioli, developed the concept of psychosynthesis (Kulmatycki 2008). In 1932 the German psychiatrist and neurologist Johannes Heinrich Schultz published his ground breaking work *Autogenic Training,* which presented the fundamental postulates and practical implications of his theory. "Autogenic" means relying on the patient's subjectivity to create a mental and physical state of relaxation. In the United States, the potential of relaxation in clinical practice was elaborated by Edmund Jacobson, the Harvard-trained physician, physiologist, and father of biofeedback. His 1929 work "Progressive Relaxation" concluded that one of the most important conditions of effective therapy was to teach patients how to relax. Jacobson was instrumental in helping organize professionals who had an interest in the field (Jacobson, 1987). The post-World War II period included thinkers from a variety of disciplines who attempted to synthesize eastern and western traditions. In medicine, the 1960's witnessed the development of sophrology, a precursor of the human potential movement. Sophrology, a personal development method designed to reduce stress and promote mental and physical well-being, was developed by the Spanish psychiatrist Alfonso Caycedo. The outreach of holistic medicine to broader social strata also characterized the work of Henry Wintrebert, a physical education teacher and neuro-psychiatrist who treated children at La Salpetriere, a rehabilitation clinic in Paris. (Kulmatycki 2004, 2007). By the end of the 60's, Polish scientists Tadeusz Pasek and Wieslaw Romanowski developed an original system of exercises based on Indian yoga (Grochmal, 1979). During the 1970's and 80's it was used in psychiatry for the treatment and rehabilitation of patients with neuroses and psychosomatic problems.

The transactional theory of Richard Lazarus and Susan Folkman found that stress was caused by certain ways of relating to the broader environment. According to Lazarus, stress occurs when a person experiences stimuli as exceeding his resources and threatening his welfare. It was generally held that if stress was repetitive and persistent, the psychological discomfort would contribute to serious, chronic somatic diseases. The primary diseases that were thought to be psychosomatic in origin had already been described by Franz Alexander during the decade of the forties. These included idiopathic hypertension, bronchial asthma, ulcerative colitis, atopic dermatitis, ulcers, hyperthyroidism, and chronic progressive rheumatism, among others.

In the 1970's, new concepts appeared in studies of the stress response. An original questionnaire for measuring stress, known as The Social Readjustment Rating Scale, was developed by Thomas H. Holmes, Richard H. Rahe, Meyer Friedman, and Ray Rosenman. They determined that the so-called "Type A Personality" was the most susceptible to stress reactions. The attempt to counteract stress gained new impetus when developments in electronic technology permitted the measurement and control of biofeedback. Other assistive techniques were developed to engage the mind, even in potentially fatal diseases. The 1978 book *Getting Well Again* by Stephanie Simonton, Carl Simonton and James Creighton noted the role that the patient's mind could play in cancer therapy. Another area of research was initiated by cardiologist Herbert Benson, who pioneered the notion of a "relaxation response." He found that this was a distinct physiological and metabolic state characterized by a reduction in heart rate, breathing, blood pressure, and brain activity. (Benson & Proctor 1974; Benson & Klipper, 1975). During the 1980's and 90's Jon Kabat-Zinn used meditation to bring his patients to a state of awareness known as "mindfulness." (Kabat-Zinn et al. 1992; Kabat-Zinn, 1995) We can complete our historical survey of Western relaxation techniques by mentioning the Mind and Life Institute, established in the mid-80's by Chilean scientist Francisco Varela. The Institute provided a venue in which Western scientists could create a dialogue between themselves and Buddhist monks, drawing on the field's ancient foundations and yielding new possibilities for experimentation and theory.

## 4. The four dimensions of relaxation – systems, methods, techniques, exercises

The scientific description of relaxation depends on the relevant methods, techniques, and exercises employed. I propose a four-dimensional typology based on the therapy's intended effect on the patient, and the specifications of a given technique. The table below summarizes this relationship. We may condense the intended effects into four categories. These are:

- Relaxation as a system or way of life usually refers to relaxation within a particular philosophy or spiritual system. In this case, applied relaxation is only one aspect of a broader set of practices or exercises (Irwin; 1999). These may be derived from a system designed to attain something other than, or more than, ordinary health and well-being as such. For example, the synoptic, coherent, and complex philosophies of Hinduism or Buddhism may lead an individual to reside in a yoga ashram or Zen meditation center. In such instances, relaxation is practiced but it may be incidental to a much broader set of practices designed to promote a holistic human awareness.

- Relaxation as an instrumental method used in the context of another activity, or as a means to attain some end other than itself. Examples of such external activities include dancing, dynamic movement, static meditation, work with the voice and work with the body. The use of technology, for example the technique of *biofeedback* relaxation for the relief of stress, might be particularly suited to this kind of practice.
- Relaxation is sought for its own sake, and a comprehensive set of protocols or techniques is followed. This approach often bears the label of a particular system, teacher, or set of exercises. Examples of such would include for example, the Alexander technique, or the breathing techniques of the "ashtanga yoga" school. A particular technique may be common to several of these systems, but will occupy a greater or lesser degree of significance, or perhaps in a different sequence, depending of the methods and goals of the particular school.
- Relaxation practiced simply as a specific exercise. This is the foundation of all the other categories of practice. It is characterized by a specific and well defined technique that prescribes and defines which body postures, breath controls, and mental images will be used. The procedure is couched in a language of moving from one step to the next, and will often require a repetition of the procedures, with some modification or variation. An example of relaxation as an exercise would be the counting of breaths that is often used in Zen meditation to concentrate the mind. The count may be increased or decreased depending on the progress of the student.

The state of relaxation can also be related to states of being in general, or dimensions of subjective consciousness. Each of these dimensions has an associated set of bodily conditions and mindsets. The following table (Table 1) shows the most characteristic features of each of the four dimensions.

| **Dimension** | **Dominant characteristics of the channel dimension and impact** | **An example of modern technique of relaxation** |
|---|---|---|
| Ecstatic | Ritual, symbol, magic, trance, movement, body | Relaxational dance of G. Roth |
| Receptional | Silence, observation, passivity, immobility, case | Relaxation Response by H. Benson |
| Imaginative | Imagination, vision, game, staging, unreality | Yoga nidra<br>S. Satyanandy |
| Training | Repeatability, control, body, physiology, rivalry, | Autogenic training by J. Schultz |

Table 1. Characteristics of the four dimensions of relaxation and sample techniques (my own description).

## 5. Nature of relaxation and nature of relaxation states

Relaxation is a complement to the normal states of wakefulness, sleep and dreaming, one which provides an opportunity for a more informed understanding of the inner world. The relaxation process involves going beyond the rational, discursive mind through the "suspension" of its activity. The ability to turn away from ingrained habits of linear, logical thinking opens the way to new associations, feelings and insights into the self. It is worth noting that the quality and depth of relaxation is related to the subject's distance from habitual thought patterns, to his ability to approach the pole of intuition. Of course, the body is a unity and life requires both logical and intuitive thought processes. But emotional disturbances can disrupt the balance of body, mind, and spirit. They may impinge on somatic processes and disrupt the balance of the neurohormonal systems. These imbalances may be experienced as disease symptoms. In turn, these symptoms may further aggravate the person's emotions in a vicious circle. This dilemma provides a rationale for the use of relaxation techniques. These can be applied methodically to treat the ailments of one individual, but they also offer a means for a more complete development and a more fulfilling life for human beings generally.

As a process, relaxation is a specific activity which is characterized by two features. The first of these is introspection, meaning a focusing of the consciousness "within oneself;" the second is mindfulness, which involves centering the consciousness on "what is" in a particular place and at a particular time. Regardless of which feature is emphasized or of which technique is used, several criteria must be met in order for the practice of relaxation to be successful.

**Dedication of a special place and time** - the process of relaxation should be a distinct reality, experienced in its own space within a delimited time of minutes or hours, an experience deliberately segregated from "ordinary" life as much as is practicable. This means that the activity should take place within a suitable setting, one free from random stimuli and distracting noise.

**Freely chosen participation** - participation in relaxation should be solely at the direction of the participant. The element of spontaneity and the person's own commitment are absolutely necessary. Relaxation should never be imposed on anyone for whatever reason, nor should a person be required to demonstrate his proficiency in any given technique. Relaxation is not a course of study and the participant is not striving to pass with honors.

**Focusing on process, not on result** - to focus on results is an obstacle to relaxation. It is essential to preserve openness to "what appears at the moment." In most techniques, it is not advisable to prescribe certain levels or goals in the process of relaxation. The unexpected is very much a part of relaxation, arising naturally in the process of observation.

**Being Receptive to the Unreal** - the experience of relaxation has a strong imaginary element. This can range from the mildly imaginary to the unreal, or even the delusional. This unreality may be perceived as relating to the physical body, sense perceptions, or imaginative images. It is inherent in the relaxation process. (see the characteristics of relaxation).

**Understanding the structure that underlies the process** - in other words, a comprehensive view of how specific methods and techniques are integrated into the overall process. In each type of relaxation there is a sequence that involves an introduction, which is followed by the prescribed exercises, and then a phase of return once the technique has finished. Normally when contact with the surrounding world is reestablished, we can expect this return phase to be one of deep tranquility and insight. Of course, the more one becomes familiar with the techniques, the less important a conceptual understanding becomes.

There is a clear difference between relaxation as a process, and relaxation as a state of being. The process consists of specific practices, including methods, techniques and relaxation exercises. It has a specific methodology and structure. However, the state of relaxation is an experience which occurs spontaneously. There is no prescribed schedule for its occurrence or inception, and it is not easy to categorize or to describe. The state of relaxation is characterized by: a feeling of balance, a muting of thoughts and sensations, a sense of slowness and non-action. It has specific features. Session participants who experienced deep relaxation (Kulmatycki 2008) commonly reported:

- Experiencing the unimportance of time. Having the feeling that time slowed down or stopped. Most of the participants who experienced deep relaxation training were not able to determine the duration of the session. They felt that much less time had elapsed than was objectively measured.
- A loss of the sense of physicality, or a distinct change in the perception of the body. In deep relaxation, one gives up control over the body. Often participants described this physical experience as total submergence or collapse, including the feeling that their body did not belong to them, that it was just "there."
- A diminished connection to one's own ego, or a loss of the sense of "I" as an identity. Relaxation session participants sometimes had the feeling that it did not matter what or who was in control, or who was the subject, or the object of consciousness. Often after the relaxation session, people were not able to describe what had happened during the session, comparing it to a half-remembered dream or to a blank space in their memory. In such cases the subjects could not have been asleep, because they had responded to the leader's instructions. Afterward, though, they could not remember these.
- An alteration of normal thought processes, losing the discursive ego or the inner monologue. This was supplanted by a more intuitive kind of thinking. Over time, and as the length of sessions increased, subjects reported that the mind suspended its usual activity in favor of spontaneously arising thoughts and insights - intuitive thinking. Often, the content of these insights were surprising or revealing to the subjects.
- Emotional changes brought about by a deeper contact with one's inner world. In the process, hidden or repressed feelings and experiences could come to the surface. Because the subject releases control, relaxation may produce a kind of spontaneous emotional catharsis which may be experienced as either positive or negative.
- Changes in sensory perception due to a reduction in sensory stimuli. When relaxation follows a reduction of external stimuli, the participant may experience illusions, visions or hallucinations.
- A more complete understanding of oneself. A sense of increased insight into one's actions and motivations, or into the nature of personality as such. These insights were sometimes described as "beyond words" or "unspeakable." In these cases, participants defined the experience as something mysterious or even as transcendentally mystical.
- A renewed feeling of vitality, energy, and a more positive attitude about oneself, others, and the world. An accompanying sense of being more open to others.

## 6. Relaxation directed to three states of being

There are three levels of a person's being which may be the object of relaxation techniques: the psychosomatic, the mental, and the transpersonal (metaphysical). This study focuses on the psychosomatic. In this case, the most important practices are aimed at harmonizing the body and mind. The second, or mental level, involves practices for the relief of physical tension, mental stress, or for the enhancement of interpersonal relationships. The third level goes beyond the interpersonal and might be termed the transpersonal, or metaphysical state, in which the subject is open to the highest levels of consciousness. For each state, relaxation has a definite goal. The first mostly involves the health of the body; the second, relief of the negative effect of mind on the body (or enhancing the positive effect), and the third, on intellectual and spiritual development. Subject and therapist may collaboratively decide to focus on one of these, or the focus may evolve spontaneously.

In most real-world situations, relaxation training begins with the first level. As experience is accumulated, it becomes easier to go on to the others. However, it should be emphasized that these levels are not a rigid hierarchy. There are situations when beginning at the second or third level may be most suitable. It depends on the needs of the subject and on the instructor's or therapist's experience and preferences. A knowledge of these levels can make the process more focused and effective.

The psychosomatic level of relaxation is characterized by three types of experience:

**Physical sensations.** The patient focuses primarily on muscular and visceral sensations, becoming aware of their location, origin, and quality.

**Experience within the mind-body.** The patient becomes aware of how his physical state, such as the quality and depth of breath or the speed of heartbeat, is linked to prevailing thoughts, emotions, and ideas.

**Experiences within the process of awareness.** The patient becomes more discerning of different types of awareness, and becomes sensitive to the intellectual, emotional and intuitive faculties of consciousness. The experience may be cognitive-rational or metaphysical.

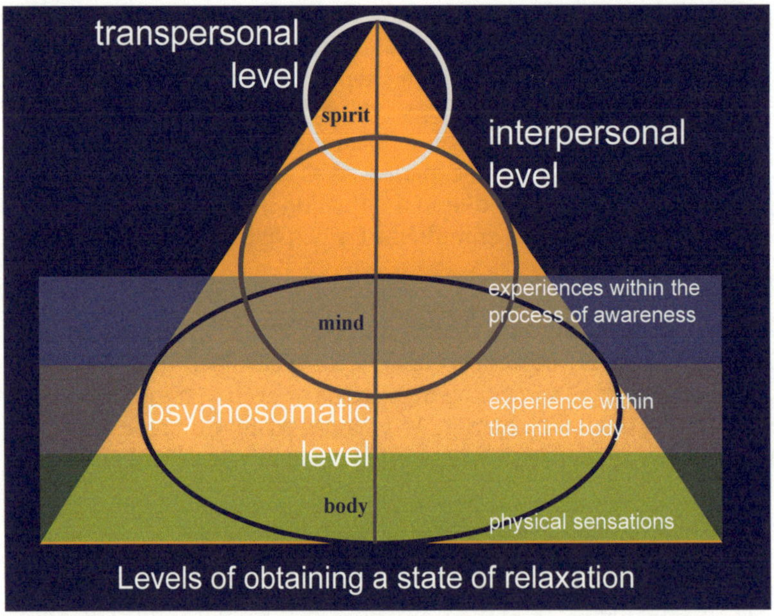


Fig. 1. Levels of obtaining a state of relaxation (my own).

## 7. Two main ways of achieving relaxation states

In the literature about relaxation techniques, researchers have formulated different classifications based on a variety of criteria. Rosemary A. Payne distinguishes two types of technique, the somatic and the cognitive (Payne, 2005). In contrast, based on their work in practical stress relief, Paul M. Lehrer and his team (Lehrer et. al. 2008) divided their techniques into seven groups: muscle relaxation, methods of hypnosis, breathing exercises and practices, methods based on Eastern meditation and therapy, cognitive methods, and miscellaneous techniques. The classification schema published by Jonathan C. Smith was also based on clinical practice (Smith, 2007). He divided techniques into six categories. He identified the bodily areas experiencing stress, and then recommended the appropriate technique. Alternatively, the Polish rehabilitation expert Stanislaw Grochmal defined methods of relaxation in light of psychotherapy. His scheme was comprehensive and has greatly influenced thinking in the field. He distinguished three classical methods of relaxation (Grochmal 1979). The first was global, viewing stress in terms of the overall development of the personality. Secondly, he cited analytical methods that were focused on particular symptoms or parts of the body. His third category was the narrowest and reflected the most common type of practice, the intuitive-physiotherapeutic approach designed to achieve relaxation quickly and with the least expenditure of effort. This intuitive method was always conditioned by the particular relationship of therapist to patient.

Increasingly, techniques are being described in terms of their reliance on physical activity or passivity. According to this approach, techniques can be divided into two main groups: ergotropic and trophotropic.[1] Ergotropic techniques involve practice-oriented activity of the body, body work and movement meditation (Fischer, 1971; Kulmatycki, 2002, 2005). Trophotropic techniques focus more on the content of consciousness while maintaining the passivity of the body and passive meditation. The two approaches occasionally use the same technique, for example in the reduction of hypertonia. They should be treated as equivalent in their potential effect on subjects. Each can be effective in improving the comfort and quality of life of the healthy, as well as that of patients in rehabilitation. The choice of one type or another depends largely on the personality and predisposition of the subject. Below there is a short description of the two groups of techniques. Thanks to functional magnetic resonance imaging or MRI (fMRI), researchers may attempt to measure activity in the brain regions associated with emotion, cognitive activity and imagination.

The Montreal experiment (Grant & Rainville, 2009) did so to investigate the phenomenon of pain. Previous studies had showed that advanced students of Zen meditation were much less sensitive to pain. Experiments in Wroclaw (Kulmatycki, 2011), Poland investigated whether and to what extent trophotropic and ergotropic activities affected the perception of pain. At a four-day short course in relation, subjects participated in the different approaches. Ergotropic techniques were presented in a trial of "movement meditation." In this group,

---

[1]This typology of scope and methods is similar to the one presented in sub-chapter 4. Here, the four dimensions of relaxation are have been modified, taking into account the specific characteristics of the participants. The division in to trophotropic and ergotropic techniques takes physical activity into greater account and also considers interaction among particpants. The terms trophotropic and ergotropic, as defined in the work of Roland Fischer.

participants had contact with each other. Each session included three recurring parts: an introduction to working with the body, practice meditation exercises, and 10 minutes of passive relaxation. At the end of each session there were questions and discussion. Trophotropic techniques at the trial took the form of "sitting meditation". Participants in this group spent the sessions in complete silence. During each session, there were three blocks of meditation: focused, analytical and receptive. Every 20 minutes there was a 7 minute break with stretching exercises. After each session there was time for questions and conversation, the same as with the ergotropic group. Afterward, the participants in both groups were randomly reassigned to two non-specific groups and their experience of pain was evaluated on the VAS (Visual Analog Scale). The results were similar for each group (within a range of 4.0-5.5). Thereafter, a proposal to participate in either "movement meditation" or "sitting meditation" was correlated with the responses to a CSQ (coping strategies questionnaire) that identified preferred methods of coping with pain. The average reading on the VAS scale was 2.7 for those who favored ergotropic techniques, and 3.8 for those who preferred trophotropic methods. The correlations achieved statistical significance. The experiment indicated that techniques of dynamic relaxation were more effective in reducing pain than passive ones. Perhaps more fundamental in the choice of relaxation techniques, however, is the individual's ability to deal with pain as such. The choice of one technique or another must be suited to each person.

| **Features of the characteristics of groups** | **Ergotropic relaxation techniques** | **Trophotropic relaxation techniques** |
|---|---|---|
| Outside-inside | Outer-directed, plosive and transformative | Inner-directed, emphasizes desensitization, acceptance. |
| Body-mind | A significant degree of movement and involvement of the body | Favors working with the mind and consciousness |
| Individually-collectively | Utilizes forms of collaboration and cooperation among participants | Focuses on individual approaches, self-directed activities in the arts. |
| Structure-improvisation | In the first phase structure is important as the basis for subsequent improvisation | Advisable to maintain a specific structure for the next phase of work |
| Sample technique | Dance relaxation, massage techniques, contact improvisation, shiatzu, tai-chi, meditation movement | Autogenic training, Alexander Technique, FeldenkreisMethod, Zen meditation, Mindfulness, Yoga Nidra |

Table 2. Characteristics of Ergotropic and Trophotropic Groups and Applicable Relaxation Techniques (my own).

## 8. Five strategic types of relaxation techniques

Based on prior experience, the various techniques were classified into five strategic groupings (Table 3, below). Each group is determined by the modality or focus of its techniques. The first two of these strategies involve techniques that focus on the physical body. The body, either at rest or in motion, is the means to achieve a state of relaxation. The third strategy links the body with the mind. Here, techniques mostly involve the flow of energy and control of the breath. The final two strategic groups involve techniques that work with the psyche, including both mental-imaginative and emotional- intuitive faculties. Each group of techniques has, in turn, been analyzed according to six criteria. The first of these assesses the importance of the body in implementing the techniques, the second assesses the importance of the mind, the third and fourth involve the degree of activity or passivity, the fifth and sixth the individual or cooperative focus. Each of these criteria is referenced to the five strategies, and a numerical value from 1 (lowest) to 4 (highest) indicates how often a given criterion occurs as part of a given strategy. For example, for the strategy that uses mental relaxation and concentration techniques, the physical body is minimally represented (1), while the mind is rated as 4.

| **Five strategic types of relaxation techniques** | | | | | |
|---|---|---|---|---|---|
| Features description of the strategy | Stretching techniques and physical postures | The techniques of body movement | Relaxation techniques and breathing energy | Techniques of mental relaxation and concen-tration | Techniques of physical expression of emotional |
| The physical body | 4 | 4 | 2 | 1 | 3 |
| Mind | 2 | 2 | 2 | 4 | 2 |
| Activity | 2 | 4 | 3 | 1 | 4 |
| Passivity | 4 | 1 | 2 | 3 | 1 |
| Own person | 4 | 3 | 4 | 4 | 2 |
| Cooperation with others | 1 | 2 | 1 | 1 | 3 |

Table 3. The five strategic types of relaxation technique, referenced by six criteria. Incidence is rated numerically on a scale of 1-4 (my own).

**Stretching techniques and physical postures**. The primary objective of this strategic group is to promote a physiological balance by affecting the nervous system. Stretching postures and physical postures interact, massage the internal organs, and especially stimulate the nerve centers and endocrine glands. The isometric extension of individual muscle groups results in an increase in blood supply and oxygenation.

**Techniques of body movement.** The purpose of this strategy is to control the fluid movements of the body with the aim of providing "physical frame" in support of the mind. Typically, those who work in this field must thoroughly understand a precise sequence of exercises or movements.

**Relaxation techniques with the energy and breath.** The purpose of this group strategy is to become aware of one's breathing and bodily energies, and to modify them in order to achieve a state of internal calm.

**Mental Relaxation techniques and concentration.** The goal here is to guide the processes associated with the mind and imagination. Mastery of this approach requires long preparation and training. In the first stage of this work, the essential part of practice is the restriction of conscious attention to what is happening at the given moment and in the given place, and the maintenance of this attention. The next phase is directed explicitly at an awareness of the workings of the mind.

**Techniques of physical emotional expression.** The aim of this strategic group is to remove or expel negative emotions through the use of physical expression or vocalization. The techniques usually rely on simple, repetitive forms of movement. An important element of work is improvisation within the overall structure. Movement serves as a pretext or vehicle to enable an individual to better contact his emotions. Typically, the first stages of practice involve an individual emotional catharsis. Later on, there may be work in groups, introducing the important element of interpersonal contact.

## 9. Therapist-patient relationship

The therapist-patient relationship sets the context in which relaxation occurs. This context involves particular people and settings, the universe of meanings and values, and other distinct features. The patient is more than a mass of symptoms or the expression of his disease. No matter what his condition, he brings his feelings, will, thought and spirit to the practice setting. The patient expects professional, technical assistance but also has a right to expect the therapist to care about him as a suffering person. The therapist-patient relationship is as much a part of the practice as any other element (Kulmatycki & Szczuka, 2007). It is an essential complement of the therapy, a link between the particular and the universal (Fig. 2).

<table>
<tr><td rowspan="5">PATIENT</td><td>UNIVERSAL</td><td rowspan="5">THERAPIST</td></tr>
<tr><td>RELAXATIONAL<br>(center, support, goal)</td></tr>
<tr><td>SUBSTANTIVE</td></tr>
<tr><td>RELAXATIONAL<br>(center, support, goal)</td></tr>
<tr><td>UNIVERSAL</td></tr>
</table>

Fig. 2. Three spheres of the patient-therapist relationship (substantive, relaxational and universal) (my own).

**The substantive sphere** is determined simply by the content of a given specialty, and by the professional competence of the practitioner. The particular practices of a physiotherapist, for example, will obviously differ from those of a social worker. Each specialty will differ in terms of professional knowledge, skills, the quality of contact with the patient and in the formal relationship (contract) with the patient.

**The universal sphere** is much broader and its effect on the patient is more diffuse. Psychological factors are very important, even if the practice is not psychotherapeutic per se. These are multifaceted, extending to the personality traits of those involved, their personal lifestyles and outlooks, and their general sociability. The most important personal traits include verbal communication skills, the ability to listen or maintain a silence, the manner of expressing emotion and the ease of establishing an emotional bond with the patient universal sphere constitutes the most fundamental relationship of therapist and patient. The given professional specialty is not important as such, nor is the particular health problem of the subject. Each relationship is unique, but they all share three basic characteristics. The first is the establishment of a relationship which is a true partnership, one that displays authenticity, acceptance, openness, and empathy. The second is the ability to gauge the nuances of the patient's health and illness, to see his problem in a broader environmental context. The third feature is the way in which the practitioner's own environmental and cultural background, his values and ethics, come into play in the therapeutic process. Thus, the universal sphere and the substantive represent the context or boundaries within which the actual technique is practiced.

**The relaxational sphere** from a therapeutic practical standpoint, is something natural, and simple to apply. Three types of relaxation due to the impact of:

- **intermediary** relaxation, as a means or a way to establish better contact with the patient both in substance and universal.
- **supportive** relaxation, support and supplement other techniques and methods of work with the patient or patient as a preparation for another phase of his treatment or rehabilitation, for example: treatment, surgery.
- **targeted** relaxation, as having end in itself, it is associated with a reduction of excessive tension and control stress levels.

## 10. Practical application of relaxation techniques

Holistic therapists view disease as having multiple determinants, and consider patients' suffering and the healing process from a broad perspective. Generally they seek to improve the resilience of patients and to rely on the self-healing properties of the organism to restore homeostasis. In coming decades, these therapies will increasingly consider the question of how to improve the quality of life in the health as well as in illness. This question raises two sets of issues: the first concerns the science involved in choosing appropriate techniques. What criteria will the average person use when deciding on a technique that suits his health problems, physical capabilities, and personality traits? Second, which techniques will provide tangible benefits?

A summary of some of the practical benefits from rehabilitation therapy and relaxation techniques is shown in Table 4. These benefits fall into four categories. The first category refers to techniques directed at somatic problems, mainly the treatment of cardiovascular disease, cancer, and traumatic injuries. The second group includes psychosomatic diseases whose origin is psychological, but which manifest in some type of somatic, functional impediment. The third group includes psychological and interpersonal problems that are most often manifest in addiction, neurosis, depression, problems of interpersonal communication, and difficulties in interpersonal relationships. The fourth and last group involves philosophical and spiritual pursuits; it is not appropriate to view these benefits as the result of particular techniques, as here, the techniques evolve from a philosophical or religious world view. The overall view or system takes precedence (see sub-chapter.3).

| Problem | Relaxation technique | Approach / Objective |
|---|---|---|
| Existential | Yoga (the system)<br>Zen (the system) | Inner transformation / change their relationship to each other and the world |
| Mental | Meditation<br>Schultz Autogenic Training (Phase II)<br>Sofrologic Meditation of Caycedo<br>Psychosyntesis of Assogioli<br>Relaxation Response of Benson<br>Gurdzijew Meditation<br>Joganidra | Personal development / self-acceptance and self-fulfillment |
| Psychosomatic | Relaxation Response of Benson<br>Schultz Autogenic Trening (phase I)<br>Feldenkrais Method<br>Relaxation Dance of Roth<br>Mindfulness Kabat-Zinn'a<br>Transcendental Meditation | Anti-stress / coping with the challenges |
| Somatic | Progressive Relaxation of Jacobson<br>Relaxation of Witrebert<br>Alexander Technique<br>Shiatsu Massage<br>Postural yoga<br>Dynamic yoga<br>Tai-chi | Physiological / back to internal balance and strengthen the positive health and welfare |

Table 4. Relaxation techniques map for the four groups of health problems

Relaxation techniques at the somatic or psychosomatic level are inseparable from a knowledge of stress, its mechanisms, and its effects on the body. A number of studies have shown that anxiety and other forms of psychological distress lead to increased strain in the sympathetic nervous system (Carlson & Hoyle, 1993; Kulmatycki et al. 2006; Kulmatycki & Supiński, 2006; Kulmatycki & Burzyński 2007; Kulmatycki & Burzyński, 2008). Studies have shown specific, positive reactions after twelve minutes of sustained relaxation (Benson & Klipper, 1975). These include, first, a reduction in the level of lactic acid in the muscles - a consequence of relaxing large areas of the body (arms, legs, trunk), a reduction in oxygen consumption of from 10% to 20%, a calming of the heart and reduction of the heart rate, and a decrease in blood pressure (Kulmatycki 1994). There is also an increase in the frequency and intensity of alpha waves, which is perhaps even more important for patients' mental tranquility (Benson & Klipper, 1975). Some studies of psychological stress have demonstrated its effects on the immune system. Sometimes these effects manifest themselves as decreased resistance, sometimes as cancer, an autoimmune reaction (a destructive immune reaction that targets the body's own tissues) (Syrjala, et al. 1995; Burish, et al. 1988). Other studies concern the effectiveness of relaxation techniques for patients in rehabilitation after stroke (Gnat, et al. 2000).

The regular use of relaxation techniques has been found to reduce the stress response, especially the level of anxiety. The techniques allow patients to better cope with aggressive behavior and negative emotions. (Borkovec & Costello 1993; Haaga et al. 1994; Andreoli et al. 1995; Scheufele 2000; Kulmatycki & Burzynski 2007; Kulmatycki & Burzynski 2008, Kulmatycki et al. 2010). Is should be noted than an increasing number of people who participate in relaxation exercises do so in order to improve their general health, rather than to treat a particular complaint. Studies of such individuals have shown that training in yoga relaxation significantly affected their subjective feeling of relaxation (Kulmatycki & Burzyński, 1999; Kulmatycki 2004). In other cases where specific complaints were present, research has shown that patients with chronic pain who used these therapies have become more active and have experienced less anxiety, anger, sadness or irritability compared to similar patients who did not. (Kulmatycki 2011). Research conducted by R.K. Wallace and H. Benson of Harvard University found that mental relaxation helped subjects perform cognitive tasks involving concentration, shifting their attention, and keeping their emotional balance. They were able to perform these tasks without tension, mental fatigue, or negative thoughts and emotions (Wallace & Benson; 1972). Research has shown the necessity of taking the subject's personality type into account when making this choice, whether the participant is healthy or ill. (Kulmatycki & Miedzinska, 1999; Kulmatycki 2011). Highly optimistic people who underwent autogenic training adopted the practice more easily compared to those who were pessimistic about the future or their ability to influence it. There is also a positive correlation between the level of dispositional optimism and the subjective level of relaxation. (Kulmatycki et al. 2006).

## 11. Conclusion

The growing popularity of alternative methods and techniques of relaxation necessitates a fair, scientific evaluation. Research design faces some difficulties, especially in dealing with the possibility of placebo effects. There is also no uniform method of assessing the physiological and psychological impact of these methods. Many studies are encouraging, but some do not show positive results. In many other cases, the methodology raises many doubts. However, we cannot emphasize enough that relaxation techniques are entirely free of side effects and that they are economical. It seems reasonable to accept them on the basis of the positive results that have in fact been demonstrated. These techniques may complement physical and somatic therapies. They can increase the effectiveness of any therapeutic relationship when personnel have been trained in their use. One of the best-known experts in relaxation techniques, Jonathan C. Smith of Roosevelt University Stress Institute, has written about a revolution that is occurring in approaches to treatment and patient care. He states that complementary methods and therapies must be used in addition to conventional approaches (Smith, 2001). These methods not only can reduce costs and increase the effectiveness of treatment; more importantly, they can increase the patient's sense of responsibility for and control over his health. It has been known for some time that patients who consider themselves healthy are treated differently than those who self-identify as sick.

*relaxing setting off on a journey in search of deep relaxation ...*
*start with no action ... sit down, lie down ...*
*then I move on to the body, its center of gravity, feeling of warm, letting go ...*
*I notice a balanced breath ...*
*regular heart rhythm ...*

*I am aware of the thoughts that arise in the mind ...*
*I realize the emotions that accompany the tension releasing ...*
*and pictures showing themselves in the imagination ...*
*I remain with all of this for some time ...*
*suspended in time and space, I feel myself ...*
*and it just so much ... at the beginning*
*the next time I try to understand the profound peace is not something specific ...*
*what can be measured ...*
*but it is ephemeral ...*
*every time something else ...*
*and will come a moment ...*
*realize that I take from that is the only way to not follow self-assessment ...*
*where there is neither the observer nor the observed or process of observation.*

## 12. References


Andreoli, A. et al. (1995). Expressive relaxation training and anxiety disorders. *New Trends in Experimental and Clinical Psychiatry*, 11, 123-129, ISSN: 0393-5310

Antonovsky, A. (1993). Complexity, Conflit, Chaos, Coherence, Coercion and Civility, *Social Science & Medicine,* 37, 969-974, ISSN: 02779536

Astin, J. A, et al. (2003). Mind-Body Medicine: State of the Science, Implications for Practice. *The Journal of the American Board of Family Practice* 16:131-147, ISSN 1557-2625

Benson, H. & Klipper, M. Z. (1975). *The Relaxation Response,* 2nd edition. Harper Collins Publishers, ISBN 0 380 00676 6, New York

Benson, H. & Proctor, W. (1974). *Beyond The Relaxation Response,* 2nd edition, William Collins Sons & Co Ltd, ISBN 0 00 626852 8, Glasgow

Blaxter, M. (2004). *Health, Key Concept.* Polity Press, ISBN 0-7456-3083-9, Cambridge

Borkovec, T. D. & Costello, E. (1993). Efficacy of applied relaxation and cognitive-behavioral therapy in the treatment of generalized anxiety disorder. *Journal of Consulting and Clinical Psychology,* 61(4): 611-619, ISSN: 0022-006X

Burish, T. G. et al. (1988). Posttreatment use of relaxation training by cancer patients. *Hospice Journal,* 4(2), pp. 1-8, ISSN: 1049-9091

Carlson, C. R., & Hoyle, R. H. (1993). Efficacy of abbreviated progressive muscle relaxation training: A quantitative review of behavioral medicine research. *Journal of Consulting and Clinical Psychology, 61*(6), 1059-1067, ISSN 0022-006X

Gnat, R. et al. (2000). Próba zwiększenia efektywności terapii pacjentów po udarach mózgu wybranymi elementami psychorelaksacji. *Fizjoterapia.* (*Physiotherapy*) Tom 8, nr 1, ISSN 1230-8323

Grant, J. A. & Rainville, P. (2009). Pain Sensitivity and Analgesic Effects of Mindful States in Zen Meditators: A Cross-Sectional Study. *Psychosomatic Medicine,* 71: 106-114, ISSN 0033-3174

Grochmal, S. (red.), (1979). *Teoria i metodyka ćwiczeń relaksowo-koncentrujących,* PZWL, ISBN 83 200 0081 5, Warszawa

Haaga, D. A. et al. (1994). Mode-specific impact of relaxation training for hypertensive men with Type A behavior pattern. *Behavior Therapy, 25,* 209-223, ISSN 0005-7894

Harrington, A. (2008). *The Cure Within: A History of Mind-Body Medicine,* W.W. Norton & Company, ISBN 978-0393-06563-3, New York

Herzlich, C. (1985). *Health and Illness*. Academic Press, ISBN 9780123441508, London

Herzlich, C. & Pierret, J. (1987), *Illness and Self in Society*, Johns Hopkins University Press, ISBN 9780801832284, Michigan

Jacobson, E. (1987). Progressive relaxation. *American Journal of Psychology*, 100 (3-4), 522-537, ISSN 0002-9556

Fischer, R. (1971). A Cartography of the Ecstatic and Meditative States. *Science*. 174 (November 26), 897-903, ISSN 0036-8075

Kabat-Zinn, J. (1995). *Właśnie jesteś. Przewodnik uważnego życia, (Wherever You go There You Are Mindfulness Meditation In Everyday Life)* Jacek Santorski &CO, ISBN 83 85386 85 8, Warszawa.

Kabat-Zinn, J. et al. (1992). Effectiveness of a meditation-based stress reduction program in the treatment of anxiety disorders. *American Journal of Psychiatry*, 149, 936–943, ISSN 0002-953X

Kulmatycki, L. (1994). Wpływ treningu relaksacyjnego jogi na obniżenie ciśnienia tętniczego. *Postępy Rehabilitacji, (Advances in Rehabilitation)* 2, pp. 55-60, ISSN 0860-6161

Kulmatycki, L. & Burzyński, Z., (1999). Opinie studentów o ćwiczeniach jogi, In: *Problemy kultury fizycznej w badaniach naukowych*, K. Zatoń (Ed.), Wydawnictwo AWF Wrocław. ISBN 83-87389-37-4, Wrocław

Kulmatycki, L. & Miedzińska, B. (1999). Podatność na relaksację a cechy osobowości, (Proneness to Relaxation and Personalisty Traits), *Postępy Rehabilitacji, (Advances in Rehabilitation)* 3/99, pp. 151-159. ISSN 0860-6161

Kulmatycki, L. (2004). *Joga nidra, sztuka relaksacji, (Yoga nidra, Art of relaxation)*, Książka i Wiedza, ISBN 85-05-13354-0, Warszawa

Kulmatycki, L. et al. (2006). Ćwiczenia relaksacyjne Schultza a poziom optymizmu, (*Schultz relaxation exercises and optimism level*), *Annales Universitatis Mariae Curie-Skłodowska*. Sectio D: Medicina, vol. 60; supl.16; 4 (375); pp. 119-121, ISSN 0066-2240

Kulmatycki, L. & Supiński, J. (2006). Influence of the Jacobson relaxation training for well-being and for anxiety level among adolescents, *Polish Journal of Environmental Studies*. vol. 15/5B, pp. 198-201, ISSN 1230-1485

Kulmatycki, L. & Szczuka, E., (2007). Rola relaksacji w pracy fizjoterapeuty z pacjentem. *Fizjoterapia*. (*Physiotherapy*) 15, 1, pp. 75-84, ISSN 1230-8323

Kulmatycki, L. (2007). Stany relaksu - wymiary i poziomy. *Postępy Rehabilitacji, (Advances in Rehabilitation)*, Tom.XXI, 3, pp. 37-43, ISSN 0860-6161

Kulmatycki, L. & Burzyński, Z., (2007). Relaksacja joga nidry i medytacji Bensona a poziom lęku oraz emocje gniewu i depresji, *Postępy Rehabilitacji, (Advances in Rehabilitation)*, Tom. XXI, nr 3, pp. 23-29, ISSN 0860-6161

Kulmatycki, L.(2008). Joga nidra - integracyjna podróż wewnętrzna (Yoga nidra - integrative inner journey), In: *Wokół psychologii analitycznej C.G. Junga. Refleksje, inspiracje, zastosowania*, K. Nieweglowska-Rzepa (Ed.), Wydawnictwo A. Marszałek. pp. 128-139, ISBN 978-83-7441-947-5, Wrocław

Kulmatycki, L. & Burzyński, Z. (2008). Trening jogi posturalnej w radzeniu sobie z negatywnymi emocjami. (Postural yoga training In doping with negative emotions), In: *Psychologiczne konteksty aktywności fizycznej człowieka*. M. Krawczyński (Ed.), Ateneum. pp. 81-92, ISBN 978-83-61079-04-0, Gdańsk

Kulmatycki, L. et al. (2010). Postural relaxation yoga versus progressive relaxation in anxiety and tension reduction. In; B. Bergier (ed) *Physical Activity in Disease Prevention and Heath Promotion*. Institute of Heath. Państwowa Wyższa Szkoła, pp. 149-158, ISBN 978-83-61044-05-5. Biała Podlaska

Kulmatycki, L. (2011). Trophotropic and ergotrophic techniques in working with the patient. In: *The Third National Scientific Conference 'Relaxation in Rehabilitation- Pain in the Reaction of the Body'*, Organized by Polish Association of Body Awareness in Rehabilitaion, Zaborek by Janow Podlaski, 05/19-22/2011 (Unpublished pilot study)

Lehrer, P.M. et al. (2007) *Principles and Practice of Stress Management*, The Guilford Press, ISBN 978-1-60623-000-8, New York

Payne, R.A. (2005). *Relaxation Techniques. A Practical Handbook for the Health Professional*, Elsevier, ISBN 0 443 07447 X, Churchill Livingstone

Pelletier, K. (1977). *Mind as Healer Mind as Slayer*, A Delta Book, ISBN 0-440-55592-2, New York

Scheufele, P. M. (2000). Effects of progressive relaxation and classical music on measurements of attention, relaxation, and stress responses. *Journal of Behavioral Medicine, 23*(2), 207-228, ISSN 0160-7715

Smith, J. C. (2007). The Psychology of Relaxation, In: (2007) *Principles and Practice of Stres Management*, Lehrer P.M., Woolfolk R.L., Sime W.E. (Ed.), The Guilford Press, pp 38-52, ISBN 978-1-60623-000-8, New York

Smith, J.C. (2001). *Advances in ABC Relaxation: application and inventories*, Springer. ISBN 978-0-82611-282-8, New York

Syrjala, K.L., et al. (1995). Relaxation and imagery and cognitive- behavioral training reduce pain during cancer treatment: A controlled clinical trial. *Pain*, 63, 189-198. ISSN: 0304-3959

Wallace, R. K & Benson, H. (1972). The physiology of meditation. Scientific American, 226: 84-90. ISSN 0036-8733

# Part 6

# Physical Therapy Interventions in Patients with Respiratory and Cardiovascular Disorders

# 14

# Effect of Inspiratory Muscle Training on Weaning Success in Critically Ill Intubated Patients

Samária Cader, Rodrigo Vale and Estélio Dantas
*Human Motricity Biosciences Laboratory of the Federal University of Rio de Janeiro State, Labimh/Unirio, Brazil*

## 1. Introduction

Approximately 50% of patients hospitalized in an intensive therapy center (ITC) rely on mechanical ventilation (MV) for an indeterminate time, depending on the reason for intubation. Once the triggering cause of acute respiratory failure (ARF) is stabilized, patients are weaned from mechanical ventilation, which is followed by extubation (Scheinhorn, Chao, & Stearn-Hassenpflug, 2000).

Weaning from MV is the gradual removal of the ventilatory prosthesis, a different process to extubation, which involves removing the endotracheal tube (MacIntyre et al., 2001). The best moment to start weaning depends on multiple factors. Premature exhubation puts the patient at risk of reintubation and airway trauma. However, prolonged weaning exposes individuals to risk of hospital infection and increases hospitalization costs (C. D. Epstein et al., 2002).

Although there are a wide range of weaning protocols, there is still no consensus on which one is the best (Butler et al., 1999). In addition to this diversity, there are also so-called predictors for ventilatory interruption and successful extubation (Chatila et al., 1996; Khamiees et al., 2001; Meade et al., 2001). Among these markers are: (a) rapid shallow breathing index or Tobin index (Li et al., 2000; Meade, et al., 2001) (b) maximal inspiratory pressure (MIP) (A. Bruton, 2002; El-Khatib et al., 1996); and (c) maximal expiratory pressure (MEP) (Salam et al., 2004) .

The Tobin index is the most commonly used predictor to indicate the possibility of weaning success (Frutos-Vivar et al., 2006; Meade, et al., 2001; Stroetz & Hubmayr, 1995), although some studies did not find a significant response with respect to effective extubation (Conti et al., 2004).

Even though these and other predictors are well controlled, during the weaning process some individuals experience spontaneous breathing difficulties, thereby prolonging dependence on the mechanical ventilator (Chao & Scheinhorn, 1998).

One of the main causes of unsuccesful weaning and extubation is failure of the respiratory muscle pump, mainly inspiratory muscles (diaphragm, external intercostals, scalene, pectoral and sterno-cleido-occipito-mastoid) (X. Capdevila et al., 1998; Cluzel et al., 2000; A. D. Martin et al., 2002; S T Venkataraman et al., 2000), which leads to prolonged dependence on MV (Vassilakopoulos & Petrof, 2004). Muscle failure is due to two factors: (1) increase in workload to which the muscle is submitted owing to carbon dioxide ($CO_2$) accumulation; greater airway resistance; reduced pulmonary compliance; increase in dead space or in respiratory stimulus; (2) lower neuromuscular capacity caused by phrenic nerve dysfunction; neuromuscular disorder due to dependence on MV and reduced respiratory muscle strength or endurance (Chao & Scheinhorn, 1998; Meade, et al., 2001; Spitzer et al., 1992)

This reduction in muscle strength results from systemic problems, such as kidney failure, poor glycemic control, history of prolonged corticosteroid use, sepsis, electrolytic and mineral abnormalities, in addition to the use of neuromuscular blockers,widely administered in cases of patients who are not successfully sedated (A. A. Bruton et al., 2002). To delimit this muscle weakness, the most observed symptom is dyspnea (McConnell & Romer, 2004; Nield, 1999), and the most reliable parameter for measuring this variable (inspiratory muscle strength) is MIP (Clanton et al., 2002; Green et al., 2002).

To prevent dyspnea from interfering in the weaning process and successful extubation, some studies suggest the need to increase inspiratory muscle strength (Caruso et al., 2005; Chang et al., 2005).

The effectiveness of inspiratory muscle strengthening has been reported in apparently healthy elderly (S A Cader et al., 2007), athletes (Inbar et al., 2000) and diverse pathologies, such as: restrictive pulmonary disease (Chatham, 2000); myotonic muscular dystrophy (de Freitas Fregonezi et al., 2006); Duchenne muscular dystrophy (Topin et al., 2002); cystic fibrosis (Enright et al., 2004), chronic obstructive pulmonary disease (COPD) (Sturdy et al., 2003), congestive heart failure (CAD) (Laoutaris et al., 2004), asthma (Weiner et al., 2002), sarcoidosis (Brancaleone et al., 2004), cancer (Bruera et al., 2000), spinal cord injury (SCI) (Liaw et al., 2000), tetraplegia (Uijl et al., 1999), ankylosing spondylitis (Van der Esch et al., 2004), osteoporosis (Cimen et al., 2003) and multiple sclerosis (Klefbeck et al., 2003). However, there are no studies on inspiratory muscle strengthening in critically ill intubated patients, hospitalized in an ITC.

One of the inspiratory muscle training (IMT) techniques used in intubated patients is a reduction of pressure support mechanical ventilation. However, no studies are available that have investigated on the threshold IMT instrument.

The present study assumes that, when aiming for strength, IMT is suitable for patients with weak inspiratory muscles, since it exhibits the same principles as limb strength training, namely: (1) as with skeletal muscles, respiratory muscle weakness occurs with disuse; (2) with increased demand on the respiratory system, patients are unable to maintain spontaneous breathing for long periods; and (3) perception of respiratory effort is inversely proportional to inspiratory muscle strength, that is, the lower the inspiratory muscle strength, the higher the effort perception (A. A. Bruton, et al., 2002; Campbell et al.Gandevia, Killian et al., 1980; Cluzel, et al., 2000; Yue & Cole, 1992).

Although IMT performed over a short time period is not sufficient to cause significant changes in distribution of muscle fiber types or in the transverse muscle section, a number of mechanisms are evident: changes in the motor program and excitability of the neuromuscular system; neural adaptation, including a decrease in coactivation in the antagonist muscle group, a synchronic increase in the motor unit and a more efficient motor program (Carolan & Cafarelli, 1992; McComas, 1994; Milner-Brown et al., 1975; Sale, 1988; Yue & Cole, 1992).

Prior explanation is needed regarding mechanical ventilation and its ventilatory modes, weaning and its predictors, as well as inspiratory muscle strengthening itself in order to better understand the proposal of this chapter.

## 2. Mechanical ventilation

The main reason for patient admittance to the ICU is the need to install MV so as to reduce respiratory work and offer adequate oxygen support (Tobin, 2001). Esteban et al. (Esteban et al., 2000) conducted a study in eight countries with a total of 1638 patients. Indications for MV included the following: acute respiratory failure (ARF - 66%); coma (15%); COPD (13%); and neuromuscular disorders (5%). ARF primarily involved pneumonia, sepsis, surgical complications and trauma.

ARF is defined as the failure of lungs and the respiratory pump to maintain adequate gas exchange between oxygen (O2) and CO2. Hypoxemic ARF (type 1) consists of intrapulmonary impairment resulting in alveolar collapse, which leads to intrapulmonary shunting and inadequate arterial oxygenation. Hypercapnic ARF (type 2) is an extrapulmonary injury, resulting in the inability of the respiratory pump to sustain sufficient alveolar ventilation to eliminate metabolically produced CO2 (Manthous et al., 1998). In this respect, the mechanical ventilator can substitute the respiratory pump until the initial disorder triggering ARF has been suitably reversed (Esteban et al., 1997; S. T. Venkataraman et al., 2000).

Chronic dependence on the mechanical ventilator is not only the main medical problem (N Ambrosino, 2005), but also an extremely uncomfortable state with important social implications (Appendini et al., 1996). Furthermore, MV is associated to greater complications, such as infection, barotrauma, cardiovascular impairment, tracheal lesion, oxygen toxicity and MV-induced pulmonary injury (Vassilakopoulos & Petrof, 2004). Some investigators have sought to determine the pathological mechanisms that lead to unsuccessful weaning in patients chronically dependent on MV (Appendini, et al., 1996).

### 2.1 Conventional ventilation modes

All patients using ventilatory prostheses are ventilated by the following modes: assist/control ( volume or pressure) and synchronized intermittent mandatory ventilation (SIMV) with pure pressure suport (PS) or pressure suport ventilation (PSV) (Esteban, Anzueto, et al., 2000). Choosing the ideal ventilation mode and adequate parameters is essential to reducing respiratory muscle overload, since there must be patient-ventilator synchronism (Tobin et al., 2001).

In the assist/control mode (the most widely used), the ventilator releases a previously established current volume (volume-controlled ventilation -VCV) or inspiratory pressure

(pressure-controlled volume - PCV) to the patient, in the presence of patient effort (assist mode) or not (control mode) (Rose & Ed, 2006). Even though VCV guarantees adequate current volume, it does not limit peak pressure, which may result in altered compliance and resistance, in addition to provoking barotrauma. However, although PCV does not ensure constant current volume, it reduces the risk of pulmonary injury by determining and maintaining airway pressure and gas exchange, decreasing respiratory work and preventing hyperdistension of healthy alveoles (Esteban, Alia, et al., 2000).

SIMV with PS was introduced to provide a gradual level of assistance. This mode intersperses mandatory cycles, with pre-established respiratory frequency and current volume, as well as spontaneous cycles, in which respiratory frequency and current volume are peculiar to the patients themselves. The spontaneous cycle offers pressure support for the patient to generate adequate current volume (Leung et al. 1997).

PSV and SIMV with PS also offer ventilatory assistance, which is gradually lowered, but consists only of spontaneous cycles. PS is reduced according to the patient's respiratory frequency and current volume (Esteban, Anzueto, et al., 2000). PSV, characterized by low pressure values (7-8 cmH2O), has been used as a weaning protocol. Kuhlen et al. (Kuhlen et al., 2003) observed that among T-piece, automatic tube compensation and PSV, the latter provided the least ventilatory overload in patients submitted to weaning, about to be disconnected from the ventilatory prosthesis. Similar data were recorded by Ezingeard et al. (Ezingeard et al., 2006). However, Haberthur et al. (2002) found no significant differences among the 3 protocols.

In a randomized prospective study, Matic et al. (Matic et al., 2007) observed that PSV weaning was more successful than theT-piece device in the extubation process ($p<0.001$) and the group submitted to T-piece also spent more time on MV and in the ICU ($p<0.001$), when compared with the PSV group.

New ventilation modes are increasingly being introduced. Recent proposals include increasing respiratory muscle rest; preventing deconditioning; improving gas exchange; avoiding pulmonary lesion and increasing patient-ventilator coordination (Tobin, 2001). Within these modalities are volume assured pressure support (VAPS), bilevel positive airway pressure (BiPAP) and airway pressure release ventilation (APRV) (Rose & Ed, 2006).

## 3. Weaning and extubation

Difficult-to-wean patients account for 20-25% of those mechanically ventilated (Vassilakopoulos & Petrof, 2004), and an even higher percentage of individuals with COPD (Matic, et al., 2007). Weaning commences with reduced ventilatory support, depending on the degree to which patients can ventilate themselves adequately, culminating with their being withdrawn from MV; extubation consists of removing the endotracheal tube (Meade, et al., 2001; Robriquet et al., 2006).

Weaning failure is a result of complications in the central respiratory stimulus or, more commonly, due to neuromuscular abnormalities, such as muscle fatigue and altered pulmonary mechanisms or gas exchange (Meade, et al., 2001). The ventilatory pattern of patients that are not successfully weaned is similar to that observed during AFR (Del Rosario et al., 1997). Thus, some investigators recommend inspiratory muscle strengthening to reduce respiratory muscle fatigue (Chang, Boots, Brown, et al., 2005) and facilitate weaning from MV (Del Rosario et al., 1997).

Extubation failure occurs when there is a need to reinstitute ventilatory support between 24 and 48 hours after tube removal. Rothaar & Epstein (Rothaar & Epstein, 2003) sumarize the main causes for extubation failure can be seen in chart 1.

Although the ideal weaning procedure has yet to be established, the literature reports three main techniques: SIMV, T-piece and PSV. Other less widely used methods are automatic tube compensation, continuous positive airway pressure (CPAP) and bilevel positive airway pressure - BiPAP (Esteban et al., 1995; Haberthur, et al., 2002).

MV must be interrupted as soon as the patient is capable of adequately sustaining spontaneous breathing and gas exchange (Farias et al., 2002). However, the benefits of rapid release from MV must be weighed against the risks of weaning and premature extubation (Meade, et al., 2001), since these have been associated to increased hospital costs, mortality, length of ICU stay and need for tracheostomy (Esteban, et al., 1997; Rothaar & Epstein, 2003), in addition to generating new trauma in the airways due to reintubation (Noizet et al., 2005).

| |
|---|
| 1- Airway obstruction (granulous tissue, inflammation, ulceration and edema) |
| 2- Excessive respiratory tract secretion |
| 3- Inability to protect airways |
| 4-Heart failure or ischemic heart disease |
| 5- Encephalopathy |
| 6- Respiratory failure (imbalance between respiratory muscle strength and the overload to which it is submitted) |
| 7- Gastrointestinal bleeding |
| 8- Sepsis |
| 9- Need for surgery |

Thus, the complex decision to extubate requires an adequate assessment of predictor accuracy in this process (S. K. Epstein, 2002).

Chart 1. Main causes of extubation failure.

### 3.1 Weaning and extubation predictors (Tobin index, MIP and MEP)

Several indices have been proposed to predict weaning from MV. These reflect respiratory mechanics, pulmonary gas exchange, respiratory control and respirastory muscle function (chart 2) (Alvisi et al., 2000; Hoo & Park, 2002; Meade, et al., 2001; Rothaar & Epstein, 2003). Despite the wide range of predictors, Vallverdú et al. (Vallverdu et al., 1998) considered the Tobin index, MIP and MEP as good predictors.

| |
|---|
| 1- Minute ventilation (MV) |
| 2- Vital capacity (VC) |
| 3- Current volume (CV) |
| 4- Respiratory rate (RR) |
| 5- Negative inspiratory force (NIF) |
| 6- Tobin index (RR/CV) |
| 7- MIP and Pimax |
| 8- Respiratory work (Wob) |
| 9- Dead space |
| 10- Central respiratory stimulus (P0.1) |
| 11- CROP index (compliance, rate, oxygenation and MIP) |
| 12- Pressure-time index |
| 13- P0.1/Pimax ratio |
| 14- Dynamic compliance |
| 15- MIP/Pimax |

Chart 2. Diversity of weaning and extubation predictors.

A common finding in patients who were not successfully weaned is a high Tobin index score (RR/CV, in liters). It is measured with the patient disconnected from MV, where minute volume is verified during a 1-minute period (Goldstone, 2002). Yang and Tobin (Yang & Tobin, 1991) observed that this index has a sensitivity of 0.97 and a specificity of 0.64.

Thus, the Tobin index is the most widely accepted of the weaning indices. However, it does not consider ventilatory force or endurance, important requirements for successful weaning (Vassilakopoulos & Petrof, 2004).

Although the cause of weak respiratory muscles is not entirely known, it is suggested that electrolytic disorders (hypophophatemia and hypomagnesemia), the use of certain medications (curarizing agents) and malnutrition, combined with muscle disuse, have contributed to the occurrence of this phenomenon. The atrophy process is rapid, occurring from 7 to 10 days after a reduction in maximum diaphragmatic pressure of 50% (Anzueto et al., 1997).

MIP is the index used to measure inspiratory muscle strength (Volianitis et al., 2001). Although there is a conventional way to check this variable (Green, et al., 2002), Marini, Smith and Lamb (Marini et al., 1986) suggest a protocol that does not require patient collaboration. Some studies applying it as a predictor observed a significant difference ($p<0.05$) between patients who were successfully weaned (higher values) and those who were not (A. Bruton, 2002; C. D. Epstein, et al., 2002). Using a cutoff point of -17 $cmH_2O$, Conti et al. (Conti, et al., 2004) found a sensitivity of 0.95, specificity of 0.42 and accuracy of 0.71.

For an effective cough, deep inspiration and expiratory force are important, since they generate high thoracic-abdominal pressure. Based on this cough mechanism, several assisted cough techniques were developed for individuals with respiratory muscle weakness. Although expiratory muscle strength is essential for an effective cough, inspiratory muscle strength is equally important in the inspiratory phase of a cough, capable of generating an ideal pulmonary volume (Jaeger et al., 1993). Nevertheless, there are few studies that correlate cough capacity with inspiratory muscle strength (Kang et al., 2006).

Chart 3 demonstrates the variability of MIP and Tobin index thresholds, as predictors of successful weaning and/or extubation.

| PREDICTORS | VALUES | REFERENCES |
|---|---|---|
| Tobin index | 60 ipm/l | X. J. Capdevila et al.(1995) |
| | 96 ipm/l | Gandia & Blanco (1992) |
| | 100 ipm/l | Vallverdu et al.(1998) |
| | 100 ipm/l | Yang (1993) |
| | 100 ipm/l | S. K. Epstein & Ciubotaru (1996) |
| | 100 ipm/l | S. K. Epstein (1995) |
| | 100 ipm/l | Afessa et al. (1999) |
| | 100 ipm/l | Del Rosario et al.(1997) |
| | 100 ipm/l | Dojat et al. (1996) |
| | 100 ipm/l | Chatila et al. ( 1996) |
| | 100 ipm/l | Leitch et al.(1996) |
| | 105 ipm/l | Krieger et al. (1997) |
| | 105 ipm/l | Mergoni et al. (1996) |
| | 105 ipm/l | Lee, Hui, & Chang ( 1994) |
| | 105 ipm/l | Yang & Tobin ( 1991) |
| MIP | 15 cmH2O | Yang (Yang, 1993) |
| | 20 cmH2O | Leitch et al. (1996) |
| | 20 cmH2O | Del Rosario et al.( 1997) |
| | 20 cmH2O | Krieger et al.(1997) |
| | 20 cmH2O | Afessa et al.( 1999) |
| | 25 cmH2O | Manthous et al.(1998) |
| | 25 cmH2O | Goldwasser (1998) |
| | 30 cmH2O | Mergoni et al. (1996) |

Chart 3. Variability of MIP and Tobin index values for successful weaning and/or extubation.

## 4. Respiratory muscle training

Malnutrition, detraining and disuse due to prolonged bedrest may induce severe skeletal muscle dysfunction. Combined with other factors, such as reduced oxygen support, metabolic acidosis, electrolytic alterations and endocrine disorders, muscle disuse may lead to muscle atrophy (N Ambrosino, 2005; A. A. Bruton, et al., 2002). Within these factors there is predominance of isolated corticosteriod action, potentialized by neuromuscular blocking agents, immobility and sepsis (Larsson et al., 2000).

Respiratory muscles are skeletal muscles and, during prolonged mechanical ventilation (more than 48 hours), (A. Bruton, 2002; Chang, Boots, Brown, et al., 2005; Chelluri et al. 2002; Douglas et al. 2002) - inspiratory muscles move passively, favoring atrophy from disuse (Angus et al., 2005; A. A. Bruton, et al., 2002). The level of this strength reduction is related to duration of MV (Chang, Boots, Brown, et al., 2005; Powers et al., 2002; Sassoon, Caiozzo et al., 2002; Sprague & Hopkins, 2003) and has been reported as one of the major determinants of weaning failure in patients receiving MV (N Ambrosino, 2005; Gayan-Ramirez & Decramer, 2002). Anzueto et al. (Anzueto, et al., 1997) observed a transdiaphragmatic pressure reduction of 45% in rats after 11 days of controlled MV.

Thus, some studies described using IMT to increase inspiratory muscle effectiveness, in order to assist with difficult weaning. The main training techniques are: (a) endurance training using eucapnic hyperventilation (Aldrich & Karpel, 1985; Aldrich et al., 1989; Aldrich & Uhrlass, 1987; Belman, 1981); (b) threshold IMT strength training (A. D. Martin, et al., 2002; U. J. Martin et al., 2005; Sprague & Hopkins, 2003) and (c) resistance training by reducing ventilatory prosthesis sensitivity (Caruso, et al., 2005). The literature also cites other less widely used IMT methods, such as the solenoid device and the weighted plunger (Bardsley et al., 1993; Johnson et al., 1996).

Inspiratory muscle endurance can be defined as sustained MIP (SMIP), which is the ability to maintain a simple contraction over a prolonged period of time. It is a dynamic maneuver, whereas MIP is considered static or quasi-static (A. A. Bruton, et al., 2002).

SMIP is measured from residual volume and the orifice must be occluded immediately at the onset of inspiration so as to generate negative intrathoracic pressure, which is verified on the gauge (Green, et al., 2002). Inspiration must last at least 3 seconds, with maximum force and time possible. This procedure is repeated three times, and the best result is used (Cimen, et al., 2003). However, it is not easy to reproduce in mechanically ventilated patients due to lack of understanding or even interaction with the examiner. As such, an alternative for this problem is the use of an expiratory unidirectional valve since it permits free expiration, progressively reaching a volume close to the residual value, whereas for inspiration the patient must exert an effort to generate sufficient negative pressure (Caruso et al., 1999; Marini, et al., 1986).

Several factors indicate an individual should be submitted to IMT, the primary one being a simple decrease in MIP. After MIP is measured, inspiratory muscle strengthening and conditioning can be initiated (S A Cader, 2006). However, a significant problem for the inspiratory muscle training program is the protocol used. Chart 4 illustrates the range of protocols, none of which are standardized because of different medical conditions.

Although IMT of patients under MV is of short duration, some studies report that muscle strength gain is due to neural adaptation (A. D. Martin, et al., 2002; Sprague & Hopkins, 2003) rather than muscle hypertrophy. The following mechanisms emerge as a result of this adaptation: increased maximum voluntary contraction ability; decreased antagonist muscle coactivation; greater motor unit synchrony or a more efficient motor program (Carolan & Cafarelli, 1992; McComas, 1994; Milner-Brown, et al., 1975; Sale, 1988; Yue & Cole, 1992). However, further studies are needed given that the exact mechanism cannot be determined from these findings.

| PROTOCOL | REFERENCES | PATHOLOGY |
|---|---|---|
| 8W/ 3X/ 10' / 20-60% | Fregonezi et al.(Fregonezi, Resqueti, Guell, Pradas, & Casan, 2005) | Miasthenia gravis |
| 8W/ 3X / 45'* / 80% | Enright et al.(Enright, et al., 2004) | Cystic Fibrosis |
| 10W/ 3X/ time?/ 60% | Laoutaris et al.(Laoutaris, et al., 2004) | CHF |
| 10W/ INT (2)/ 3s-10x/ 40-60% | Klefbeck & Hamrah Nedjad (Klefbeck & Hamrah Nedjad, 2003) | Multiple Sclerosis |
| 8W/ 3X/ 20'-2' + 1'/ 70% | Sturdy et al.(Sturdy, et al., 2003) | COPD |
| 5W/ 5X/ 30'/ 40-50% | Ramirez-Sarmiento et al. (Ramirez-Sarmiento et al., 2002) | COPD |
| 12W/ 3X(2)/ 5'/ 5-7% de increase every 2 S | de Freitas Fregonezi, Resqueti, & Casan (de Freitas Fregonezi, et al., 2006) | Myotonic muscular dystrophy |
| 24W/ 6X/ 30'/ 60-70% | Sanchez Riera et al. (Sanchez Riera et al., 2001) | COPD |
| 6W/7x(2)/10'/30% | Topin et al. (Topin, et al., 2002) | Duchenne muscular dystrophy |
| 16W/ 7X/ 30'/ 30-60% | Covey et al. (Covey et al., 2001) | COPD |
| 6W/ 5X/ 20'/ > 40% | de Jong et al. (de Jong, van Aalderen, Kraan, Koeter, & van der Schans, 2001) | Cystic fibrosis |
| 6W/ 7X (2)/ 15-20'/? | Liaw et al.(Liaw, et al., 2000) | SCI |
| 6W/ 7X (2)/ 15'/ 70% | Uijl et al. (Uijl, et al., 1999) | Tetraplegia |
| 60W/ 7X (2)/ 15'/ 15% | Sudo et al.(Sudo et al., 1997) | COPD |
| 10W/ 6X/ 30'/ 30% | Lisboa et al.(Lisboa et al., 1997) | Chronic airflow limitation |
| 8W/ 7X (3)/ 5-15'/ 20% | Cahalin, Semigran, & Dee (Cahalin, Semigran, & Dee, 1997), | CHF (indication for transplantation) |

W: weeks; X: times per week; (2): twice a day; (3): three times a day; %: percentage of MIP.

Chart 4. Diversity of inspiratory training protocols.

### 4.1 Effects of inspiratory muscle training on the weaning process in critically ill tracheostomized patients – Results of previous studies

Imbalance between increased ventilatory overload and reduced ventilatory endurance is an important determinant of MV dependence. Thus, respiratory muscle weakness is nearly always present as a contributing factor for weaning failure. Even so, there is little information on the use of the respiratory muscle training program for difficult-to-wean patients (Caruso, et al., 2005; A. D. Martin, et al., 2002; U. J. Martin, et al., 2005). Although some studies recommend IMT during MV (Chang, Boots, Brown, et al., 2005; Del Rosario, et al., 1997; Sprague & Hopkins, 2003), Chang et al. (2005) observed a post-weaning reduction in inspiratory muscle strength in patients submitted to long-term MV.

Hawkes et al. (2007) used electromyographic analysis to examine the effect of submaximal inspiratory overload on MIP and diaphragm and intercostal muscles. Results obtained 15 minutes after the test revealed an increase in MIP ($p=0.032$). During MIP, an increase was recorded in the range of diaphragm ($p=0.032$) and intercostal ($p=0.016$) muscle action ($p=0.039$). These data support the hypothesis of greater muscle strength after submaximal inspiratory effort or IMT, resulting in a reduction in perception of respiratory effort (dyspnea) (McConnell & Romer, 2004; Volianitis, et al., 2001).

A.D. Martin et al. (2002) carried out a study in which IMT was performed using threshold IMT in 10 patients (tracheostomized) dependent on MV (34±33 days), in order to facilitate weaning from the ventilator by improving inspiratory muscle strength. In addition to a significant increase in strengthening workload used (7±3 cmH2O for 18±7 cmH2O, $p<0.05$) in threshold IMT, after 44±43 days of IMT, 9 of 10 patients submitted to training were successfully weaned.

The effect of IMT was observed in a randomized prospective study, evidenced by reduced ventilator sensitivity and shorter weaning and reintubation time in critically ill patients (Caruso, et al., 2005). These authors observed no significant difference between training and control groups with respect to weaning time ($p=0.24$), reintubation frequency ($p=0.39$) and final MIP value ($p=0.34$). However, it is important to underscore that, in addition to the small sample size, the fact that some patients did not have an ideal level of consciousness to determine MIP may have caused its underestimation. Consequently, the load applied might have been too low. This fact, along with short training time, could have influenced the final results.

Sprague & Hopkins (2003) aimed at weaning tracheostomized MV-dependent patients by increasing inspiratory muscle strength using threshold IMT. The protocol consisted of: 6-7 days per week, once a day, 4 series of 6-8 repetitions, with a 5-10-minute rest period (under MV) between series. Training load (approximately 50% of MIP) was modified according to Borg's scale. After IMT, patients remained under spontaneous breathing via T-piece, with a gradual increase over time. Results showed a rise in training load (9.3 cmH2O for 27.5 cmH2O; 195%); increase in MIP (22.0 to 54.0 cmH2O; 140%); and successful weaning in all patients over a period of 9 to 28 days.

Frutos-Vivar et al. (2006) observed that pneumonia, as the initial trigger of AFR (with consequent intubation), is one of the primary causes of intubation. Based on this information, the study conducted by Kang et al. (2006) is very important to the present investigation, since these authors aimed to assess the relationship between inspiratory muscle strength and cough capacity, but in patients with SCI. They obtained intermediate correlation between VC and voluntary cough capacity ($r=0.583$; $p<0.001$), intermediate-high between VC and MIP ($r=0.749$; $p<0.001$) and intermediate between VC and MEP ($r=0.438$; $p<0.001$); MIP also exhibited more significant correlations ($p<0.001$, with intermediate to intermediate-high correlations) with peak cough flow compred to MEP ($p<0.05$, with low-intermediate and intermediate correlations).

Another investigation, conducted in patients with spinal cord injury, underscored the importance of respiratory muscle training, increasing endurance and strength, in order to wean these individuals (Gutierrez et al., 2003). Comparative pre- and post-training data demonstrated a significant rise in MIP ($p<0.001$; mean of -18 cmH2O), MEP ($p<0.001$; mean of 21.6 cmH2O) and VC ($p<0.001$; mean of 340cc) in patients with lower lesions, favoring their weaning.

Few studies have assessed the magnitude and extent of generalized muscle weakness in patients with prolonged MV dependence. In a retrospective analysis of 49 patients submitted to whole-body rehabilitation (with strength and resistance training of upper (UULL) and lower limbs (LLLL), inspiratory muscles with threshold IMT, as well as the torso and gait), U. J. Martin et al. (2005) recorded a significant response ($p<0.001$) in functional status and inspiratory muscle, UULL and LLLL strength, in addition to a significant ($r=0.720$; $p<0.001$) and inverse correlation between UULL strength at admission and weaning time.

Additional research, conducted in patients with no artificial airway, is also relevant for the present study (Baker et al., 2003; Mathers-Schmidt & Brilla, 2005; Sapienza et al., 1999). Some post-extubation impairments occur in the upper airways (UAW), with only glottic edema and vocal cord lesions causing stridor. These alterations restrict air flow in the upper airways, resulting in dyspnea and eventual reintubation (Esteban, et al., 1997). Baker et al. (Baker, et al., 2003) studied in patients with upper airway limitation (glottis) resulting from bilateral vocal cord paralysis (abductors). The authors aimed to determine whether inspiratory muscle strengthening increases inspiratory muscle ability, generating greater MIP and reducing dyspnea sensation during exercise and speech. Threshold IMT was used for training, with the following protocol: load of 75% of MIP, 5 times per week, 5 series of 5 repetitions, for 5 weeks. Results revealed an increase of 47% in MIP, reduced dyspnea during exercise and speech of 2-4 points on Borg's scale and improved airway opening (glottis), visualized by endoscopy.

In a study performed with healthy individuals, nuclear magnetic resonance showed acute and chronic responses of the UAWs to ITM (How et al., 2007). Results demonstrated significant alterations in the experimental group, as follows: increase in UAW dilator muscles (genioglossus and geniohyoid) ($p<0.001$); increase in transverse section area ($p=0.025$) and lateral diameter ($p=0.003$) of the UAWs; increase in nasopharynx ($p<0,001$) and laryngopharynx ($p=0.031$).

Another advantage of IMT was the reduction in central respiratory stimulus ($P_{0.1}$) (Huang et al., 2003). In this investigation, the intervention was performed in 23 apparently healthy individuals (28.8±2.3 years). The protocol used an increasing inspiratory load of 30% to 75% of MIP; 5 days a week; 4 series of 6 repetitions; for 4 weeks. The significant increase in MIP ($p<0.001$) had a significant negative correlation ($p<0.01$) with the significant reduction in $P_{0.1}$ ($p<001$). These data suggest that once $P_{0.1}$ is reduced, the occurrence of dyspnea decreases (N. Ambrosino et al., 1994; Ferrari et al., 1997).

### 4.2 Effects of inspiratory muscle training on the weaning process of critically ill intubated patients – Results of a pioneering study

Cader et al.(S. A. Cader et al., 2010) investigated elderly patients to detemine whether inspiratory muscle training improved maximal inspiratory pressure, breathing pattern and length of weaning time from mechanical ventilation. Of the 198 patients screened during the recruitment, the 67 eligible individuals were monitored daily to assess readiness to start weaning. Twenty of the subjects were tracheostomized, 5 died and 1 was transferred to another center before the start of weaning. The remaining 41 were randomized: 21 to the experimental group and 20 to the control group. Four participants in each group died before extubation. Three participants in the experimental group and two in the control group were tracheostomized before extubation.

Group data for all outcomes at the start of weaning and at extubation for the experimental and control groups are presented in Table 1. Maximal inspiratory pressure increased significantly more in the treatment group than in the control group (MD 7 $cmH_20$, 95% CI 4 to 10). The Tobin index decreased significantly more in the treatment group (73.6 br/min/l to 81.5 br/min/l) than among controls (81.5 br/min/l to 95.9 br/min/l). In those who did not die or receive a tracheostomy, time to weaning was significantly shorter in the treatment group than in the control group (MD 2 days, 95% CI 0 to 3).

Among uncensored participants, inspiratory muscle training significantly reduced the spontaneous ventilation (ie, weaning) period. Inspiratory muscle training shortened weaning time by 1.7 days (95% CI 0.4 to 3.0), as shown in Table 2.

| Outcome | Groups | | | | Within-group difference | | Between-group difference |
|---|---|---|---|---|---|---|---|
| | Pre-test | | Post-test | | Post-test minus pre-test | | Post-test minus pre-test |
| | Exp (n = 14) | Con (n = 14) | Exp (n = 14) | Con (n = 14) | Exp | Con | Exp minus Con |
| MIP ($cmH_2O$) | 15.1 (2.6) | 15.3 (2.2) | 25.0 (3.9) | 17.6 (1.9) | 9.9 (2.5) | 2.3 (2.1) | 7.6 (5.8 to 9.4) |
| TI (*br/min/L*) | 73.6 (8.8) | 81.5 (6.9) | 79.7 (11.2) | 95.9 (12.4) | 6.1 (3.6) | 14.4 (9.1) | -8.3 (-13.7 to -2.9) |

TI = Tobin index (ipm/l); MIP = maximal inspiratory pressure ($cmH_2O$); Exp = experimental group; Con = control

Table 1. Mean (SD) outcomes for each group, mean (SD) difference within groups and mean difference (95% CI) between groups.

| Outcome | Groups | | Mean inter-groupdifference |
|---|---|---|---|
| | Exp (n=14) | Con (n=14) | Exp minus Con |
| Total ventilation period | 10.4 (4.0) | 11.0 (3.5) | 0.6 (-3.5 to 2.4) |
| Spontaneous ventilation period | 3.6 (1.5) | 5.3 (1.9) | 1.7 (0.4 to 3.0) |

Exp = experimental group, Con = control group

Table 2. Mean (SD) duration (days) of total ventilation and spontaneous ventilation periods among uncensored participants.

The receiver operating characteristic curve (sensitivity and specificity) was used to evaluate weaning predictors and extubation success. Beginning with the Tobin index, the receiver operating characteristic curve reveals an area below the curve of 0.877 (SD 0.06). Sensitivity and specificity of the Tobin index was 1.00 and 0.36, respectively. Maximum inspiratory pressure was 0.86 for sensitivity and 0.72 for specificity. The area below the receiver operating characteristic curve shows a value of 0.845 (SD 0.07).

Thus, it was concluded that inspiratory muscle training in the experimental group contributed to a significant increase in maximum inspiratory pressure (between and within groups) and a reduction in the Tobin index (within groups). These are considered good weaning predictors , which is consistent with our finding that inspiratory muscle training significantly reduces the weaning period in patients who did not die or receive a tracheostomy.

## 5. Conclusions

This chapter presented an important alternative for inspiratory muscle training in critically ill patients. The technique aims at shortening weaning time and increasing extubation success rates.

IMT was found to improve some weaning predictors such as TI, MIP and MEP, favoring a reduction in weaning time. This outcome is essential in the ICU, since it decreases physiological alterations and complications from ventilator use, such as barotrauma, tracheal lesion, sarcopenia, repeated infections, scabs and surgical invasions. These findings are relevant because they contribute to shorter hospital stays and lower costs, resulting in fewer social repercussions and enhanced pre-and post-hospitalization quality of life.

Recommendations of the present study are divided into two categories: applicability and continuity.

With respect to continuity, the authors recommend that randomized controlled studies be carried out to compare IMT using threshold IMT with other interventions, such as reduced sensitivity and mechanical ventilator support pressure in intubated subjects. We further suggest that prospective studies be applied to assess the effects of training and its continued use in hospital patients, following up these individuals after discharge from the intensive therapy center (ITC) to a semi-intensive unit, infirmary or room.

In relation to the applicability of the present study, it is hoped that the knowledge acquired will be made available to physical therapists so they may intervene as early as possible, thereby avoiding future complications previously experienced in the ICU, mainly when elderly patients are involved. It is suggested that this intervention (IMT) be implemented in ITCs in order to reduce the number of tracheostomies, increase extubation sucess rates and decrease hospital stays, all of which will improve patient well-being.

## 6. References

Afessa, B., Hogans, L., & Murphy, R. . (1999). Predicting 3-day and 7-day outcomes of weaning from mechanical ventilation. . *Chest, 116*, 456-461.

Aldrich, T. K., & Karpel, J. P. (1985). Inspiratory muscle resistive training in respiratory failure. *Am Rev Respir Dis, 131*(3), 461-462.

Aldrich, T. K., Karpel, J. P., Uhrlass, R. M., Sparapani, M. A., Eramo, D., & Ferranti, R. (1989). Weaning from mechanical ventilation: adjunctive use of inspiratory muscle resistive training. *Crit Care Med, 17*(2), 143-147.

Aldrich, T. K., & Uhrlass, R. M. (1987). Weaning from mechanical ventilation: successful use of modified inspiratory resistive training in muscular dystrophy. *Crit Care Med, 15*(3), 247-249.

Alvisi, R., Volta, C. A., Righini, E. R., Capuzzo, M., Ragazzi, R., Verri, M., . . . Milic-Emili, J. (2000). Predictors of weaning outcome in chronic obstructive pulmonary disease patients. *Eur Respir J, 15*(4), 656-662.

Ambrosino, N. (2005). Weaning and respiratory muscle dysfunction: the egg-chicken dilemma. *Chest, 128*(2), 481-483.

Ambrosino, N., Opasich, C., Crotti, P., Cobelli, F., Tavazzi, L., & Rampulla, C. (1994). Breathing pattern, ventilatory drive and respiratory muscle strength in patients with chronic heart failure. *Eur Respir J, 7*(1), 17-22.

Angus, D., Ishizaka, A., Matthay, M., Lemaire, F., Macnee, W., & Abraham, E. (2005). Critical care in AJRCCM 2004. *Am J Respir Crit Care Med, 171*(6), 537-544.

Anzueto, A., Peters, J. I., Tobin, M. J., de los Santos, R., Seidenfeld, J. J., Moore, G., . . . Coalson, J. J. (1997). Effects of prolonged controlled mechanical ventilation on diaphragmatic function in healthy adult baboons. *Crit Care Med, 25*(7), 1187-1190.

Appendini, L., Purro, A., Patessio, A., Zanaboni, S., Carone, M., Spada, E., . . . Rossi, A. (1996). Partitioning of inspiratory muscle workload and pressure assistance in ventilator-dependent COPD patients. *Am J Respir Crit Care Med, 154*(5), 1301-1309.

Baker, S E, Sapienza, C M, Martin, D, Davenport, P. D., Hoffman-Ruddy, B, & Woodson, G. (2003). Inspiratory Pressure Threshold training for upper airway limitation: a case of bilateral abductor vocal fold paralysis. *Journal of Voice, 17*(3), 384-394.

Bardsley, P. A., Bentley, S., Hall, H. S., Singh, S. J., Evans, D. H., & Morgan, M. D. (1993). Measurement of inspiratory muscle performance with incremental threshold loading: a comparison of two techniques. *Thorax, 48*(4), 354-359.

Belman, M. J. (1981). Respiratory failure treated by ventilatory muscle training (VMT). A report of two cases. *Eur J Respir Dis, 62*(6), 391-395.

Brancaleone, P., Perez, T., Robin, S., Neviere, R., & Wallaert, B. (2004). Clinical impact of inspiratory muscle impairment in sarcoidosis. *Sarcoidosis Vasc Diffuse Lung Dis, 21*(3), 219-227.

Bruera, E., Schmitz, B., Pither, J., Neumann, C. M., & Hanson, J. (2000). The frequency and correlates of dyspnea in patients with advanced cancer. *J Pain Symptom Manage, 19*(5), 357-362.

Bruton, A. (2002). A pilot study to investigate any relationship between sustained maximal inspiratory pressure and extubation outcome. *Heart Lung, 31*(2), 141-149.

Bruton, A. A., Conway, J. H., & Holgate, S. T. (2002). Inspiratory muscle dysfunction after prolonged periods of mechanical ventilation. *Physiotherapy, 88*(3), 131-137.

Butler, R., Keenan, S. P., Inman, K. J., Sibbald, W. J., & Block, G. (1999). Is there a preferred technique for weaning the difficult-to-wean patient? A systematic review of the literature. *Crit Care Med, 27*(11), 2331-2336.

Cader, S A. (2006). Força Muscular Respiratória: como aferi-la e fortalecê-la? *Fitness e Performance Journal, 5*(1), 57.

Cader, S A, Silva, E B, Vale, R G S, Bacelar, S C, & Dantas, E H M. (2007). Efeito do treino dos músculos inspiratórios sobre a pressão inspiratória máxima e a autonomia funcional de idosos asilados. *Motricidade, 3*(1), 279-288.

Cader, S. A., Vale, R. G., Castro, J. C., Bacelar, S. C., Biehl, C., Gomes, M. C., . . . Dantas, E. H. (2010). Inspiratory muscle training improves maximal inspiratory pressure and may assist weaning in older intubated patients: a randomised trial. [Randomized Controlled Trial]. *Journal of physiotherapy, 56*(3), 171-177.

Cahalin, L. P., Semigran, M. J., & Dee, G. W. (1997). Inspiratory muscle training in patients with chronic heart failure awaiting cardiac transplantation: results of a pilot clinical trial. *Phys Ther, 77*(8), 830-838.

Campbell, E. J., Gandevia, S. C., Killian, K. J., Mahutte, C. K., & Rigg, J. R. (1980). Changes in the perception of inspiratory resistive loads during partial curarization. *J Physiol, 309*, 93-100.

Capdevila, X. J., Perrigault, P. F., Perey, P. J., Roustan, J. P., & d'Athis, F. (1995). Occlusion pressure and its ratio to maximum inspiratory pressure are useful predictors for successful extubation following T-piece weaning trial. *Chest, 108*(2), 482-489.

Capdevila, X., Perrigault, P. F., Ramonatxo, M., Roustan, J. P., Peray, P., d'Athis, F., & Prefaut, C. (1998). Changes in breathing pattern and respiratory muscle performance parameters during difficult weaning. *Crit Care Med, 26*(1), 79-87.

Carolan, B., & Cafarelli, E. (1992). Adaptations in coactivation after isometric resistance training. *J Appl Physiol, 73*(3), 911-917.

Caruso, P., Denari, S. D., Ruiz, S. A., Bernal, K. G., Manfrin, G. M., Friedrich, C., & Deheinzelin, D. (2005). Inspiratory muscle training is ineffective in mechanically ventilated critically ill patients. *Clinics, 60*(6), 479-484.

Caruso, P., Friedrich, C., Denari, S. D., Ruiz, S. A., & Deheinzelin, D. (1999). The unidirectional valve is the best method to determine maximal inspiratory pressure during weaning. *Chest, 115*(4), 1096-1101.

Chang, A. T., Boots, R. J., Brown, M. G., Paratz, J., & Hodges, P. W. (2005). Reduced inspiratory muscle endurance following successful weaning from prolonged mechanical ventilation. *Chest, 128*(2), 553-559.

Chang, A. T., Boots, R. J., Henderson, R., Paratz, J. D., & Hodges, P. W. (2005). Case report: inspiratory muscle training in chronic critically ill patients--a report of two cases. *Physiother Res Int, 10*(4), 222-226.

Chao, D. C., & Scheinhorn, D. J. (1998). Weaning from mechanical ventilation. *Crit Care Clin, 14*(4), 799-817, viii.

Chatham, K. (2000). Individualised fixed load inspiratory muscle training responses in a patient with severe restrictive lung disease and an élite sportsman. *Physiotherapy, 86*(1), 28-31.

Chatila, W., Jacob, B., Guaglionone, D., & Manthous, C. A. (1996). The unassisted respiratory rate-tidal volume ratio accurately predicts weaning outcome. *Am J Med, 101*(1), 61-67.

Chelluri, L., Rotondi, A., Sirio, C., & al., et. (2002). 2-Month mortality and functional status of critically ill adult patients receiving prolonged mechanical ventilation. *Chest, 121*, 549-558.

Cimen, O. B., Ulubas, B., Sahin, G., Calikoglu, M., Bagis, S., & Erdogan, C. (2003). Pulmonary function tests, respiratory muscle strength, and endurance of patients with osteoporosis. *South Med J, 96*(5), 423-426.

Clanton, T, Calvery, P M, & Celli, B R. (2002). Tests of respiratory muscle endurance. *American Journal of Respiratory and Critical Care Medicine, 166*(4), 559-570.

Cluzel, P., Similowski, T., Chartrand-Lefebvre, C., Zelter, M., Derenne, J. P., & Grenier, P. A. (2000). Diaphragm and chest wall: assessment of the inspiratory pump with MR imaging-preliminary observations. *Radiology, 215*(2), 574-583.

Conti, G., Montini, L., Pennisi, M. A., Cavaliere, F., Arcangeli, A., Bocci, M. G., . . . Antonelli, M. (2004). A prospective, blinded evaluation of indexes proposed to predict weaning from mechanical ventilation. *Intensive Care Med, 30*(5), 830-836.

Covey, M. K., Larson, J. L., Wirtz, S. E., Berry, J. K., Pogue, N. J., Alex, C. G., & Patel, M. (2001). High-intensity inspiratory muscle training in patients with chronic obstructive pulmonary disease and severely reduced function. *J Cardiopulm Rehabil, 21*(4), 231-240.

de Freitas Fregonezi, G. A., Resqueti, V. R., & Casan, P. (2006). [Domiciliary respiratory muscle training in myotonic dystrophy]. *Arch Bronconeumol, 42*(11), 605-607.

de Jong, W., van Aalderen, W. M., Kraan, J., Koeter, G. H., & van der Schans, C. P. (2001). Inspiratory muscle training in patients with cystic fibrosis. *Respir Med, 95*(1), 31-36.

Del Rosario, N., Sassoon, C. S., Chetty, K. G., Gruer, S. E., & Mahutte, C. K. (1997). Breathing pattern during acute respiratory failure and recovery. *Eur Respir J, 10*(11), 2560-2565.

Dojat, M., Harf, A., Touchard, D., Laforest, M., Lemaire, F., & Brochard, L. (1996). Evaluation of a knowledge-based system providing ventilatory management and decision for extubation. *Am J Respir Crit Care Med, 153*(3), 997-1004.

Douglas, S.L., Daly, B.J., & Gordon, N. (2002). Survival and quality of life: short-term versus long-term ventilator patients. *Critical Care Medicine, 30*, 2655-2662.

el-Khatib, M. F., Baumeister, B., Smith, P. G., Chatburn, R. L., & Blumer, J. L. (1996). Inspiratory pressure/maximal inspiratory pressure: does it predict successful extubation in critically ill infants and children? *Intensive Care Med, 22*(3), 264-268.

Enright, S., Chatham, K., Ionescu, A. A., Unnithan, V. B., & Shale, D. J. (2004). Inspiratory muscle training improves lung function and exercise capacity in adults with cystic fibrosis. *Chest, 126*(2), 405-411.

Epstein, C. D., El-Mokadem, N., & Peerless, J. R. (2002). Weaning older patients from long-term mechanical ventilation: a pilot study. *Am J Crit Care, 11*(4), 369-377.

Epstein, S. K. (1995). Etiology of extubation failure and the predictive value of the rapid shallow breathing index. *Am J Respir Crit Care Med, 152*(2), 545-549.

Epstein, S. K. (2002). Decision to extubate. *Intensive Care Med, 28*(5), 535-546.

Epstein, S. K., & Ciubotaru, R. L. (1996). Influence of gender and endotracheal tube size on preextubation breathing pattern. *Am J Respir Crit Care Med, 154*(6 Pt 1), 1647-1652.

Esteban, A., Alia, I., Gordo, F., de Pablo, R., Suarez, J., Gonzalez, G., & Blanco, J. (2000). Prospective randomized trial comparing pressure-controlled ventilation and volume-controlled ventilation in ARDS. For the Spanish Lung Failure Collaborative Group. *Chest, 117*(6), 1690-1696.

Esteban, A., Alia, I., Gordo, F., Fernandez, R., Solsona, J. F., Vallverdu, I., . . . Goldwasser, R. S. (1997). Extubation outcome after spontaneous breathing trials with T-tube or pressure support ventilation. The Spanish Lung Failure Collaborative Group. *Am J Respir Crit Care Med, 156*(2 Pt 1), 459-465.

Esteban, A., Anzueto, A., Alia, I., Gordo, F., Apezteguia, C., Palizas, F., . . . Tobin, M. J. (2000). How is mechanical ventilation employed in the intensive care unit? An international utilization review. *Am J Respir Crit Care Med, 161*(5), 1450-1458.

Esteban, A., Frutos, F., Tobin, M. J., Alia, I., Solsona, J. F., Valverdu, I., . . . et al. (1995). A comparison of four methods of weaning patients from mechanical ventilation. Spanish Lung Failure Collaborative Group. *N Engl J Med, 332*(6), 345-350.

Ezingeard, E., Diconne, E., Guyomarc'h, S., Venet, C., Page, D., Gery, P., . . . Zeni, F. (2006). Weaning from mechanical ventilation with pressure support in patients failing a T-tube trial of spontaneous breathing. *Intensive Care Med, 32*(1), 165-169.

Farias, J. A., Alia, I., Retta, A., Olazarri, F., Fernandez, A., Esteban, A., . . . Sheehan, G. (2002). An evaluation of extubation failure predictors in mechanically ventilated infants and children. *Intensive Care Med, 28*(6), 752-757.

Ferrari, K., Goti, P., Duranti, R., Iandelli, I., Misuri, G., Mancini, M., . . . Scano, G. (1997). Breathlessness and control of breathing in patients with COPD. *Monaldi Arch Chest Dis, 52*(1), 18-23.

Fregonezi, G. A., Resqueti, V. R., Guell, R., Pradas, J., & Casan, P. (2005). Effects of 8-week, interval-based inspiratory muscle training and breathing retraining in patients with generalized myasthenia gravis. *Chest, 128*(3), 1524-1530.

Frutos-Vivar, F., Ferguson, N. D., Esteban, A., Epstein, S. K., Arabi, Y., Apezteguia, C., . . . Anzueto, A. (2006). Risk factors for extubation failure in patients following a successful spontaneous breathing trial. *Chest, 130*(6), 1664-1671.

Gandia, F, & Blanco, J. (1992). Evaluation of indexes predicting the outcome of ventilator weaning and value of adding supplemental inspiratory load. *Intensive Care Med, 18*, 327.

Gayan-Ramirez, G., & Decramer, M. (2002). Effects of mechanical ventilation on diaphragm function and biology. *Eur Respir J, 20*(6), 1579-1586.

Goldstone, J. (2002). The pulmonary physician in critical care. 10: difficult weaning. *Thorax, 57*(11), 986-991.

Goldwasser, R.S. (1998). *Estudo da eficácia de dois diferentes testes de tolerância de interrupção da ventilação mecânica para predizer o êxito da extubação em pacientes portadores de doença pulmonar obstrutiva crônica.* Tese de Mestrado., Universidade Federal do Rio de Janeiro., Rio de Janeiro.

Green, M, Road, J, Sieck, G C, & Similowski, T. (2002). Tests of respiratory muscle strength. *American Journal of Respiratory and Critical Care Medicine, 166*(4), 528-547.

Gutierrez, C. J., Harrow, J., & Haines, F. (2003). Using an evidence-based protocol to guide rehabilitation and weaning of ventilator-dependent cervical spinal cord injury patients. *J Rehabil Res Dev, 40*(5 Suppl 2), 99-110.

Haberthur, C., Mols, G., Elsasser, S., Bingisser, R., Stocker, R., & Guttmann, J. (2002). Extubation after breathing trials with automatic tube compensation, T-tube, or pressure support ventilation. *Acta Anaesthesiol Scand, 46*(8), 973-979.

Hawkes, E. Z., Nowicky, A. V., & McConnell, A. K. (2007). Diaphragm and intercostal surface EMG and muscle performance after acute inspiratory muscle loading. *Respir Physiol Neurobiol, 155*(3), 213-219.

Hoo, G.W.S., & Park, L. (2002). Variations in the measurement of weaning parameters. *Chest, 121*, 1947-1955.

How, S C, McConnell, A K, Taylor, B J, & Romer, L M. (2007). Acute and chronic responses of the upper airway to inspiratory loading in hellthy awake humans: An MRI study. *Respiratory Physiology & Neurobiology, 157*, 270-280.

Huang, C H, Martin, A D, & Davenport, P. W. (2003). Effect of inspiratory muscle strength training on inspiratory motor drive and RREP early peak components. *J Appl Physiol, 94*, 462-468.

Inbar, O., Weiner, P., Azgad, Y., Rotstein, A., & Weinstein, Y. (2000). Specific inspiratory muscle training in well-trained endurance athletes. *Med Sci Sports Exerc, 32*(7), 1233-1237.

Jaeger, R. J., Turba, R. M., Yarkony, G. M., & Roth, E. J. (1993). Cough in spinal cord injured patients: comparison of three methods to produce cough. *Arch Phys Med Rehabil, 74*(12), 1358-1361.

Johnson, P. H., Cowley, A. J., & Kinnear, W. J. (1996). Evaluation of the THRESHOLD trainer for inspiratory muscle endurance training: comparison with the weighted plunger method. *Eur Respir J, 9*(12), 2681-2684.

Kang, S. W., Shin, J. C., Park, C. I., Moon, J. H., Rha, D. W., & Cho, D. H. (2006). Relationship between inspiratory muscle strength and cough capacity in cervical spinal cord injured patients. *Spinal Cord, 44*(4), 242-248.

Khamiees, M., Raju, P., DeGirolamo, A., Amoateng-Adjepong, Y., & Manthous, C. A. (2001). Predictors of extubation outcome in patients who have successfully completed a spontaneous breathing trial. *Chest, 120*(4), 1262-1270.

Klefbeck, B., & Hamrah Nedjad, J. (2003). Effect of inspiratory muscle training in patients with multiple sclerosis. *Arch Phys Med Rehabil, 84*(7), 994-999.

Krieger, B. P., Isber, J., Breitenbucher, A., Throop, G., & Ershowsky, P. (1997). Serial measurements of the rapid-shallow-breathing index as a predictor of weaning outcome in elderly medical patients. *Chest, 112*(4), 1029-1034.

Kuhlen, R., Max, M., Dembinski, R., Terbeck, S., Jurgens, E., & Rossaint, R. (2003). Breathing pattern and workload during automatic tube compensation, pressure support and T-piece trials in weaning patients. *Eur J Anaesthesiol, 20*(1), 10-16.

Laoutaris, I., Dritsas, A., Brown, M. D., Manginas, A., Alivizatos, P. A., & Cokkinos, D. V. (2004). Inspiratory muscle training using an incremental endurance test alleviates dyspnea and improves functional status in patients with chronic heart failure. *Eur J Cardiovasc Prev Rehabil, 11*(6), 489-496.

Larsson, L., Li, X., Edstrom, L., Eriksson, L. I., Zackrisson, H., Argentini, C., & Schiaffino, S. (2000). Acute quadriplegia and loss of muscle myosin in patients treated with nondepolarizing neuromuscular blocking agents and corticosteroids: mechanisms at the cellular and molecular levels. *Crit Care Med, 28*(1), 34-45.

Lee, K H, Hui, K P, & Chang, T. (1994). Rapid shallow breathing did not predict extubation outcome. *Chest, 105*, 540.

Leitch, E. A., Moran, J. L., & Grealy, B. (1996). Weaning and extubation in the intensive care unit. Clinical or index-driven approach? *Intensive Care Med, 22*(8), 752-759.

Leung, P., Jubran, A., & Tobin, M. J. (1997). Comparison of assisted ventilator modes on triggering, patient effort, and dyspnea. *Am J Respir Crit Care Med, 155*(6), 1940-1948.

Li, Y., He, G., & Chen, R. (2000). [Clinical study of weaning predictors in COPD patients with prolonged mechanical ventilation]. *Zhonghua Jie He He Hu Xi Za Zhi, 23*(4), 217-220.

Liaw, M. Y., Lin, M. C., Cheng, P. T., Wong, M. K., & Tang, F. T. (2000). Resistive inspiratory muscle training: its effectiveness in patients with acute complete cervical cord injury. *Arch Phys Med Rehabil, 81*(6), 752-756.

Lisboa, G., Villafranca, C., Leiva, A., Cruz, E., Pertuze, J, & Borzone, G. (1997). Inspiratory muscle training in chronic airflow limitation: effect on exercise performance. *Eur Respir J, 10*(3), 537-542.

MacIntyre, N. R., Cook, D. J., Ely, E. W., Jr., Epstein, S. K., Fink, J. B., Heffner, J. E., . . . Scheinhorn, D. J. (2001). Evidence-based guidelines for weaning and discontinuing ventilatory support: a collective task force facilitated by the American College of Chest Physicians; the American Association for Respiratory Care; and the American College of Critical Care Medicine. *Chest, 120*(6 Suppl), 375S-395S.

Manthous, C. A., Schmidt, G. A., & Hall, J. B. (1998). Liberation from mechanical ventilation: a decade of progress. *Chest, 114*(3), 886-901.

Marini, J J, Smith, T C, & Lamb, V. (1986). Estimation of inspiratory muscle strength in mechanically ventilated patients: The measurement of maximal inspiratory pressure. *Journal of Critical Care, 1*(1), 32-38.

Martin, A. D., Davenport, P. D., Franceschi, A. C., & Harman, E. (2002). Use of inspiratory muscle strength training to facilitate ventilator weaning: a series of 10 consecutive patients. *Chest, 122*(1), 192-196.

Martin, U. J., Hincapie, L., Nimchuk, M., Gaughan, J., & Criner, G. J. (2005). Impact of whole-body rehabilitation in patients receiving chronic mechanical ventilation. *Crit Care Med, 33*(10), 2259-2265.

Mathers-Schmidt, B. A., & Brilla, L. R. (2005). Inspiratory muscle training in exercise-induced paradoxical vocal fold motion. *J Voice, 19*(4), 635-644.

Matic, I., Danic, D., Majeric-Kogler, V., Jurjevic, M., Mirkovic, I., & Mrzljak Vucinic, N. (2007). Chronic obstructive pulmonary disease and weaning of difficult-to-wean patients from mechanical ventilation: randomized prospective study. *Croat Med J, 48*(1), 51-58.

McComas, A. J. (1994). Human neuromuscular adaptations that accompany changes in activity. *Med Sci Sports Exerc, 26*(12), 1498-1509.

McConnell, A. K., & Romer, L. M. (2004). Dyspnoea in health and obstructive pulmonary disease : the role of respiratory muscle function and training. *Sports Med, 34*(2), 117-132.

Meade, M., Guyatt, G., Cook, D., Griffith, L., Sinuff, T., Kergl, C., . . . Epstein, S. (2001). Predicting success in weaning from mechanical ventilation. *Chest, 120*(6 Suppl), 400S-424S.

Mergoni, M., Costa, A., Primavera, S., Salvadori, A., Saccani, A., & Zuccoli, P. (1996). [Assessment of various new predictive parameters of the outcome of mechanical ventilation weaning]. *Minerva Anestesiol, 62*(5), 153-164.

Milner-Brown, H. S., Stein, R. B., & Lee, R. G. (1975). Synchronization of human motor units: possible roles of exercise and supraspinal reflexes. *Electroencephalogr Clin Neurophysiol, 38*(3), 245-254.

Nield, M. A. (1999). Inspiratory muscle training protocol using a pressure threshold device: effect on dyspnea in chronic obstructive pulmonary disease. *Arch Phys Med Rehabil, 80*(1), 100-102.

Noizet, O., Leclerc, F., Sadik, A., Grandbastien, B., Riou, Y., Dorkenoo, A., . . . Leteurtre, S. (2005). Does taking endurance into account improve the prediction of weaning outcome in mechanically ventilated children? *Crit Care, 9*(6), R798-807.

Powers, S. K., Shanely, R. A., Coombes, J. S., Koesterer, T. J., McKenzie, M., Van Gammeren, D., . . . Dodd, S. L. (2002). Mechanical ventilation results in progressive contractile dysfunction in the diaphragm. *J Appl Physiol, 92*(5), 1851-1858.

Ramirez-Sarmiento, A., Orozco-Levi, M., Guell, R., Barreiro, E., Hernandez, N., Mota, S., . . . Gea, J. (2002). Inspiratory muscle training in patients with chronic obstructive pulmonary disease: structural adaptation and physiologic outcomes. *Am J Respir Crit Care Med, 166*(11), 1491-1497.

Robriquet, L., Georges, H., Leroy, O., Devos, P., D'Escrivan, T., & Guery, B. (2006). Predictors of extubation failure in patients with chronic obstructive pulmonary disease. *J Crit Care, 21*(2), 185-190.

Rose, L., & Ed, A. (2006). Advanced modes of mechanical ventilation: implications for practice. *AACN Adv Crit Care, 17*(2), 145-158; quiz 159-160.

Rothaar, R. C., & Epstein, S. K. (2003). Extubation failure: magnitude of the problem, impact on outcomes, and prevention. *Curr Opin Crit Care, 9*(1), 59-66.

Salam, A., Tilluckdharry, L., Amoateng-Adjepong, Y., & Manthous, C. A. (2004). Neurologic status, cough, secretions and extubation outcomes. *Intensive Care Med, 30*(7), 1334-1339.

Sale, D. G. (1988). Neural adaptation to resistance training. *Med Sci Sports Exerc, 20*(5 Suppl), S135-145.

Sanchez Riera, H., Montemayor Rubio, T., Ortega Ruiz, F., Cejudo Ramos, P., Del Castillo Otero, D., Elias Hernandez, T., & Castillo Gomez, J. (2001). Inspiratory muscle training in patients with COPD: effect on dyspnea, exercise performance, and quality of life. *Chest, 120*(3), 748-756.

Sapienza, C. M., Brown, J., Martin, D., & Davenport, P. (1999). Inspiratory pressure threshold training for glottal airway limitation in laryngeal papilloma. *J Voice, 13*(3), 382-388.

Sassoon, C. S., Caiozzo, V. J., Manka, A., & Sieck, G. C. (2002). Altered diaphragm contractile properties with controlled mechanical ventilation. *J Appl Physiol, 92*(6), 2585-2595.

Scheinhorn, D. J., Chao, D. C., & Stearn-Hassenpflug, M. (2000). Approach to patients with long-term weaning failure. *Respir Care Clin N Am, 6*(3), 437-461;vi.

Spitzer, A. R., Giancarlo, T., Maher, L., Awerbuch, G., & Bowles, A. (1992). Neuromuscular causes of prolonged ventilator dependency. *Muscle Nerve, 15*(6), 682-686.

Sprague, S S, & Hopkins, P D. (2003). Use os inspiratory strength training to wean six patients who were ventilator-dependent. *Physical Therapy, 83*(2), 171-181.

Stroetz, R. W., & Hubmayr, R. D. (1995). Tidal volume maintenance during weaning with pressure support. *Am J Respir Crit Care Med, 152*(3), 1034-1040.

Sturdy, G., Hillman, D., Green, D., Jenkins, S., Cecins, N., & Eastwood, P. (2003). Feasibility of high-intensity, interval-based respiratory muscle training in COPD. *Chest, 123*(1), 142-150.

Sudo, E., Ohga, E., Matsuse, T., Teramoto, S., Nagase, T., Katayama, H., . . . Ouchi, Y. (1997). [Duration of effect of pulmonary rehabilitation in elderly patients with chronic obstructive pulmonary disease]. *Nippon Ronen Igakkai Zasshi, 34*(9), 739-742.

Tobin, M. J. (2001). Advances in mechanical ventilation. *N Engl J Med, 344*(26), 1986-1996.

Tobin, M. J., Jubran, A., & Laghi, F. (2001). Patient-ventilator interaction. *Am J Respir Crit Care Med, 163*(5), 1059-1063.

Topin, N., Matecki, S., Le Bris, S., Rivier, F., Echenne, B., Prefaut, C., & Ramonatxo, M. (2002). Dose-dependent effect of individualized respiratory muscle training in children with Duchenne muscular dystrophy. *Neuromuscul Disord, 12*(6), 576-583.

Uijl, S G, Houtman, S, Folgering, H T M, & Hopman, M T E. (1999). Training of the respiratory muscles in individuals with tetraplegia. *Spinal Cord, 37*(8), 575-579.

Vallverdu, I., Calaf, N., Subirana, M., Net, A., Benito, S., & Mancebo, J. (1998). Clinical characteristics, respiratory functional parameters, and outcome of a two-hour T-piece trial in patients weaning from mechanical ventilation. *Am J Respir Crit Care Med, 158*(6), 1855-1862.

van der Esch, M., van 't Hul, A. J., Heijmans, M., & Dekker, J. (2004). Respiratory muscle performance as a possible determinant of exercise capacity in patients with ankylosing spondylitis. *Aust J Physiother, 50*(1), 41-45.

Vassilakopoulos, T., & Petrof, B. J. (2004). Ventilator-induced diaphragmatic dysfunction. *Am J Respir Crit Care Med, 169*(3), 336-341.

Venkataraman, S T, Khan, N, & Brown, A. (2000). Validation of predictors of extubation success and failure in mechanically ventilated infants and children. *Crit Care Med, 28*(8), 2991-2996.

Venkataraman, S. T., Khan, N., & Brown, A. (2000). Validation of predictors of extubation success and failure in mechanically ventilated infants and children. *Crit Care Med, 28*(8), 2991-2996.

Volianitis, S., McConnell, A. K., Koutedakis, Y., & Jones, D. A. (2001). Specific respiratory warm-up improves rowing performance and exertional dyspnea. *Med Sci Sports Exerc, 33*(7), 1189-1193.

Weiner, P., Magadle, R., Beckerman, M., & Berar-Yanay, N. (2002). The relationship among inspiratory muscle strength, the perception of dyspnea and inhaled beta2-agonist use in patients with asthma. *Can Respir J, 9*(5), 307-312.

Yang, K. L. (1993). Inspiratory pressure/maximal inspiratory pressure ratio: a predictive index of weaning outcome. *Intensive Care Med, 19*(4), 204-208.

Yang, K. L., & Tobin, M. J. (1991). A prospective study of indexes predicting the outcome of trials of weaning from mechanical ventilation. *N Engl J Med, 324*(21), 1445-1450.

Yue, G., & Cole, K. J. (1992). Strength increases from the motor program: comparison of training with maximal voluntary and imagined muscle contractions. *J Neurophysiol, 67*(5), 1114-1123.

# 15

# Lipedema

Győző Szolnoky
*Department of Dermatology and Allergology, University of Szeged, Hungary*

## 1. Introduction

Lipedema is a sparsely recognized clinical entity almost entirely affecting the female population. It poses a significant importance as being one of the most common disorder to be mistaken with lymphedema and obesity [1,2,3,4].

## 2. Definition

Lipedema is a disproportional, bilateral, symmetrical fatty swelling of the legs whereas arms are also commonly involved [1,2,3,4,5]. Generally, first symptoms appear around puberty [2,3,4]. Almost solely women are affected and males usually develop lipedema on the basis of hormonal disturbance [6]. The general incidence of lipedema among women is reported to be up to 11% [2]. Lipedema is presumably due to endocrinological and genetic background [2,7,8]. Two prominent hallmarks are the frequent spontaneous or minor trauma-induced bruising and spontaneous or palpation-induced pain [1,2,3,4]. Lipedema, especially in untreated cases, is commonly associated with lymphatic and/or venous insufficiency [2,3,4]. Lipedema is commonly combined with obesity making differential diagnosis considerably difficult [2,3,4]

## 3. Clinical diagnosis

In most cases the diagnosis of lipedem is simply set using patient history and clinical examination [2,3,4]. There is no absolutely unambiguous pathognomonic diagnostic test for lipedema.

### 3.1 Classification

In stage I, the skin looks flat, but the subcutis is thicker and on palpation feels like 'styrofoam balls in a plastic bag'. In stage II walnut- to apple-sized indurations appear and the overlying skin has an irregular surface ('mattress phenomenon'). Stage III shows large, sometimes lobular fat deposits. The location-based classification concurs with the area of the fat deposits: mainly buttocks (type I), buttocks to knees (type II), buttocks to ankles (type III), mainly arms (type IV) and mainly lower legs (type V) [9].

### 3.2 Differential diagnosis

The differential diagnosis of lipedema embraces obesity, various forms of lipohypertrophy and phleb- or lymphedema.

### 3.3 Instrumental diagnosis

These tests comprise waist-to-height ratio as an anthropomethric measure [10], Streeten test for capillary permeability [11] and vacuum suction method with Parrot's angiosterrometer for capillary fragility assessments [2,3,4,12], evaluation of aortic stiffness [13], subjective pain perception [14,15,16]. Accurate techiques as high resolution ultrasonography [17,18], computer tomography [19], magnetic resonance imaging [20,21], lymphoscintigraphy [22,23], fluorescent microlymphography [24] are also useful in diagnosis or making differential diagnosis.

## 4. Clinical management

### 4.1 Evidences in lipedema care

#### 4.1.1 The conservative approach in the intensive therapeutical phase

corresponds to complex decongestive physiotherapy (CDP) consisting of manual lymph drainage (MLD), intermittent pneumatic compression (IPC) as a possible supplementary treatment, physical exercise, multilayered and multicomponent compression bandaging and meticulous skin care [2]. The first observational study disclosed a maximally achieved reduction by CDP as nearly 10% of the original leg girth [25]. In an other clinical setting MLD-based CDP was compared with MLD+IPC along with compression. Each treatment modality turnd out significant limb volume reductions, however no significant difference was proven between the two regimens in that pilot study [26]. Further controlled trials showed that MLD+IPC-based CDP drastically decreased capillary fragility and pain perception of lipedema patients [12].

Gentle forms of liposuction give reliable benefit to lipedema patients without proven relevant damage of lymphatics [27].

#### 4.1.2 Maintenance phase

includes skin care, daily wear of support garments (stockings or hosieries). Nocturnal bandaging might be applied. [28,29,30,31,32].

### 4.2 General description of physiotherapy and compression material

#### 4.2.1 Manual lymph drainage

Although there is a relatively strong empirical body advocating the benefits of MLD, research dat barely support its use in lymphatic insufficiency [33,34,35,36,37] or in lipedema. Beyond its well-known benefits, MLD is able to improve blood microcirculation blood flow, has peripheral analgesic, central sedative, analgesic, vagotonic reaction [38,39,40,41] and improves muscular recovery after physical exercise [42].

#### 4.2.2 Steps of manual lymph drainage after Dr. Vodder in various forms of leg lipedema [33]

**Central treatment:** All proximal lymphatic drainage pathways must be cleared before any attempt to move lymph into them from a more distal area.

Supine position:

Contact with the neck region

Deep abdominal treatment
Treatment of axillar region and inguino-axillar anastomoses of both sides
Decongestion of both lower edematous body quadrants to the direction of axillary lymph nodes of identical sides
Treatment of lower body quadrants (emptying the quadrant)

Prone position:

Treatment of inguino-axillar anastomoses of both sides
Decongestion of gluteal region to the direction of axillary lymph nodes of identical sides
Treatment of lower body quadrants (emptying the healthy quadrant)

**Leg treatment**

Supine position:

Stimulation of inguinal lymph nodes
Decongestion of the lateral side of the thigh
Drainage from the lateral to the medial side of the thigh
Drainage from the distal to the proximal part of the thigh
Stimulation of the knee region
Drainage from the distal to the proximal part of the calf
Drainage from the distal to the proximal part of the foot

Prone position:

It is consistent with the previous steps in supine position, however the stimulation and decongestion of retromalleolar region is also mandatory.
Leg treatment steps should be repeated so as to treat all regions several time

### 4.3 Intermittent pneumatic compression

IPC is assumed to reduce oedema by decreasing capillary filtration, rather than by accelerating lymph flow. IPC alone is particularly effective in nonobstructive edemas like lipedema. Clinical trials prefer the utilization of multi-chambered pumps to single-chambered ones [43,44], however pressures should be adjusted according to individual response. In general, pressures of 30-60 mmHg are mostly applied, however higher pressures also improve limb edema and lower pressures (20-30 Hgmm) are advised in palliative care and a duration of 30 minutes to two hours daily is recommended [45,46]. IPC is outstandingly efficacious in the edema treatment of immobile patient [47]. It is able to squeeze the water content of a lymphedematous extremity without improving the lymphatic drainage to an adequate amount leading to an increase of the oncotic tissue pressure necessitating a continuation of compression therapy [48]. IPC may exacerbate or cause congestion at the noncompressed root of a treated limb and also in the adjacent, genital region [49]. Recent examinations showed remarkable induction of lymph return in lymphedema [50].

### 4.4 Compression

Compression therapy is recognized as one of the most effective treatment modalities in the management of venous and lymphatic disease. Until recently, evidence of efficacy was based mostly on empirical study. Experimental data on the effect of conventional compression therapy on lymphedema are sparse. Effects of compression include shape and volume restoration and remodelling, improvement of skin changes, elimination of lymphorrhoea, subcutaneous tissue softening. There is a strong recommendation for the management of acute deep vein thrombosis, chronic venous insufficiency and lymphoedema with compression therapy [51,52,53]. Patients with lower limb lymphoedema with reduced ankle-brachial pressure index (ABPI ) of 0.5-0.8 should not receive sustained compression exceeding 25 mmHg. Patients with ABPI <0.5 can mostly receive only intermittent compression. [54] Compression therapy comprises the use of compression bandages or stockings.

#### 4.4.1 Multilayer compression

It is a treatment of choice when we use bandages to achieve appropriate compression. The principles are seen as follows:

1. Skin care prevents dryness and consequent microinjuries of the epidermis
2. Finger or toe bandaging may prevent or reduce toe swelling
3. Tubular cotton bandage provides a protective layer between the skin and other bandages
4. Subbandage padding (soft synthetic wool, foam roll or sheet) protects the skin and subcutaneous tissues, normalizes shape, protects bony prominences and equalizes the distribution of pressure produced by outermost bandage layers.
5. Dense foam is applied locally to soften hard areas of tissue thickening and fibrosis or areas particularly vulnerable to oedema (ankles).
6. Inelastic bandages (non-adhesive, adhesive or cohesive bandages) should be applied at nearly full extension and 50% overlap. Elastic bandages should be applied at 50% extension and overlap. Several pieces of bandages are used to achieve desired pressure.
7. Adhesive tapes or cohesive bandages fix compression bandages and stabilize the whole compression system

To achieve optimal volume reduction, high initial interface pressures are necessary to compensate pressure decrease. The pressure drop is already significant after 2 hours and mainly caused by volume reduction explaining the need for a more frequent bandage change in the beginning of lymphedema therapy compared to the current practice where change of bandage system is recommended once a day in the initial phase [55].

In general, inelastic compression depending on the strength of tensile pressure and stiffness can be worn overnight without major influence on microcirculation, hence subbandage pressure does not significantly interfere with capillary function in supine position.

Unlike inelastic compression, the nocturnal wear of elastic bandages is normally not recommended due to the relatively high interface pressure in supine position and in case of diminished arterial influx severe occlusion can occur.

When compression with bandages is applied, exerted pressure can be calculated using Laplace's law: $P=T \times N / C \times W$ (P = sub-bandage pressure (mmHg), T = bandage tension (kilograms force), N = number of layers, C = limb circumference (cm), W =bandage width (cm).

Pressure, LAyers, Components, and Elastic properties (P-LA-C-E) are the main factors that have to be taken into consideration when a compression bandage is applied.

Pressure measured in vivo in the medial gaiter area in the supine position for training purposes may be classified into the following categories:

mild (<20 mmHg), moderate (20–40 mmHg), strong (40–60 mmHg), or very strong (>60 mmHg).

Layers: A double-layer bandage is characterized by an overlap of at least 50%. Components of a bandage consist of same or different materials that may have different functions (padding, protection, retention). Elasticity: The elastic properties of a single bandage may be inelastic (rigid bandages or short-stretch bandages) or elastic (long-stretch bandages). Inelastic bandage extensibility is under 100% of the original length. Several layers of material (either identical or different materials) have the tendency to make the bandage system stiffer. In the lympholgy practice we encourage the application of inelastic bandages because they exert high working pressure in active physical exercises and depending on tension forces during pulling of bandages evoke relatively low resting pressure at lying position.

Multilayer bandage systems may behave as inelastic systems even though the individual layers act as elastic materials due to the friction generated between bandage layers. Therefore, in case of multilayer bandage systems and kits, the terms "high or low stiffness" should be used to describe the behavior of the final bandage. Stiffness is defind as the increase of interface pressure measured in the gaiter area when standing up from the supine position. A raise in the pressure of more than 10 mmHg measured in the gaiter area is characteristic of a stiff bandage system [53].

### 4.4.2 Use of elastic bandages

In some situations (ineffective calf muscle pump, phlebolymphedema, large volume loss is predicted, extremely large adipose tissue deposition and low potential of muscle pump), the inelastic bandages may be replaced with elastic ones. The stiffness produced by multiple layers produces high working pressure. However, the resting pressure is usually higher than with inelastic systems at the same strength of application of both bandage systems.

### 4.5 Medical compression stockings

Compression garment utilization comprises the the long-term management of lipedema in maintenance phase. Limbs with relatively normal shape require round-knitted stockings while flat-knitted stockings better fit to limbs with unusual shape or remarkable distorsion than round-knitted ones. Appropriate and well-fitting medical compression stockings in lipedema are mostly custom-made because lipedematous legs usually do not belong to average sizes.

In general, compression stockings have a lower stiffness index than inelastic bandages, especially when these bandages are worn in a multilayered fashion. MCSs drop their pressure to a much less degree compared to compression bandages [56,57,58].

### 4.6 Exercise

Exercise/movement should be tailored to the patient ability and disease status. Compression should be worn during exercise whenever it is possible. Walking, swimming, cycling and low impact aerobics are recommended.

## 5. Prognosis

Lipedema is a progressive disorder but in few cases may remain self-limited. Early diagnosis and treatment are cornerstones for this disorder otherwise gradual enlargement of massive fatty tissue causes severely compromised mobility, debilitating condition and further co-morbidities like arthrosis and lymphatic insufficiency. Interlobar areas may become susceptible for infections that may further progress to cellulitis or septicaemia. Lipedema has remarkable psychological impact ranging from mild upset to severe anxiety, depression or even anorexia [2,3,4].

## 6. References

[1] Allen, EV, EA Hines. Lipedema of the legs: a syndrome characterized by fat legs and orthostatic edema. Mayo Clin. Proc 1940:15:184-7.
[2] Földi E, Földi M. [Das Lipödem] in Földi M, Kubik S eds, Lehrbuch der Lymphologie, 5nd edn, Chapt 9. München-Jena:Gustav Fischer, 2002:449-58.
[3] Langendoen SI, Habbema L, Nijsten TEC, Neumann HAM. Lipoedema: from clinical presentation to therapy. A review of the literature. Br J Dermatol 2009:161:980–6.
[4] Todd M. Lipoedema: presentation and management. Br J Community Nurs 2010:15:S10-6.
[5] Meier-Vollrath I, Schneider W, Schmeller W: Lipödem: [Verbesserte Lebensqualität durch Therapiekombination]. Dtsch Ärtzteblatt 2005:102:1061-7.
[6] Chen S, Hsu SD, Chen TM, Wang HJ. Painful fat syndrome in a male patient. Br J Plast Surg 2004: 57:282–6.
[7] Child AH, Gordon KD, Sharpe P, Brice G, Ostergaard P, Jeffery S, et al. Lipoedema: an inherited condition. Am J Med Genet Part A:152A:970–6.
[8] Szolnoky G, Kemeny L. Lipoedema: from clinical presentation to therapy: Further aspects. Br J Dermatol 2010:162:889.
[9] Meier-Vollrath I, Schneider W, Schmeller W: [Das Lipödem: neue Möglichkeiten der Therapie]. Schweiz Med Forum 2007:7:150–5.
[10] Herpertz U. [Adipositas-Diagnostik in der Lymphologie]. LymphForsch 2009:13:90-3
[11] Streeten DH. Idiopathic edema. Pathogenesis, clinical features, and treatment. Endocrinol Metab Clin North Am 1995:24:531-47.
[12] Szolnoky G, Nagy N, Kovács RK, Dósa-Rácz E, Szabó A, Bársony K, et al. Complex decongestive physiotherapy decreases capillary fragility in lipedema. Lymphology 2008:41:161-6.
[13] Nemes A, Gavallér H, Csajbók É, Forster T, Csanády M. Obesity is associated with aortic enlargement and increased stiffness: an echocardiographic study. Int J Cardiovasc Imaging 2008:24:165-71.
[14] Schmeller W, Meier-Vollrath I. [Schmerzen beim Lipödem: Versuch einer Annäherung]. LymphForsch 2008:12:7-11.
[15] Hamner, JB, Fleming MD. Lymphedema therapy reduces the volume of edema and pain in patients with breast cancer. Ann. Surg. Oncol. 2007:14:1904-8.
[16] Wong DL, Baker CM: Smiling faces as anchor for pain intensity scales. Pain 2001:89:295-300.
[17] Marshall M, Schwahn-Schreiber C. [Lymph-, Lip- und Phlebödem. Differenzialdiagnostische Abklärung mittels hochauflösender Duplexsonographie]. Gefässchirurgie 2008:13:204–12.
[18] Naouri M, Samimi M, Atlan M, Perrodeau E, Vallin C, Zakine G et al. High resolution cutaneous ultrasonography to differentiate lipoedema from lymphoedema. Br J Dermatol 2010 Apr 16 [Epub ahead of print].
[19] Vaughan BF. CT of swollen legs. Clin Radiol 1990:41:24–30.
[20] Dimakakos PB, Stefanopoulos T, Antoniades P, . MRI and ultrasonographic findings in the investigation of lymphedema and lipedema. Int Surg 1997:82:411–6.

[21] Lohrmann C, Foeldi E, Langer M. MR imaging of the lymphatic system in patients with lipedema and lipo-lymphedema. Microvasc Res 2009:77:335-9.
[22] Brauer WJ. [Altersbezogene Funktionslymphszintigraphie beim Lipödem und Lipolymphödem]. LymphForsch 2000:4: 74-7.
[23] Harwood CA, Bull RH, Evans J, Mortimer PS. Lymphatic and venous function in lipedema. Br J Dermatol.1996:134:1–6.
[24] Amann-Vesti BR, Franzeck UK, Bollinger A. Microlymphatic aneurysms in patients with lipedema. Lymphology 2000:134:170-5.
[25] Deri G, Weissleder H: [Vergleichende prä- und posttherapeutische Volumenmessungen in Beinsegmenten beim Lipödem]. Lymph Forsch 1997: 1: 35–7.
[26] Szolnoky G, Borsos B, Bársony K, Balogh M, Kemény L. Complete decongestive physiotherapy of lipedema with or without pneumatic compression: a pilot study. Lymphology 2008:41:50-2.
[27] Schmeller W, Meier-Vollrath I. Tumescent liposuction: a new and successful therapy for lipedema. J Cutan Med Surg 2006:10:7-10.
[28] Wozniewski M, Jasinski R, Pilch U, Dabrowska G: Complex physical therapy for lymphoedema of the limbs. Physiotherapy 2001:87:252-6.
[29] Wozniewski M, Pilch U, Jezierski C: Physiotherapy for patients with lower extremity lymphoedema. Phlebology 1995: suppl 1:1051-3.
[30] Lymphoedema Framework. Best Practice for the Management of Lymphoedema. International consensus. London: MEP Ltd, 2006:3-52.
[31] Ko TS, Lerner R, Klose G, Cosimi AB: Effective treatment of lymphedema of the extremities Arch Surg 1998:133: 452-8.
[32] Vignes S: Physical therapy in limb lymphedema La kinésithérapie dans le traitement des lymphoedémes des membres). Ann Dermatol Venereol 2005:132:185-7.
[33] Strössenreuther RHK. [Hinweise zur Durchführung der ML/KPE bei primären und sekundären Lymphödemen sowie weiteren ausgewahlten Krankheitsbildern] in Földi M, Kubik S eds, Lehrbuch der Lymphologie, 5nd edn, Chapt 19. München-Jena:Gustav Fischer, 2002:621-658.
[34] Badger C, Preston N, Seers K, Mortimer P. Physical therapies for reducing and controlling lymphoedema of the limbs. Cochrane Database Syst Rev 2004: 4: CD003141.
[35] McNeely ML, Magee DJ, Lees AW, Bagnall KM, Haykowsky M, Hanson J. The addition of manual lymph drainage to compression therapy for breast cancer related lymphedema: a randomized controlled trial. Breast Cancer Res Treat 2004:86: 95-106.
[36] Johansson K, Albertsson M, Ingvar C, Ekdahl C. Effects of compression bandaging with or without manual lymph drainage treatment in patients with postoperative arm lymphedema. Lymphology 1999:32:103-10.
[37] Vignes S, Coupé M, Baulieu F, Vaillant L;Groupe Recommandations de la Société Francaise de Lymphologie. Limb lymphedema: Diagnosis, Exploration, complications. J Mal Vasc 2009:34:314-22.
[38] Hutzschenreuter P, Brümmer H, Ebberfeld K. Experimental and clinical studies of the mechanism of effect of manual lymph drainage therapy. Z Lymphol 1989:13:62-64.
[39] Trettin H. Manual lymph drainage in migraine treatment - a pathophysiologic explanatory model. Z Lymphol 1989:13:48-53.
[40] Hutzschenreuter P, Ehlers R. Effect of manual lymph drainage on the autonomic nervous system. Z Lymphol 1986:10:58-60.
[41] Kurz W, Wittlinger G, Litmanovitch YI, Romanoff H, Pfeifer Y, Tal E, et al. Effect of manual lymph drainage massage on urinary excretion of neurohormones and minerals in chronic lymphedema. Angiology 1978:29:764-72.

[42] Schillinger A, Koenig D, Haefele C, Vogt S, Heinrich L, Aust A, et al. Effect of manual lymph drainage on the course of serum levels of muscle enzymes after treadmill exercise. Am J Phys Med Rehabil 2006:85:516-520
[43] International Society of Lymphology. The diagnosis and treatment of peripheral lymphedema.Consensus document of the International Society of Lymphology. Lymphology 2009:42:51-60.
[44] Bergan JJ, Sparks S, Angle N. A comparison of compression pumps in the treatment of lymphedema. J Vasc Surg 1998:32:455-62.
[45] Partsch H, Mostbeck A, Leitner G. Experimentelle untersuchungen zur Wirkung einer Druckwellenmassage (Lymphapress) beim Lymphödem. Phlebol. U Proktol 1980:9:124-8.
[46] Szuba A, Achalu R, Rockson SG. Decongestive lymphatic therapy for patients with breast carcinoma-associated lymphedema. A randomised, prospective study of a role of adjunctive pneumatic compression. Cancer 2002:95:2260-7.
[47] Partsch H: Intermittent pneumatic compression in immobile patients. Int Wound J 2008:5:389-97.
[48] Miranda F Jr, Perez MC, Castiglioni ML, Juliano Y, Amorim JE, Nakano LC, et al. Effect of sequential intermittent pneumatic compression. Lymphology 2001:34:135-41.
[49] Boris M, Weindorf S, Lasinski BB. The risk of genital edema after external pump compression for lower limb lymphedema Lymphology 1998:31:15-20.
[50] Olszewski WL, Cwikla J, Zaleska M, Domaszewska-Szostek A, Gradalski T, Szopinska S. Pathways of lymph and tissue fluid flow during intermittent pneumatic massage of lower limbs with obstructive lymphedema.Lymphology. 2011 Jun;44(2):54-64.
[51] Badger CM, Peacock IL, Mortimer P. A randomised controlled parallel group clinical trial comparing multilayer bandaging followed by hosiery *versus* hosiery alone in the treatment of patients with lymphoedema of the limb. Cancer 2000:88:2832-7.
[52] Badger C, Seers K, Preston N, Mortimer P. Physical therapies for reducing and controlling lymphoedema of the limbs. The Cochrane Library, Issue 2, 2003. Oxford: Update Software.
[53] Partsch H, Clark M, Mosti G, Steinlechner E, Schuren J, RN, Abel M, et al. Classification of Compression Bandages: Practical Aspects. Dermatol Surg 2008:34:600-9.
[54] Marston W, Vowden K. Compression therapy: a guide to safe practice. In: European Wound Management Association (EWMA). Position Document: Understanding compression therapy. London: MEP Ltd, 2003: 11-7.
[55] Damstra RJ, Brouwer E, Partsch H. Controlled, comparative study of relation between volume changes and interface pressure under short-stretch bandages in leg lymphedema patients. 2008:34:773-9.
[56] Partsch H, Partsch B, Braun W. Interface pressure and stiffness of ready made compression stockings: Comparison of in vivo and in vitro measurements. J Vasc Surg 2006:44:809-14.
[57] Vignes S. Elastic garment in management of lymphedema (Interet de la compression élastique dans le traitement des lymphoedémes). Ann Dermatol Venereol 2005:132:290-1.
[58] Larsen AM, Futtrup I. Watch the pressure - it drops! EWMA Journal 2004:4:8-12.

# Part 7

# Physical Therapy for Patients with Stroke

# 16

# Running a Complementary Stroke Rehabilitation and Aftercare Program: Experiences of Four European Centres

Wilfried Schupp[1,3,4], Napasakorn Komaratat[1,2] and Simon Steib[3]
[1]*m & i Fachklinik Herzogenaurach, Department of Neurology,*
[2]*Sirindhorn National Medical Rehabilitation Centre,*
[3]*Institute for Sports Science and Sports,*
*Friedrich-Alexander University Erlangen-Nuremberg,*
[4]*On Behalf of the CERISE Study Group*
[1,3,4]*Germany*
[4]*Belgium*
[4]*Switzerland*
[4]*United Kingdom*
[2]*Thailand*

## 1. Introduction

Stroke is one of the major public health concerns throughout the world. It is a leading cause of long-term disability (Shah, 2006), consuming a large amount of health care resources. In Western Europe, the incidence varies between 1.0 and 3.75 per 1000 habitants (Weir et al., 2001). The proportion of patients who die or are dependent at six months after stroke ranges from 20 - 41% (Brainin et al., 2000). The consequences after stroke are not only persistent neurological impairment which lead to physical,functional and psychosocial problems, but also life-time disability that requires rehabilitation to enable optimal function. There is strong evidence that specialized interdisciplinary stroke rehabilitation units are correlated with improved functional outcomes, mortality reduction, shorter lengths of hospital stay, and reduced need for long term institutionalization in moderate to severe stroke patients (Teasell et al., 2009). The benefits of specialized interdisciplinary stroke rehabilitation can be attributed to the fact that they emphasize specialized training for stroke rehabilitation inpatients, while an emphasis on working as a complementary rehab team ensures that diverse skills can be applied to deal with the complex needs of the stroke patient. Specialized interdisciplinary stroke rehabilitation units currently represent the "gold standard" in stroke rehabilitative care (Teasell et al., 2009). There are many specialized interdisciplinary stroke rehabilitation units, running throughout the European countries. Regarding the concepts of Health care provision which is composed of structures of care, processes of care, and outcomes (Donabedian, 1988), each organized stroke rehabilitation centre has its own practice ( structure of care and process of care) which will effect on its outcome and make a difference to other centres. Therefore, comparing different stroke rehabilitation systems (multicentre study) in fields of practice and outcome from various centres between European countries allows for greater appreciation of alternative approaches and will assist in determining optimal strategies for improving outcomes and rehabilitation services (Markus, 2004).

The CERISE project (Collaborative Evaluation in Rehabilitation of Stroke across Europe) was a multicentre study that compared stroke care practices and their outcomes among 4 European centres.

## 2. Objectives

The European project CERISE (Collaborative Evaluation in Rehabilitation of Stroke across Europe) is a prospective longitudinal multicentre cohort study which mainly aims to gain insight into various clinical and organizational aspects of stroke rehabilitation and care by comparing inpatient stroke outcome, recovery patterns, process and structure of stroke rehabilitation and care during 6 months after stroke between four European countries: United Kingdom, Switzerland, Germany and Belgium.

This project consisted of 2 main components:

1. To document multidimensional clinical profile including motor, functional, emotional recovery and socioeconomic aspects of stroke patients until 6 months after stroke and to compare the results between the rehabilitation centres. This category comprised of 4 sub- studies.

   1.1 The Use of a Biplot in Studying Outcomes After Stroke (De Wit et al.,2009)
   1.2 Motor and Functional Recovery After Stroke (De Wit et al.,2007)
   1.3 Anxiety and depression in the first six months after stroke (De Wit et al.,2008)
   1.4 Effect of socioeconomic status on functional and motor recovery after stroke (Putman et al.,2007)

2. To document and compare the results between the centers in the provision of stroke services and in subject of the organizational characteristics of the different rehabilitation units. This includes consideration of following sub- studies.

   2.1 The use of time by stroke patients (De Wit et al.,2005)
   2.2 Amount of time spent by therapy ,emphasized on physiotherapy and occupational therapy (Putman et al.,2006)
   2.3 The content of physiotherapy & occupational therapy (De Wit et al.,2006)
   2.4 Admission criteria to stroke rehabilitation units (Putman et al.,2007)
   2.5 Variation in follow- up services after inpatient stroke rehabilitation (Putman et al.,2009)

## 3. Material and methods

### 3.1 Main study design: Longitudinal multicentre cohort study

#### 3.1.1 Settings and subjects

The CERISE project was conducted in 4 European countries. Data was collected in 6 stroke rehabilitation units (SRUs) as followed:

- The Fachklinik, Herzogenaurach, Germany (SRU-DE)
- Queen's Medical Centre and City Hospital, Nottingham, UK (analysed together, SRU-UK)
- 2 SRUs at the University Hospital Pellenberg, Belgium (analysed together, SRU-BE )
- The Rehab Clinic, Zurzach Switzerland (SRU-CH)

Each stroke rehabilitation unit has the provision of inpatient multidisciplinary care. Patients were transferred to the units from an acute setting. Between March 2002 and September 2004, all consecutive patients fitting the following inclusion criteria were recruited:

1. First-ever stroke as defined by WHO (WHO MONICA project,1988)
2. 40 to 85 years of age
3. Score on Gross Motor function of the Rivermead Motor Assessment (Lincoln & Leadbitter, 1979) (RMA-GF) ≤ 11 and/or a score on Leg and Trunk function (RMA-LT) ≤ 8 and/or a score on Arm function (RMA-AR) ≤ 12 on admission to the rehabilitation centre

The exclusion criteria were:

1. Other neurological impairments with permanent damage
2. Stroke-like symptoms attributable to subdural hematoma, tumor, encephalitis, or trauma
3. Pre-stroke Barthel Index (BI) (Mahoney & Barthel,1965) < 50
   (To be able to distinguish between pre- existing disabilities and disabilities resulting from the stroke)
4. Admitted to the rehabilitation centre more than 6 weeks after stroke
5. No informed consent
   The study was approved by the ethic committee for each centre.

### 3.1.2 Data collection

The basic documented data of all studies in the CERISE project were collected in similar pattern. These data comprised of patients'age, gender, time between stroke onset and admission assessment (TSOA), prestroke disability (assessed by Barthel Index), type of stroke & side of impairment (determined on the basis of NMR/CT and clinical examination), severity of neurological deficit (NIHSS : range 0-42), comorbidities, urinary incontinence and swallowing problems. Also the occurrence of dysarthria and dysphasia using items from the National Institutes of Health Stroke Scale and length of stay (LOS) were documented. Educational level, equivalent income, place of residence were recorded in the study as parameters of socioeconomic status on functional and motor recovery after stroke and the study of variations in follow-up services after inpatient stroke rehabilitation.

### 3.1.3 Outcome measurement and data analysis

Each of the sub-studies has different study design, outcome measurement and analysis method. All data and outcome were analyzed and corrected by reliable statistic methods which were suitable for each individual study and variations in case-mix were corrected using multiple regression analysis. They were summarized in Table 1.

| Outcomes | Measurement /Assessment methods or Study design | Time of measurement/ assessement | Details/Description |
|---|---|---|---|
| **Motor recovery pattern** | RMA (Rivermead Motor Assessment)<br>• RMA - GF (Gross function)<br>• RMA - LT (Leg/Trunk)<br>• RMA- A (Arm) | On admission, at discharge, 2 ,4, 6 months after stroke | The RMA-GF into 5 classes (0 -2,3-5, 6-7, 8-9, 10-13)<br>RMA-LT into 4 classes (0 -3, 4-6, 7-9,10)<br>RMA-A into 5 classes (0 -1, 2-5, 6-9, 10-12, 13-15) |
| **Self care and function recovery pattern** | Barthel Index (BI)<br>The Nottingham Extended Activities of Dailay Living (NEADL) (Nouri & Lincoln,1978) | On admission, at discharge, 2 ,4, 6 months after stroke | BI into 5 classes (0 -20, 25-40, 45-60, 65-80, 85-100)<br>NEADL into 6 classes (0 -2, 3-5, 6-8, 9-11, 12-16, 17-22) |
| **Emotional recovery pattern** | Hospital Anxiety and Depression Scale (HADS) (Zigmond & Snaith,1983) | 2, 4, 6 months after stroke | Measuring the symptoms of anxiety and depression, not provided by a psychiatric diagnosis<br>The score ≥ 8 were to be considered anxious or depressed |
| **Quality of life** | • Euroqol ( EQ ) (The EuroQol Group,1990)<br>• Euroqol-Health State ( EQ- HS)<br>• Euroqol-Visual Analogue Scale ( EQ - VAS ) | 2, 4, 6 months after stroke | |
| **The burden on the patients' principal caregiver** | Caregiver Strain Index ( CSI ) (Robinson,1983) | 2, 4, 6 months after stroke | |
| **Socioeconomic** | • Education level International Standard Classification of Education ( ISCED) (OECD,1999)<br>• Equivalent income : monthly household income and household composition | At discharge by a structured interview | • Education level : 2 classes Low education : ISCED classification of 0-2 High education : ISCED classification of 3 or higher<br>• Income: 3 categories low, moderate ,high |

Table 1. Outcome measurement methods in each sub-studies.

| Outcomes | Measurement /Assessment methods or Study design | Time of measurement/ assessement | Details/Description |
|---|---|---|---|
| **Time in therapy** | 60 randomly selected stroke patients were observed at 10-minute intervals using behavioral mapping in each centre. | 30 randomly selected days, equally distributed over the 5 weekdays to cover regular activities | Behavioral mapping was used to document patients' activity, location, and interaction |
| **Task characteristics of physiotherapists and occupational therapists** | Documented diary by PT and OT :Activity,number of patients,number of stroke patients, involvement of other people,location and frequency of each activity. | The therapists documented their activities in 15 min periods. | Task characteristics of physiotherapists and occupational therapists |
| **Contents of therapy** | Recording individual 15 PT and 15 OT sessions using a mobile camera. Five experienced PT and Five experienced OT verified that the list contained all activities ( 53 activities in 12 categories) performed when treating stroke patients. | | |
| **Admission criteria to inpatient rehabilitation unit** | Each medical consultants do Questionnaires (To document the impact of clinical and non-clinical factors on the admission ) followed by a qualitative round of semi-structured interviews | | Questionnaires: 3 main categories<br>1. Factors related to the patient: physical condition, cognition,psychological behavioral, socioeconomic conditions and patient's network.<br>2. Factors related to the network between facilities<br>3. Factors related to the referring hospital<br>The medical consultants were asked to score the impact of each identified factor on the admission on a 4-point scale, ranging from no effect ,low effect,high effect and very high effect. |

Table 1. Outcome measurement methods in each sub-studies. (Continuation)

| Outcomes | Measurement /Assessment methods or Study design | Time of measurement/ assessement | Details/Description |
|---|---|---|---|
| **Follow-up sevices after inpatient stroke rehabilitation** | Structured interview patients or carers to document:<br>• place of residences<br>• the services post discharge | At 2,4, 6 months post- stroke | Place of residences 4 categories: living at home<br>1. In rehabilitation centre<br>2. In acute hospital<br>3. Institutionalized<br>4. ( long-term facility)<br>The services 6 types:<br>1. PT<br>2. OT<br>3. nursing care<br>4. speech & language therapy<br>5. psychological therapy<br>6. general practitioner (GP) |

Table 1. Outcome measurement methods in each sub-studies. (Continuation)

## 4. Results

Between March 2002 and September 2004, 1297 stroke patients were admitted to the 4 centres. Of these, 765 patients were excluded due to inclusion and exclusion criteria and the remaining 532 were included in the study. Patients' characteristics and prognostic data were compared between 4 centres (Table 2.)

There were some significant data differences. Patient in the UK and Swiss centres were significantly older than those in the Belgian and German centres. In the German centre, there were significantly more male patients compared with the other centres. Patients in the UK centre have shorter time between stroke onset and admission assessment and length of stay when compared with the other centres, while length of stay was significantly longer in the Belgian Centre compared to the other centres. No significant differences were found between centres for side of impairment,type of stroke, or prestroke BI. Comparison of prognostic data between centres revealed that patients in the Swiss and German centres had significantly less severe stroke compared with the others,reflected in higher initial BI, RMA-GF,RMA-LT, and RMA-A scores. There were significantly more patients with urinary incontinence in the Belgian and UK centres. More swallowing problems in UK centre were presented, whereas dysarthria occurred significantly more in the Belgian centre and dysphagia was found more often in the Swiss centre.

According to missing data at the each period of assessment , the number of eligible patients were adjusted and analysed by each of study. The study of using a Biplot in studying outcomes after stroke and the study of motor and functional recovery after stroke showed 69 patients were lost to follow-up 6 months after stroke; 18 died, 46 refused to participate, and 5 could not be assessed (missed assessment, poor medical condition). Hence, assessments were carried out for 463 patients. For the study of anxiety and depression in the first six months after stroke, 27 patients did not complete the HADS at any time, resulting in 505 remaining patients but complete HADS data could be collected for 435 ( 86%) patients. 113 patients were excluded from the analyses in the studies of effect of socioeconomic status and on functional and motor recovery after stroke.

| | All centre n = 532 | Belgian Centre n = 127 | UK Centre n = 135 | Swiss Centre n = 135 | German Centre n = 135 | P |
|---|---|---|---|---|---|---|
| Characteristics | | | | | | |
| Age,y : mean | 69.5 | 67.8 | 72.0 | 71.7 | 66.5 | <0.0001* |
| SD | 10.3 | 11.2 | 9.5 | 9.6 | 9.8 | |
| Men: n (%) | 283 (53) | 57 (45) | 66 (49) | 72 (53) | 88 (65) | 0.006† |
| **Side of impairment: n (%)** | | | | | | |
| Left | 285 (54) | 69 (54) | 81 (60) | 67 (50) | 68 (50) | 0.28† |
| Right | 227 (43) | 51 (40) | 53 (39) | 62 (46) | 61 (45) | |
| Both | 20 (3) | 7 (6) | 1 (1) | 6 (4) | 6 (4) | |
| **Type of stroke : n (%)** | | | | | | |
| Hemorrhage | 77 (14) | 24 (19) | 15 (11) | 24 (18) | 14 (10.5) | 0.33ŧ |
| Ischemic infarcts | 445 (84) | 101 (79.5) | 117 (87) | 108 (80) | 119 (88) | |
| Unspecified | 10 (2) | 2 (1.5) | 3 (2) | 3 (2) | 2 (1.5) | |
| **Prestroke Barthel : median (IQR)** | 100 (100-100) | 100 (100-100) | 100 (100-100) | 100 (100-100) | 100 (100-100) | 0.09§ |
| **Urinary incontinence: n (%)** | 149 (28) | 49 (39) | 27 (35) | 24 (18) | 29 (21) | 0.0002† |
| **Swallowing problems: n (%)** | 106 (20) | 23 (18) | 41 (30) | 19 (14) | 23 (17) | 0.004† |
| **Dysarthria: n (%)** | 223 (42) | 66 (52) | 46 (34) | 52 (39) | 59 (44) | <0.0001† |
| **Dysphasia: n (%)** | 178 (33) | 36 (28) | 35 (26) | 74 (55) | 33 (24) | 0.02† |
| **TSOA (days): median (IQR)** | | 23 (17-31.3) | 12 (8-19) | 19 (15-22.5) | 20 (16-27) | <0.0001§ |
| **LOS (days): median (IQR)** | | 66 (39.5-98) | 44.5 (20-78) | 52.5 (29-77) | 49 (35-71) | 0.0004§ |
| **Initial BI: median (IQR)** | | 40 (25-60) | 45 (25-80) | 70 (30-90) | 75 (50-90) | <0.0001§ |
| **Initial RMA-GF: median (IQR)** | | 2 (1-6) | 2 (1-5) | 6 (1-9) | 8 (4-10) | <0.0001§ |
| **Initial RMA-LT:median (IQR)** | | 4 (1-7) | 4 (1-7) | 6 (3-9) | 7 (5-9) | <0.0001§ |
| **Intial RMA-A : median (IQR)** | | 2 (0-9) | 3 (0-11) | 7 (1-11) | 7 (1-11) | 0.0005§ |
| *ANOVA, †$x^2$, ŧexact,and §Kruskal-Wallis test with post-hoc analysis. | | | | | | |
| IQR indicates,interquartile ranges; LOS, length of stay; TSOA, time between stroke onset and assessment on admission | | | | | | |
| Barthel Index; (Q1- Q3) = (Quartile 1- Quartile3) | | | | | | |

Table 2. Comparison of Demographic and Prognostic Data Between Centres.

### 4.1 Using a Biplot in studying outcomes after stroke (De Wit et al., 2009)

Use of a Biplot aimed to obtain a global picture of the relation between different multidimentional 6 months poststroke outcomes. This study revealed 2 clusters (Figure1): The first cluster showed the high association between the 3 sections of the RMA (Rivermead Motor Assessment ; RMA -GF, RMA - LT, RMA- A) and the Extended Activities of Daily Living (EADL) which means that a strong association exists between motor function and dependency in activities of daily life beyond personal self care. A second cluster was formed by HADS-A,HADS-D,and EQ-VAS which showed strong association between anxiety, depression, and quality of life. The results also indicated that at 6 months after stroke, patients with BI ( Barthel Index ) < 45 had lower RMA-GF,RMA-LT,RMA-A,EQ-HS, and EQ-VAS scores but higher HADS-A and HADS-D scores compared to the others and also higher Caregiver Strain Index ( CSI ) scores when compared to the other caregivers. In contrast, the patients with initial BI score >60 had higher RMA-GF,RMA-LT,EQ-HS, and EQ-VAS sores whereas lower CSI scores and large variation of the HADS-A and HADS-D scores. This result concluded that patients with a low functional status could have higher

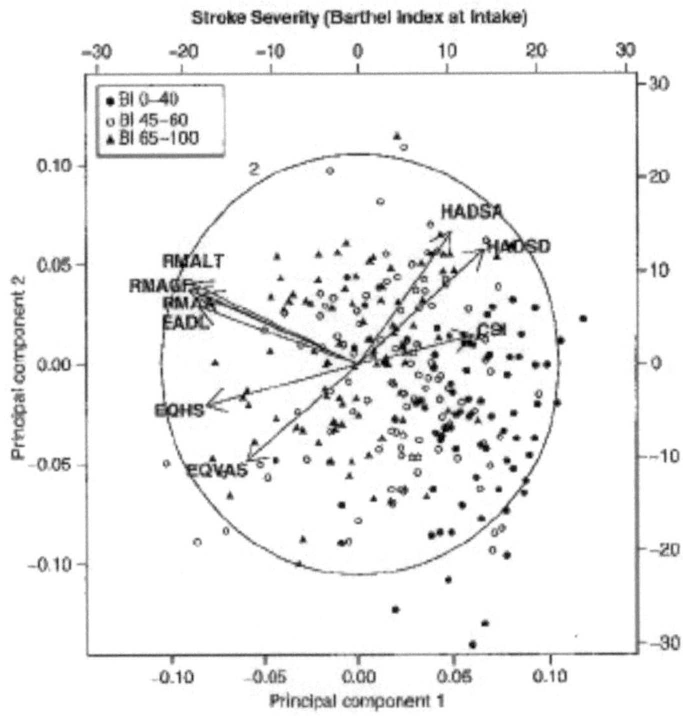


Abbreviations: BI, Barthel Index; CSI, Caregiver Strain Index; EQ-HS, EuroQol–Health State; EQ-VAS, EuroQol–Visual Analogue Scale; HADS-A, Hospital Anxiety and Depression Scale-Anxiety; HADS-D, Hospital Anxiety and Depression Scale–Depression; RMAGF, RMALT, and RMAA, Rivermead Motor Assessment-Gross Function, -Leg and Trunk, and -Arm.

Fig. 1. Biplot without CSI and with projected HADS-A and HADS-D vectors

anxiety and depression scores at 6 months after stroke, while there were large variations in anxiety and depression in patients with higher functional status. Mood disorders like anxiety and depression can occur in all stroke patients with less regard to their motor and/or functional status. EQ-HS, EQ-VAS and CSI are more related to remaining motor and functional disabilities. Patients with initial lower motor and ADL functions also have a poor prognosis for good recovery and therefore need more effort by the rehab team to avoid long term restrictions. EQ-HS and EQ-VAS at the end of the inpatient rehabilitation process seems to be additional prognostic factors for the long term perspective as it was revealed in another patient sample of the German centre (Grässel et al., 2005).

### 4.2 Anxiety and depression in the first six months after stroke (De Wit et al., 2008)

Prevalence of anxiety ranging between 22 -25 % and depression between 24 - 30% on HADS were found in this study. Median severity of both emotional disorders ranged between 4 and 5 at each time point. There were no significant differences between centres in the prevalence and severity of both disorders. The prevalence of both disorders remained stable between two and six months after stroke onset, but different patients were anxious/ depressed at subsequent stages. This may be explained by the result which showed that about 40% of the patients with initial anxiety remained anxious up to six months, while 11% and 7% of those initially not anxious became anxious for the first time at four and six months after stoke, respectively. This pattern was similar for depression. The time course of severity of both disorders also were considered in this study. Severity of anxiety decreased between four and six months. Meanwhile severity of depression remained stable.

This study also compared severity of those who were anxious/depressed throughout the whole follow- up period to severity of those who were anxious/depressed at only two time points. The result showed that severity of both emotional disorders was significantly more severe in the group who was continuously anxious/depressed compared to those who were anxious/depressed at one or two time points. These findings from 2 studies involving in emotional disorders after stroke may indicate that anxiety was almost as common as depression in this subacute phase. Patients may become anxious and/or depressed at any time during 6 months. Those patients who remain anxious or depressed throughout the subacute period suffered from more severe emotional disorders. Thus, early screening for both emotional disorders is recommended for all stroke patients, even in those patients who had mild functional deficit. Additionally, the screening should be continued for an extended period of time after stroke.

### 4.3 Motor and functional recovery 6 months after stroke (De Wit et al., 2007)

The result of motor and functional recovery after stroke in 4 European centre revealed that half of the patients of all centres reached a score of at least 70% of the maximum RMA-GF score and at least 85 % of the maximum BI score at discharge. This meant that most recovery time took place during inpatient rehabilitation phase so it is important to prescribe proper therapeutic interventions to patients in this period. Among the 4 centres, the gross motor function(RMA-GF) and functional recovery using EADL were better in the Swiss and German centre than the UK centre.

On the other hand, personal self-care recovery on basic ADL functions (BI) was significantly higher in the UK compared with the German centre. The last fact may reflect the high input of the nurses in UK with a more compensatory than functional training. The results are possibly explained by outcome findings of the other studies in the same CERISE project ,time in therapy of stroke patients , use of time by physiotherapists & occupational therapists.The result of this 2 studies showed that higher percentage of time was spent in the therapeutic activities in the German and Swiss centre than in UK centre. More details of this 2 studies were reviewed on the following paragraphs. Thus, these reasonable findings led to better gross motor and NEADL recovery in these 2 centres. Despite these methodological issues, these findings are in line with the results of a meta-analysis (Kwakkel et al., 2004), indicating that more intensive rehabilitation results in better recovery. Other meta-analysis and papers reflect on the content and specificity of the rehabilitation interventions. In common we can state that more intensive and more specific (goal directed) rehabilitative treatment may improve outcome even in initially severely disabled stroke patients.

### 4.4 Use of time by stroke patients and use of time by physiotherapists and occupational therapists (De Wit et al., 2005; Putman et al., 2006)

Considering about the use of time by stroke patients in 4 rehabilitation centres,the results were revealed as shown in Table 3, using percentage time and Figure 2a using absolute time. The percentage of time spent in non therapeutic activites were higher (>72 % of the time) than therapeutic activities in all centres. In all 4 centres, physical and occupational therapies together accounted more than half the total therapeutic time, in which physiotherapy was comprised about 40% of therapeutic time, where as occupational therapy composed of 20-30% excepts for the United Kingdom, which spent least time on this kind of therapy, while they used more than 35 % of therapy time for nursing care. The authors of this study discussed this result that may be explained by the high availability of nurses which is in line to the previous study in UK.

The absolute time spent in therapeutic measures is shown in Figure 2a. Patients in Switzerland spent more time in therapy (2 hours and 46 minutes) compared with those in Germany (2 hours and 20 minutes), Belgium (1 hour and 59 minutes), or the United Kingdom (1 hour and 0 minutes). This showed that the United kingdom spent the least time in therapeutic intervention, even they had highest overall time available from all professional groups, taking into account the staffing levels and number of working hours per week, but most of their time ( about 35% of therapy time) were for nursing care, instead of other professional therapeutic interventions. Moreover, higher percentage of non therapeutic time was found in the United Kingdom and Belgium centres,patient spent > 2 hours per days lying or sleeping, this was more than in Switzerland or Germany centres as shown in Figure 2b.

This study result showed that patients in the United Kingdom were significantly less likely to be in therapy than patients in the other centres. Differences in therapy time between 4 centres seemed to be related to differences in task characteristics of therapists and the organization of the rehabilitation programs. In addition, patients in the Swiss and German centres spent more time in therapy, possibly due to strictly-timed rehabilitation program and organization (consistent time tables for patients and therapists). The ward rounds and the team conferences monitoring the patients' progress and limitations and adapting the amount & content of therapeutic interventions have been more clearly structured in the German and Swiss centres. In general, more efficient use of human resources in both centres may be taken into the account.

| | Belgium, % (n=1200) | United Kingdom, % (n=1200) | Switzerland, % (n=1200) | Germany, % (n=1200) |
|---|---|---|---|---|
| Activity | | | | |
| Therapy time | **19.8** | **10.1** | **27.7** | **23.4** |
| Percentage of therapy time | | | | |
| Physiotherapy | 40.8 | 40.5 | 37.7 | 36.7 |
| Occupational therapy | 20.2 | 11.6 | 29.2 | 24.9 |
| Speech therapy | 5.0 | 1.7 | 5.1 | 1.8 |
| Neuropsychological training | 6.3 | 0.0 | 3.6 | 9.3 |
| Nursing care | 10.1 | 35.5 | 5.4 | 4.6 |
| Medical care | 0.8 | 2.5 | 5.7 | 5.7 |
| Sports | 1.7 | 0.0 | 0.0 | 0.7 |
| Autonomous exercising | 2.9 | 4.1 | 0.6 | 6.4 |
| Other therapeutic activity | 12.2 | 4.1 | 12.7 | 10.0 |
| Nontherapy time | **80.2** | **89.9** | **72.3** | **76.6** |
| Percentage of non-therapy time | | | | |
| Sitting | 16.7 | 29.1 | 21.1 | 22.9 |
| Lying/sleeping | 28.3 | 23.8 | 19.9 | 10.7 |
| Leisure | 10.1 | 7.8 | 9.4 | 15.9 |
| Communication | 14.1 | 14.6 | 14.5 | 14.4 |
| Transport | 8.4 | 2.1 | 13.5 | 12.8 |
| Dressing/hygiene | 9.3 | 10.8 | 7.5 | 6.4 |
| Eating | 8.0 | 6.2 | 10.3 | 11.8 |
| Other nontherapy | 4.2 | 5.5 | 3.8 | 5.2 |

Table 3. Distribution of Activities in Percentage (%) of Observations (n=1200) Between 7 AM and 5 PM in Four Rehabilitation Centres.

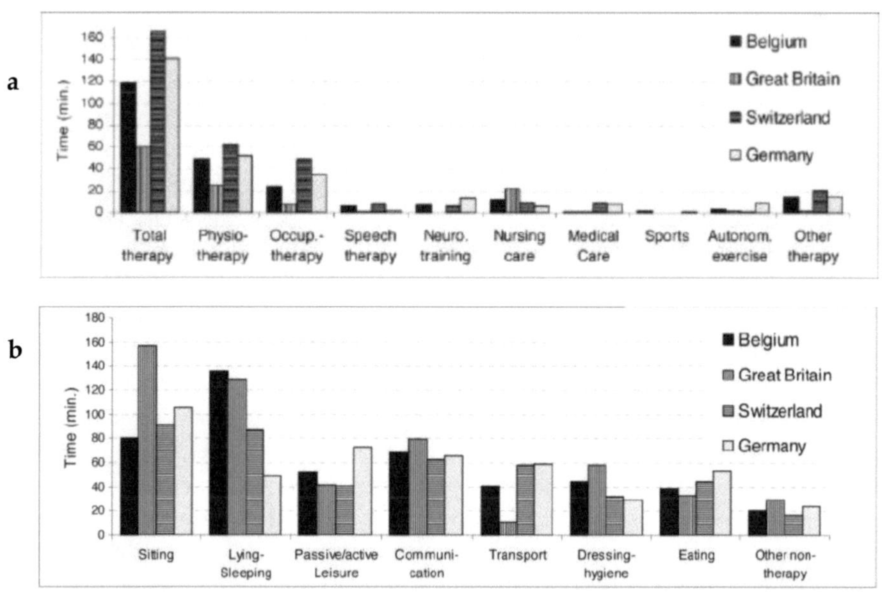


Fig. 2. Absolute time spent in therapeutic activities (a), nontherapeutic activities (b).

**Use of time by physiotherapist (PT) and occupational therapist (OT)**

This study focused on 2 professions, physiotherapists and occupational therapists, who accounted for highest proportion of therapy services in in-patient stroke rehabilitation. The aim of this study was to compare the time allocated to therapeutic activities (TA) and non-therapeutic activities of this 2 professions. Therapeutic acitivities defined all professional interventions when PT's and OT's were in personal contact with the patients(e.g. single and group therapy). Non-therapeutic activities of PT's and OT's were divided into patient linked co-ordination (e.g. patient administration, organizing technical aids, counseling care givers), unit linked co-ordination (e.g. team conferences, staff organization) and others. Table 4 showed the result of this study that the highest percentage of time spent in therapeutic activities was found in German centre (PT= 66.1 %;OT= 63.3%). The lowest was in the United Kingdom centre (PT= 45.9 %;OT = 32.9%). This was in corresponding to the finding that more than half for both PT and OT in the United Kingdom were used in non-therapeutic activities such as administration work ( PT= 54.1%;OT=67.1%) This led to less time therapy for patient.

| | SRU-GB | | SRU-CH | | SRU-DE | | SRU-BE | |
|---|---|---|---|---|---|---|---|---|
| | PT<br>n = 2476 | OT<br>n = 1284 | PT<br>n = 3883 | OT<br>n = 2033 | PT<br>n = 4548 | OT<br>n = 2157 | PT<br>n = 2545 | OT<br>n = 1495 |
| *Therapeutic activities, %* | *45.9* | *32.9* | *53.7* | *45.2* | *66.1* | *63.3* | *61.7* | *50.2* |
| mobility training | 34.6 | 9.4 | 26.7 | 14.6 | 36.1 | 31.3 | 36.9 | 13.0 |
| ADL-training | 8.0 | 19.5 | 14.6 | 22.1 | 15.0 | 18.4 | 20.2 | 23.5 |
| neuropsychological training | 0.1 | 1.2 | 1.8 | 6.9 | 4.6 | 8.4 | 0.4 | 12.8 |
| other training | 3.2 | 2.6 | 10.8 | 1.6 | 10.4 | 5.1 | 4.2 | 0.7 |
| *Non-therapeutic activities, %* | *54.1* | *67.1* | *46.3* | *54.8* | *33.9* | *36.7* | *38.2* | *49.8* |
| patient linked co-ordination | 23.1 | 32.6 | 14.0 | 24.5 | 16.7 | 12.9 | 9.7 | 13.1 |
| unit linked co-ordination | 14.6 | 22.3 | 14.9 | 13.8 | 10.0 | 15.8 | 13.4 | 15.7 |
| other | 16.4 | 12.3 | 17.4 | 16.6 | 7.2 | 8.0 | 15.1 | 21.0 |

Table 4. Percent (%) of activities of PT and OT in the four stroke rehabilitation centers

### 4.5 Contents of therapy: Physical Therapy (PT) & Occupational Therapy (OT) (De Wit et al., 2006)

Contents of therapy were categorized into 12 groups : (1) mobilization,(2) selective movements,(3) exercises and balance in lying,(4) exercises and balance in sitting,(5) exercises and balance in standing,(6) sensory,perceptual training and cognition,(7) transfers,(8) ambulatory,(9) personal activities of daily living (ADL),(10) domestic ADL, (11) leisure and work- related activities,(12) miscellaneous techniques. The comparison between the content of PT and OT sessions revealed a different emphasis of each profession in 8 of the 12 categories. PT sessions significantly more often were composed of ambulatory exercises, transfers, exercises and balance in standing and lying, whereas ADL, sensory, perceptual training, cognition, domestic and leisure activities occured significantly more in the OT sessions. No statistically significant differences were found for selective movement, mobilization, and exercises and balance in sitting. The category of miscellaneous contained too few observations for statistical analysis. This result revealed that PT and OT are distinct professions with clear demarcation of roles. When comparing the contents of PT and OT between 4 centres, there were only 2 from 12 categories that showed significant differences between the centres which were ambulatory exercises and selective movements. Ambulation exercises occurred more in the United Kingdom and Belgian centres than in the German and Swiss centres. Selective movements occurred less for both PT and OT sessions in the United Kingdom compared with the Swiss centre and the German centre, respectively.

### 4.6 Effect of socioeconomic status on functional and motor recovery after stroke (Putman et al., 2007)

Socioeconomic status ( SCS ) should be considered as one essential part for holistic stroke approach. Previous study (Jakovlijevic et al., 2001) also showed impact of socioeconomic status on morbidity and mortality. In one part of the CERISE project, the study aimed to explore the impact of education and equivalent income on motor and functional recovery between the period in the stroke rehabilitation unit and the period after discharge. The patient's educational level was converted to the International Standard Classification of Education (ISCED) (OECD,1999) for valid comparison between countries. Equivalent income was based on the monthly household income and househole composition,and calculated according to the modified Organization for Economic Cooperation and Development (OECD) scale. The Barthel index (BI) and the Rivermead Motor Assessment (RMA) were used for functional outcome measurement performed on admission, at discharge and at 6 months after stoke. There was significantly higher urinary incontinence in the low education group. The high income group had consistently higher BI outcome score compared with other subgroups at all three assessment points. This was similar to the RMA -GF which was significant higher in both the high education and high income groups. When comparing between the SES groups, BI and RMA scores, the results revealed that patients with a low educational level were less likely to improve on the functional status (BI) and the RMA of the arm section,while no significant effect of equivalent income on motor and functional recovery during inpatient rehabilitation period were found. After discharge until 6 months after stroke,neither equivalent income nor education had a significant effect on the functional recovery ( BI ) but patients with a low equivalent income were less likely to improve further on motor function, represented by RMA ( RMA- GF, RMA-Leg and trunk, RMA-arm) compared with the high income group. This may conclude that education level was a determinant of recovery during inpatient rehabilitation, while after discharge, equivalent income was the main impact of additional recovery.

### 4.7 Admission criteria to inpatient stroke rehabilitation units among the 4 European stroke rehabilitation centres (Putman et al., 2007)

Previous literature (Langhorne & Duncan,2001) reviewed that post-acute rehabilitation is an essential part of the recovery process. There was no clear evidence when it would be suitable to tranfer patients to post-acute stroke rehabilitation units. Even though there is limited evidence that early admission to stroke rehabilitation directly results in improved functional outcomes,stroke patients should be admitted to stroke rehabilitation units as soon as they are medically stable (Teasell et al.,2009).

According to the difference in the organization and financing of healthcare systems across Europe, this might have an impact on admission policies and the selection of patients for post-acute inpatient rehabilitation. This study focuses on comparing the admission procedures in 6 SRUs across 4 European countries reflecting the impact of clinical and non-clinical factors in the decision making process on admission. The finding of clinical and non-clinical factors for admission were not quite explicit and were evaluated differently between the European rehabilitation units. The factors related to the patients were identified as followed; in the SRU-UK, diagnosis of stroke was the only criterion for admission, in the Belgian, German and Swiss units, pre-morbid conditions were taken into account in admission decisions. For example, patient in SRU-DE with pre-morbid disability who required support were more or less automatically transferred to geriatric rehabilitation or nursing care. Presence of pre-morbid cognitive disability or depression reduced the likelihood of being admitted in all SRUs excepts SRU-UK. Severe behavioral problems after stroke also are considered in both SRU-UK and SRU-BE. Admission to SRU-CH was delayed if the patient showed rehabilitation potential but did not yet have the stamina required for intensive rehabilitation. Also the availability of home support was a decisive factor in the decision to admit a patient in SRU-CH. Highly motivated patients or their relatives sometimes influenced to be admitted in SRU-DE. Patient's socioeconomic status did not really influence any on a decision to admit a patient. Patient's network was concerned by the medical consultants in SRU-CH.

Factors related to the network between facilities showed that the affiliations between the SRU and other healthcare settings were highly important for the admission to SRU-BE,SRU-CH and SRU-DE. In SRU-CH, the medical consultants who worked at the acute stroke units also were responsible for referral process to other SRU. In SRU-DE, the medical consultants was not involved in the selection of stroke patients because the association of different legal (by social laws) or private cost bearers (health or personal insurances) with the rehabilitation centre was appraised as having a very high effect on the admission policy. External stakeholders decided, based on the referral letters from the medical consultants in the acute unit, whether the patient was suitable for further rehabilitation in SRU-DE.

This study concluded that clinical characteristics of stroke patients were significantly different between European rehabilitation units admitted. Decision - making process on admission was effected not only by clinical factors, but also by non-clinical factors which seemed to determine whether and where patients are referred to inpatient stroke rehabilitation services.

### 4.8 Discharge program & follow up services after inpatient stroke rehabilitation (Grässel et al., 2005, 2006; Putman et al., 2009)

An effective stroke rehabilitation program must also involve comprehensive discharge planning and give recommendations for the post discharge rehabilitation phase which follow - up services are needed and should be prescribed or ordered. Several studies on different models for discharging and follow-up for stroke patients have been performed with varying effect, and there is still no consensus concerning care of stroke patients after discharge (Holmqvist et al., 1998; Anderson et al., 2000; Ytterberg et al.,2000). "Early supported discharge" is one of the settings studied in other countries.

In CERISE study,the majority of the patients were still in the rehabilitation centre ( 50-70%) at 2 months after stroke. But at 4 months, the majority of patients were at home,with the lowest percentage for SRU-BE (55%) and the highest for SRU-DE ( 75%). At 6 months, 66 -76 % of the patients were at home. The patients in SRU-BE and SRU-UK had significantly lower scores on all motor and functional assessments compared with the other 2 centres as shown in the previous studies in this project. Patients from SRU-BE also had the lowest score of the quality of life (using EuroQuoL VAS).

The data above showed that the majority of the patients in each rehabilitation centre were discharged to their home at 4 months after stroke. However, the decision making for discharge planning such as suitable time to discharge may depend on a number of variable factors including stroke severity, family support, progression in rehabilitation, availability of outpatient resources, incentives to discharge, and discharge guidelines or policies in each unit. Like for the admission also non-clinical, mainly health care system related factors had partially great impact on the discharge.

For examples, Kalra and Walker (2009) noted that in the United Kingdom, the rehabilitation goals of professionals often do not reflect the patients' priorities, which can lead to unnecessarily prolonged rehabilitation stays. Putman and De Wit (2009) noted that in Germany distinct phases for rehabilitation have been developed so that patients receive the targeted therapy that they need, with the BI being used as help to determine which phase is right for which patient.

Besides the consideration of suitable time to discharge, the effective discharge planning program was also an essential part. This program aimed to prepare the patients to live with proper care and minimize problems occurring at home. Family caregivers of stroke patients also play a major role in this phase. Some studies revealed that familiy caregivers of stroke patients were to suffer from increased emotional distress (Dennis et al.,1998) and required support from professionals, both as regards caring itself and also in helping them to cope with emotional and psychological problems (Dennis et al.,1998; Wyller et al.,2003).

The intensified transitional concept was an example that concerned and emphasized on the effective discharge planning program and post discharge phase for stroke patients. This concept was developed by the Rehabilitation Clinic Herzogenaurach, Germany which is one of the stroke rehabilitation units in this CERISE project. This concept consisted of a standard transitional program with an additional therapeutic weekend care at home, individual training course and telephone counselling 3 months after discharge for the family caregivers in order to minimize the problems at home , decrease physical and emotional stress of family caregivers ( mostly the spouse). This concept was also studied by comparing with the standard transitional program in aspects of effect on the functional status of stroke patients and the physical and emotional burden of their caregivers at 2 time periods (6 months and about 2 and a half years after discharge) (Grässel et al.,2005,2006). The result revealed that the intensified transitional program had no effect either on the functional status of the stroke patients or the health status of the care givers at 6 months after discharge but this intensified program led to earlier visits of family physicians & therapists and more uptake of outpatient services. Additionally,significantly fewer patients in this intensified transitional program were institutionalized or had died, comparing to the standard transitional group at two and a half years after discharge. This concluded that the effect of an intensified transition program can occur and persist over a long term period.

Follow-up services can be an effective mean to alleviate the functional disabilities of stroke at home (Andersen et al., 2002). Indicators or determining factors which patients receive follow-up services remain uncertain. Patient's needs and functional abilities at discharge seem to be logical factors. However, Asplund et al. (2003) found that healthcare routines rather than the patient's condition were the major determinants of the extent to which resources were used. Therefore, differences in discharge policies especially involving in proportions of patients receiving services after discharge may influence follow - up comparisons in outcome between centres.

In Europe, wide variations are observed in healthcare provision after stroke (McKevitt et al.,2000; Wolfe et al.2004). This leads to great variations in follow-up services after inpatient stroke between the 4 European centres which were founded in the CERISE study. This study revealed that the Belgian patients were most likely to receive physiotherapy (53 -82%) but least likely to receive occupational therapy (9-11%) . German patients were least likely to receive nursing care (2-4%) and the UK patients were least likely to receive medical care from general practitioners (family doctors) (43- 61%) but most likely to receive OT (16-27 %) compared with the other patient groups. (Figure 3.) The follow-up services for stroke patients in UK is more focused on so called " stroke nurses" who did not exist as a profession of health care system in the other countries. In other countries, all the medical services must be prescribed by the family doctors who must be contacted. Low frequencies in both speech and language therapy and psychological therapy were found in all countries and could not be analysed further.

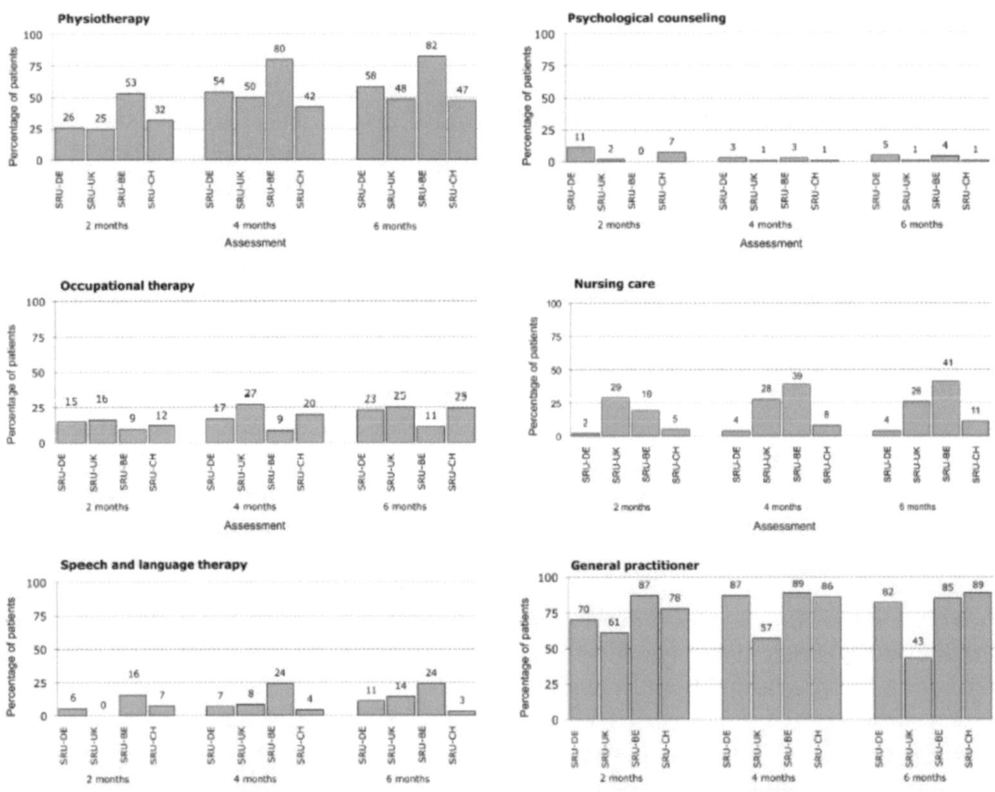


Fig. 3. Percentage of patients at home receiving services at 2, 4 and 6 months post-stroke onset. SRU-DE: German stroke rehabilitation unit; SRU-UK: United Kingdom stroke rehabilitation unit; SRU-BE: Belgian stroke rehabilitation unit; SRU-CH: Swiss stroke rehabilitation unit.

The probability of receiving services at home correlated only in part with clinical characteristic of patients. According to statistic analysis in this study, patients under 70 years, patients with initial swallowing problems or low score on the RMA-A were more likely to receive PT programs. Age under 70 years, being male and low RMA-A score were factors significantly associated with a higher likelihood of receiving OT programs.

In this study, the results showed that age of patient and motor function ( RMA-A) may influence any follow-up services provided especially in PT and OT programs 6 months after discharge. This findings were comparable to the results from the other study by U.Hoess, et al. (2008). They also investigated factors that may effect the frequency of prescription and performance of the follow -up services,predominantly PT,at 6 months and even in long term period (2 and a half years) after discharge. They found that younger stroke patients, who had more physical but less mental impairment were most likely to receive more therapies. Additionally,in the period of 6 months after discharge, stroke patients who visited the outpatient clinic more frequently,were more likely to get multiple medication and physical therapy prescriptions.

However, non-clinical factors such as cost bearing regulations in the national health care systems, also influence the variations in follow-up services after discharge from inpatient stroke rehabilitation. As follow-up services can reduce the long-term dependency after stroke, variation of patients receiving services after discharge may influence follow-up comparisons in outcome between centres. Thus, services provided after discharge from inpatient rehabilitation should be better documented to facilitate a more precise comparison on the effectiveness of rehabilitation programs and aftercare.

Despite to the results which showed that physical therapy and occupational therapy remained the common therapy services that stroke patients received after discharge in outpatient settings, in the present, there are unclear standards or guidelines for the prescription of these two therapies in follow up services.

A summarizing review of the literature on these two therapy programs ( PT and OT), their contents, prescription criteria and effects on stroke patients in outpatient service setting was recently performed in Germany (Steib & Schupp,2011). The results showed that the physical therapy programs such as progressive strength training, endurance training, gait training were effective to improve physical performance, resulting in benefits on gait, mobility and ADL. Also, occupational therapy showed an improved performance in activities of daily living (ADL / EADL), increase of participation & leisure activities and reducing risk of "poor outcome" (deterioration of condition of patients, progression of disability, death). Frequency and duration of the therapy in outpatient service usually took two to three times a week for 4 - 12 weeks, respectively. The duration of each treatment varied from 20 minutes to two hours.

## 5. Conclusion

The European project CERISE was a longitudinal multicentre and multiprofessional collaboration that studied multidimentional aspects of stroke rehabilitation. Comparison between European stroke rehabilitation centres in field of structure of care and process of care should provide insight into the aspects of stroke rehabilitation that are crucial for patients' outcome and also provide important information to improve efficiency of stroke rehabilitation programs and organization. Although differences in service provision and patient selection in each centre limit direct comparisons of outcome measures, this can be considered as an advantage to show the real situation in existing settings. However, some limitations of this project need to be addressed. Only a limited sample of each centre was included, even the centres were very experienced in stroke rehabilitation with a good reputation, but they cannot be representative for the different countries. Therefore, the variations found in this study cannot be attributed entirely to differences in national healthcare services, but may also reflect differences in local healthcare structure and policies. Study designs that need to collect data from observations by blind participants may not reflect a real situation. Data from interviews may not be a real data due to participants' conditions such as memory or behavioral status.

From all our experiences revealed by the studies cited above we can conclude that:

Stroke patients should be referred to a comprehensive rehabilitation program. The admission should be mainly guided by medical reasons and by personal and environmental context factors according to the ICF. Financial aspects of cost bearers should not interfere with the admission to such stroke rehab programs. The stroke rehabilitation units should be strictly organized and managed, that the therapists can spend the majority of their working time in direct therapeutic interactions with the patients. The therapeutic activities must be emphasized against every tendency for more bureaucracy. Physiotherapists (PT's) and occupational therapists (OT's) are responsible for the main therapeutic input ,but both professions have to clarify their different approaches and responsibilities for training in ADL, mobility, EADL, vocational and leisure activities. On the other side they must repeatedly discuss and adapt their therapeutic interventions which are to be based on neurobiological scientific research and modified by the patients' psychosocial context in at least weekly team conferences along with the patients' recovery process. Most patients have much more time available during the day in any inpatient rehabilitation program, which could be filled with additional therapy supporting activities and exercises, supervised and guided by PT's and OT's. A new role of therapists should include this task to give every patients recommendations for such self administered therapeutic activities and exercises. Such recommendations are also very important for the phase after inpatient or outpatient rehabilitation setting (aftercare and long term rehabilitation). Mood disorders like anxiety and depression often develop after stroke and negatively influence the long term outcome. Therefore, screening for such comorbidity, and if any suspicion can be derived, definite diagnosis and treatment should be performed.

The structures and performance of further rehabilitation services and treatment should be improved in every country and their health care systems. Socioeconomic parameters should not any longer impact on the long term outcome and quality of life (QoL) after stroke as we had seen in our studies. This fact is in contrast to the human rights of disabled people.

Many research studies until now in the field of stroke rehabilitation largely focus on acute or subacute stroke rehabilitation phase and single specific interventions. In the present, studies in a long term stroke care are less evident,despite the fact that the need to gain more knowledge on this aspect is increased. Further studies on long term stroke care are needed based on holistic aspects and multicentre setting to prevent further impairment, disability and improve quality of life.

## 6. Acknowledgement

The CERISE project (Collaborative Evaluation of Rehabilitation in Stroke across Europe) was supported and funded by the European Commission and Sekretariat fuer Bildung und Furschung SBF (CH) via contracts: Quality of life, key action 6,2001-2005, contract number QLK6-CT-2001-00171. It does not necessarily reflect its views and in no way anticipates the Commission's future policy in this area.

This project was conducted by I. Baert (B),P.Berman (GB), H. Beyens (B), N.Brinkmann (D),L. Connell (GB), E. Dejaeger (B), W. De Weerdt (B),L. De Wit (B),H. Feys (B),W. Jenni (CH),J. Jurkat (D),H. Kamsteegt (B),C. Kaske (CH), M. Leys (B), N.B. Lincoln (GB),F. Louckx (B),K. Putman (B),B. Schuback (CH), W. Schupp (D) and B. Smith (GB).

The authors of the project thank Dr Gaynor for her contribution. All reseachers are independent from funders.

The authors would like to thank for all researchers and participants in this projects.

## 7. References

Anderson C, Rubenach S, Mhurchu CN, Clark M, Spencer C,& Winsor A.(2000) Home or hospital for stroke rehabilitation? Result of a randomized controlled trial. 1: Health outcomes at 6 months. *Stroke.* 2000,Vol.31,pp.1031.

Andersen HE, Eriksen K, Brown A, Schultz-Larsen K, & Forchhammer BH.(2002) Follow-up services for stroke survivors after hospital discharge– a randomized control study. *Clin Rehabil* 2002,Vol.16,pp. 593–603.

Asplund K, Ashburner S, Cargill K, Hux M, Lees K, & Drummond M.(2003) Health care resource use and stroke outcome. Multinational comparisons within the GAIN International trial. *Int J Technol Assess Health Care 2003,*Vol.19,pp. 267–277.

Brainin M, Bornstein N, Boysen G, & Demarin V. (2000). Acute neurological stroke care in Europe: Results of the European Stroke Care Inventory. *Eur J neurol* 2000, Vol.7(1),pp.5-10.

De Wit L, Putman K, Dejaeger E, Baert I, Berman P,Bogaerts K, et al.(2005) Use of time by stroke patients:A comparison of four European rehabilitation centers. *Stroke.* 2005,Vol.36(9),pp.1977–1983.

De Wit L, Putman K, Lincoln N, Baert I, Berman P,Beyens H, et al.(2006). Stroke rehabilitation in Europe: What do physiotherapists and occupational therapists actually do? *Stroke.* 2006,Vol.37(6),pp.1483–1489.

De Wit L, Putman K, Schuback B, Komarek A, Angst F,Baert I, et al.(2007) Motor and functional recovery after stroke:A comparison of 4 European rehabilitation centers. *Stroke.* 2007,Vol.38(7),pp.2101–2107.

De Wit L, Putman K, Baert L, Berrice N, Angst F, Beyens H, et al. (2008) Anxiety and depression in the first six months after strokes. A longitudinal multicentre study. *Disability & Rehabilitation.*2008,Vol.30(24),pp.1858-1866.

De Wit L, Molas M, Dejaeger E, Weerdt W, Feys H, Jenni W,et al. (2009). The Use of a Biplot in Studying Outcomes After Stroke. *Neurorehabilitation and Neural Repair.*2009,Vol. 23,No.8,( Oct 2009),pp. 825-830.

Donabedian A. (1988). The quality of care. How can it be assessed? *JAMA.*1988, Vol.260(12), pp.1743–1748.

Grässel E,Biehler J,Schmidt R,& Schupp W (2005). Intensification of the transition between inpatient neurological rehabilitation and home care of stroke patients. *Clin Rehab* 2005,Vol.19,pp.725-736.

Grässel E,Biehler J,Schmidt R,& Schupp W (2006). Long-term outcome of a new concept for the transition between inpatient neurological rehabilitation and home care of Stroke patients. *Clin Rehab* 2006,Vol.20,pp.577-583.

Hoeβ U, Schupp W, Schmidt R, & Gräβel E. (2008) Home care of stroke patients with Remedies and Aids after Inpatient Neurological Rehabilitation – A Longitudinal Study. *Phy Med Rehab Kuror* 2008,Vol.18,pp.115-121.

Holmqvist L, Widen RPT, von Koch L, Kostulas V, Holm M, Widsell G,Tegler H, Johansson K, Almazan J, & de Pedro-Cuesta J.(1998) A randomized controlled trial of rehabilitation at home after stroke in southwest Stockholm. *Stroke.* 1998,Vol.29,pp.591–597.

Jakovlijevic D, Sarti C, Sivenius J,et al. (2001) Socioeconomic status and ischemic stroke: The FINMONICA Stroke Register. *Stroke* 2001,Vol.32,pp.1492-8.

Kalra L, Walker MF.(2009) Stroke rehabilitation in the United Kingdom. *Top Stroke Rehabil.* 2009,Vol.16(1),pp.27–33.

Langhorne P, Duncan P.(2001)Does the organization of postacute stroke care really matter? *Stroke* 2001,Vol.32,pp.268-174.

Lincoln N,Leadbitter D. (1979) Assessment of motor function in stroke patients. *Physiotherapy.* 1979,Vol.65,pp.48-51.

Mahoney FI, Barthel DW. (1965) Functional evaluation: the Barthel Index.*Md State Med J.* 1965,Vol.14,pp.61-65.

Markus H. (2004).Variations in care and outcome in the first year after stroke: A Western and Central European perspective. *J Neurol Neurosurg Psychiatry* .2004, Vol.75(12),pp.1660–1661.

McKevitt CJ, Beech R, Pound P, Rudd AG, & Wolfe CDA.(2000) Putting stroke outcomes into context – Assessment of variations in the processes of care. *Eur J Public Health* 2000, Vol.10, pp.120–126.

Nouri FM, Lincoln NB.(1978) An extended activity of daily living scale for stroke patients. *Cli Rehabil.* 1978, Vol.1, pp.301-305.

Organization for Economic Co-operation and Development.(1999) Classifying Educational Programmes, Manual for ISCED-97 Implementation in OECD Countries,1999 Edition. http://www.oecd.org/dataoecd/41/42/1831853.pdf

Putman K, De Wit L, Schoonacker M, Baert I, Beyens H, Brinkmann N, et al. (2007) Effect of socioeconomic status on functional and motor recovery after stroke: a European multicentre study. *J Neurol Neurosurg Psychiatry.*2007, Vol.78, pp. 593-599, jnnp.2006.094607.

Putman K, De Wit L, Schupp W, Baert I, Berman P,Connell L, et al.(2006). Use of time by physiotherapists and occupational therapists in a stroke reha bilitation unit:A comparison between four European rehabilitation centres. *Disabil Rehabil.* 2006,Vol.28(22),pp.1417–1424.

Putman K, De Wit L, Schupp W, Beyens H,Dejaeger E, De Weerdt W, et al.(2007) Inpatient stroke rehabilitation:A comparative study of admission criteria to stroke rehabilitation units in four European centres. *J Rehabil Med.* 2007, Vol.39 (1), pp.21–26.

Putman K, De Wit L, Schupp W, Baert I, Brinkmann N, Dejaeger E,et al. (2009) Variations in follow-up services after inpatient stroke rehabilitation: A multicentre study. *J Rehabil Med.*2009,Vol.41,pp.646-653.

Putman K, De Wit L.(2009) European comparison of stroke rehabilitation. *Top Stroke Rehabil.*2009,Vol.16(1),pp.20–26.

Robinson BC. (1983) Validation of a Caregiver Strain Index.*J Gerontol.* 1983, Vol.38, pp.344-348.

Shah MV.(2006).Rehabilitation of the older adult with stroke. *Clin Geriatr Med* 2006,Vol.22, pp.469-489,ISSN 0749-0690.

Steib S, Schupp W. (2011) Inhalte und Effekte von Therapiestrategien in der Schlaganfallnachsorge. *Nervenarzt* 2011, accepted for publication.

Teasell R, Foley N., Salter K, Bhogal S,Jutai J, & Speechley M (2009). Evidence-Based Review of Stroke Rehabilitation: Executive Summary, 12th Edition.*Topics in stroke rehabilitation,*Vol.16,No.6,( Nov-Dec 2009), pp.463-488.

Teasell R,Meyer J.M. , McClure A , Pan C,Fernandez M. M, Foley N,& Salter K. (2009) Stroke Rehabilitation:An International Perspective. *Topics in stroke rehabilitation,* Vol.16,No.1, (Jan-Feb 2009),pp.44-56.

The EuroQol Group.(1990) EuroQol: a new facility for the measurement of health-related quality of life. *Health Policy.* 1990,Vol.16,pp.199-208.

Weir NU, Sandercock PA, Lewis SC, Signorini DF,& Warlow CP. (2001). Variations between countries in outcome after stroke in the International Stroke Trial (IST). *Stroke* 2001,Vol.32(6),pp.1370-1377,ISSN 1524-4628.

WHO MONICA project Principal Investigators.(1988) The World Health Organization MONICA project (monitoring trends and determinants in cardiovascular disease): a major international collaboration. *Clini Epidemiol.*1988,Vol.41,pp. 105 - 114.

Wolfe CDA, Tilling K, Rudd A, Giroud M,& Inzitari D.(2004) Variations in care and outcome in the first year after stroke: a Western and Central European perspective. *J Neurol Neurosurg Psychiat* 2004,Vol.75,pp.1702–1706.

Wyller TB, Thommessen B, Sødring KM, Sveen U, Pettersen AM, Bautz-Holter E,& Laake K. (2003) Emotional well-being of close relatives to stroke survivors. *Clin Rehabil* 2003,Vol.17,pp.410-17.

Ytterberg C, Malm SA, & Britton M.(2000) How do stroke patients fare when discharged straight to their homes? A controlled study on the significance of hospital follow-up after one month. *Scand J Rehabil Med.* 2000,Vol.32,pp.323–396.

Zigmond AS, Snaith RP.(1983) The hospital anxiety and depression scale. *Acta Psychiatr Scand.*1983,Vol.67,pp.361-370.

# 17

# Group Circuit Class Therapy for Stroke Survivors – A Review of the Evidence and Clinical Implications

Coralie English[1], Ingrid van de Port[2] and Elizabeth Lynch[3]
[1] *International Centre for Allied Health Evidence, University of South Australia,*
[2]*Centre of Excellence in Rehabilitation,*
*Rehabilitation Centre De Hoogstraat & University Medical Centre Utrecht,*
[3]*Hampstead Rehabilitation Centre,*
[1,3]*Australia*
[2]*Netherlands*

## 1. Introduction

Concerns about rising costs of health care are felt by health care professionals around the world (Rosen & Karlberg 2002; Bovier et al., 2005; Robertson et al., 2011) leading to pressure to deliver services in the most cost effective way possible. As new diagnostic tests, surgical procedures or more effective drugs are developed, there is, at least in developed countries, an imperative to make these things available to all, inevitably at an increased cost to the system. This places a downward pressure on all aspects of the health care system to deliver services for the most efficient means to counteract this increased cost. In the case of physiotherapists providing rehabilitation therapy to people who have suffered a stroke, there is an expectation that the current standard of therapy provision will remain stable or in fact be enhanced, with the same or less funding. At the same time, research evidence suggests that providing more therapy time - maximising the time stroke survivors spend in active task practice each day - will maximise their functional outcomes (Kwakkel et al., 2004). So therapists are being asked to do more with less on a daily basis.

Usual practice for stroke rehabilitation around the world for the past several decades has been the provision of therapy in individual, one-to-one therapy sessions. However, group circuit class therapy is an alternate method of therapy service delivery. One study showed that providing physiotherapy in circuit classes rather than in individual therapy sessions led to a significant increase in therapy time without increasing cost (English et al., 2007). While group circuit class therapy for stroke survivors is often touted as 'novel' - there is nothing particularly new or unusual about the concept of providing therapy to groups of people. In the research literature, the first paper written about group circuit class therapy was published by Dean et al in 2000. This small pilot study was the first to describe the effectiveness of group circuit class therapy with regard to improved motor function in people after stroke. Since that time the evidence base has continued to grow and there are now two published systematic reviews on the topic (English, et al., 2007; Wevers et al., 2009).

So what is group circuit class therapy exactly? What are the theoretical underpinnings to this mode of therapy service delivery? What exactly is the current evidence base for group circuit class therapy and where do the gaps exist? This chapter will provide some answers to these questions, as well as providing tools for clinicians who may be interested in implementing this form of therapy delivery.

## 2. What is circuit class therapy?

Circuit class therapy, in its most basic interpretation, is progressive task-specific therapy provided in a group setting, as opposed to therapy provided with a one therapist to one patient ratio. As it is a relatively new concept in stroke rehabilitation, there is as yet no consensus opinion regarding the definition of circuit class therapy. The only published, specific description of circuit class therapy states: "circuit class therapy is therapy (that is) provided to more than 2 participants, involving a tailored intervention program with a focus on practice of functional tasks received within a group setting, provided to participants with similar or different degrees of functional ability and involving a staff to patient ration of no greater than 1:3" (English, et al., 2007).

Other authors have described circuit class therapy as:

- "Training, organized in a circuit with a series of workstations designed to strengthen affected muscles and provide the opportunity for task practice" (Dean et al., 2000).
- "A mode of exercise training using a series of systematically progressed workstations" (Rose et al., 2010)
- "A model of therapy delivery that utilizes active exercises and activities which are task-specific (practicing the functional task itself or parts thereof) and is provided in an intensive manner. The key components of circuit class therapy are that therapy is provided in a group setting with more than 2 participants per therapist and there is a focus on repetitive practice of functional tasks and continual progression of exercises" (English & Hillier, 2010).

Therefore, circuit class therapy is:

- *Not* recipe driven or one size fits all approach
- *Does* include tailoring and progression of exercises to suit individual participants
- *Does* include one-on-one time with individual participants
- *Does* allow for correction of movement patterns/quality of movement
- *Does* include variety in practice

The group nature of circuit class therapy is integral to the concept, as it is the mechanism by which potential cost savings can be made. However, several researchers have described implementation of circuit class therapy with a one staff member to one participant ratio (Salbach et al., 2004; Yang et al., 2006; Rose et al., 2010). In these cases participants rotated through a program of progressive, task-specific exercises, but rather than performing the exercises independently or under distant supervision, participants were under the constant supervision of a therapist. In the interests of inclusiveness these studies will be discussed in this chapter. To highlight the importance of the group nature of circuit classes, and to avoid any confusion between circuit training in individual sessions and circuit class therapy provided in a group setting, we will refer to group circuit class therapy throughout this chapter.

While group therapy can be used for a variety of aims, including improving speech and language (Simmons-Mackie & Elman 2011) or providing education (Mudge et al., 2009; Harrington et al., 2010; Marsden et al., 2010), this chapter is devoted to the concept of using circuit class therapy for improving motor function. This can include motor function of the arm or leg; however the majority of research to date has focused on outcomes relating to leg function, such as standing balance and walking.

### 2.1 Who can deliver group circuit class therapy?

To date, the majority of studies investigating the effectiveness of group circuit class therapy have involved physiotherapists delivering or overseeing the implementation of the therapy sessions. However, there is no reason why other professionals with exercise or movement-based training cannot also deliver group circuit class therapy. Health professionals who deliver exercise or movement based therapy include physiotherapists, occupational therapists, exercise physiologists, sports therapists and people trained as assistants or aides to these professions. The roles these professionals play can differ from country to country, or even within countries and between different hospitals or rehabilitation centres within the same city or area. For example in some settings in Australia the role of providing arm therapy sits solely with occupational therapists, whereas in other settings it is shared between physiotherapists and occupational therapists. While there is no evidence as to which health profession may be better able to provide group circuit class therapy, it is essential that the therapist has the skills and qualifications to be able to clinically reason the issues the stroke survivor is having with his/her mobility, design appropriate exercises to address those issues and be able to progress, update and modify the prescribed exercise program. In order to do this effectively, the therapist must also have a detailed understanding of the clinical condition of stroke.

### 2.2 In what settings can group circuit class therapy be delivered?

Group circuit class therapy can be delivered by a variety of health professionals, and it can also be delivered in a variety of settings. In this chapter, inpatient settings refer to both acute and rehabilitation hospitals in which patients spend the majority of their days and nights, with the exception of day leave or overnight leave. Outpatient settings refer to sites where therapy is provided to people who are living at home or elsewhere. We have chosen to define periods of time post stroke as either 'early' (less than 6 months since stroke onset) or 'late' (6 months or more post stroke onset).

*Group circuit class therapy provided early after stroke*

The majority of people who have a stroke are admitted to an acute hospital. There is one published paper reporting on the use of group circuit class therapy for people in acute hospitals. Zanker et al., (2007) reported on the feasibility and sustainability of group circuit class therapy provided to people with mixed diagnoses on a stroke, aged care and neurological ward in an acute hospital in Australia. In a more recent study, Rose et al (2010) described the implementation of circuit class therapy for people very early (on average 10-11 days) after stroke, but these sessions were provided with a participant to staff ratio of 1:1. Table 1 summarises published papers reporting on the use of group circuit class therapy in the acute hospital setting including the reported aims of the classes, types of participants included and the duration and frequency of classes.

| Author | Country | Aims of classes | Duration and frequency | Diagnoses of participants (mean time since stroke in days) | Participant mix (minimum level of functional ability) | Adjunct or sole method of physiotherapy delivery | Participant to staff ratio |
|---|---|---|---|---|---|---|---|
| Zanker et al 2007 | Australia | Increasing activity. Mix of mobility related activities and arm activities | Twice weekly, 40-50 minutes per session | Stroke and other neurological disorders and elderly medical patients with physical deconditioning | Sufficient sitting balance to sit in a wheelchair for at least one hour. Able to follow simple instructions. | Adjunct | Maximum of 12 participants and two staff |
| Rose et al 2010 | United States of America | Improving mobility | One 60 minute CCT session and one 30 minute individual session, five days a week | Stroke (10) | Able to follow simple commands. | Sole method | 1:1 ratio* |

*personal communication with the author

Table 1. Details of published papers describing circuit class therapy provided in acute hospital settings.

Therefore, while group circuit class therapy can theoretically be provided to people in acute hospital settings very early after stroke, and there are anecdotal reports of this occurring, there is only two published studies supporting its use in this context. This may be due to the additional barriers to providing group therapy in this setting, such as patient drowsiness precluding tolerance of longer therapy sessions, competing appointments for requisite medical investigations and other therapies, difficulties finding adequate space for the therapy, as well as difficulties transporting patients to and from the therapy area.

Following a period in an acute hospital that may last anywhere between a few days and several weeks, there are, broadly speaking, two primary models of providing rehabilitation care to people within the first month or two following stroke. In many places in Australia, New Zealand and Europe, people stay as inpatients in rehabilitation wards where they receive daily, multi-disciplinary therapy. The length of time people stay in rehabilitation hospitals varies between one week and one, two or several months. Other models of care include early supported discharge where people go home very early after stroke, and receive therapy either in their own homes or at outpatient therapy centres some or all days of the week. Furthermore, in some countries bouts of intensive, inpatient therapy is offered to people months or years later after stroke. Group circuit class therapy can be implemented in all of these scenarios, with the exception of rehabilitation in the home (where group therapy is not possible).

During this early, intensive rehabilitation period, group circuit class therapy can be delivered as either an adjunct to usual care therapy, or as the sole method of therapy service delivery. There are three published studies investigating the use of circuit class therapy in this early rehabilitation phase, and another, as yet unpublished trial (van de Port et al., 2009). Table 2 summarises the key elements of group circuit class therapy as defined in these papers.

These papers describe quite distinct models of providing group circuit class therapy. The trial by English et al (2007) investigated the use of group circuit class therapy as the sole method of physiotherapy service delivery to people receiving inpatient rehabilitation after stroke. Compared to the other group circuit class therapy trials, the exercises and structure of the classes in this trial were less strictly defined. The classes were tailored to address multiple aims including arm therapy, and included participants with a range of levels of ability, not just those who were able to walk independently. Therefore, rather than rotating around a set number of pre-determined 'stations', participants undertook their own individualised exercise program within a group setting. Classes also included group activities and activities in pairs.

In the two other published trials conducted in an inpatient rehabilitation setting for people within the first 6 months of stroke, (Blennerhassett & Dite 2004; Outermans et al., 2010) participants were offered daily group circuit class therapy sessions in addition to usual care therapy. In both of these trials participants were required to walk independently to be eligible for inclusion and the structure of the classes was more formally defined. Details of the content of therapy provided in this and other published group circuit class therapy trials is outlined in Section 2.5. In the trial by Blennerhassett & Dite (2004), participants were randomised to receive either group circuit class therapy sessions aimed at improving mobility and balance, or group circuit class therapy aimed at improving arm function. All participants in the trial by Outermans et al (2010) received group circuit class therapy aimed at improving walking and walking-related function, with the difference between groups being the intensity of exercise.

| Author | Country | Aims of classes | Duration and frequency | Diagnoses of participants (mean time since stroke in days) | Participant mix (minimum level of functional ability) | Adjunct or sole method of physiotherapy delivery | Setting | Participant to staff ratio |
|---|---|---|---|---|---|---|---|---|
| Blennerhasset and Dite 2004 | Australia | Improve walking and walking-related function OR arm function | 60-minute session, once per day, 5 days a week | Stroke (43) | Walk 10 m without physical assistance | Adjunct | Inpatient | 1:4 |
| English et al 2007 | Australia | Improve walking and walking-related function AND arm function | 90-minute sessions twice daily, 5 days a week | Stroke (39) | Able to sit unsupported | Sole | Inpatient | Between 1:3 and 1:6 (mean 1:4) |
| van de Port et al 2009 | Netherlands | Improve walking and walking-related function | 90-minute session twice a week for 12 weeks | Stroke (not known) | Walk 10 m without physical assistance | Sole | Outpatient | Between 1:4 and 1:6 |
| Outermans et al 2010 | Germany | Improve walking and walking-related function | 45-minute sessions 3 times a week for 4 weeks | Stroke (23) | Walk 10 m without physical assistance | Adjunct | Inpatient | Not stated |

Table 2. Details of published papers describing circuit class therapy provided early after stroke in settings other than acute hospitals.

The other trial conducted in the early rehabilitation period was the FIT-Stroke trial recently completed in the Netherlands (van de Port, 2009). In this study, participants were recruited on completion of their inpatient rehabilitation if they were less than six months post-stroke and were able to walk independently. Participants attended an outpatient setting twice a week and received group circuit class therapy as the sole method physiotherapy service delivery, with the exception that arm therapy was also provided to some participants outside the group circuit class therapy sessions.

*Group circuit class therapy provided later after stroke*

The majority of group circuit class therapy trials have been conducted with participants later (typically at least six months, and up to several years) post stroke. It is this group of trials that provide the greatest evidence for the effectiveness of group circuit class therapy for people after stroke. Most papers describe group circuit class therapy being provided with the primary aim of improving mobility while the control group received either 'sham' therapy, or 'usual care' therapy. The trial by Pang et al (Pang et al., 2005; Pang et al., 2006) compared people receiving either circuit class therapy aimed at mobility or at improving arm function. However, the inclusion criteria for the trial (independent walking ability and ability to raise heart rate to at least 60% of maximum heart rate) suggested that the primary research question was centred upon mobility, as there was no inclusion criterion related to arm function. The settings in which group circuit class therapy was conducted included local community halls, local hospitals and rehabilitation centres. Table 3 describes the key elements of these papers.

It is clear that the majority of trials of group circuit class therapy have included participants several months after their stroke. In these trials, circuit class therapy was provided one to three times a week for between four and 19 weeks. The staff to patient ratio ranged from 1:1 to up to seven participants per therapist (Mead et al., 2007). With the exception of two trials (Harrington et al., 2010; Marsden et al., 2010), participants in all trials were able to walk independently at enrolment. All trials indicated that group circuit class therapy was being provided with the intent to improve walking ability, although additional aims including balance, community reintegration and arm function were mentioned in some of the trials. All trials reported positive outcomes for at least some of the outcomes measured. An in-depth reporting of the evidence base for group circuit class therapy based on all published trials is reported in Section 2.4.

### 2.3 What are the theoretical underpinnings of group circuit class therapy?

It is well established that active and repetitive practice of movement is required to maximise neuroplasticity after stroke. Furthermore, physical activity is vital to general health and has been shown to reduce the risk of all stroke types (Lee et al., 2003). This section will review what is known about the levels of physical activity in people both early and later after stroke, and will present current knowledge regarding neuroplasticity and motor learning as it pertains to group circuit class therapy. There are two themes to this section - maximising physical activity to improve health and maximising neuroplasticity and recovery of function after stroke.

| Author | Country | Aims of classes | Duration and frequency | Diagnoses of participants (mean time since stroke in days) | Participant mix (minimum level of functional ability) | Setting | Participant to staff ratio |
|---|---|---|---|---|---|---|---|
| Dean, et al., 2000 | Canada | Improve walking ability | 1 hour 3 times a week for 4 weeks | Stroke (658) | Able to walk 10m independently | Outpatient (rehabilitation clinic) | 2:5 |
| Salbach et al., 2004 | Canada | Improve walking ability | 3 times a week for 6 weeks (duration not stated) | Stroke (228) | Able to walk 10m independently | Outpatient | 1:1* |
| Pang et al., 2006/2005 | Canada | Improve leg function/walking OR improve arm function | 60-minute sessions, 3 times a week for 19 weeks | Stroke (1881) | Able to walk 10 m independently, able to raise HR to at least 60% HR max | Outpatient (community hall) | Between 1:3 and 1:4 |
| Marigold et al., 2006 | Canada | Improve walking ability and standing balance | 1 hour 3 times a week for 10 weeks | Stroke (1351) | Able to walk 10m independently | Outpatient (community hall) | 1:3 |
| Yang et al., 2006 | Taiwan | Improve leg strength and walking ability | 30-minute session 3 times a week for 4 weeks | Stroke (1927) | Able to walk 10m independently | Not stated | 1:1 |
| Mead et al., 2007 | United Kingdom | Improve cardiovascular fitness and walking ability | 75-minute sessions 3 times a week for 12 weeks | Stroke (159) | Able to walk independently (minimum distance not specified) | Outpatient (rehabilitation centre) | Up to 1:7 |
| Mudge et al., 2009 | New Zealand | Improve walking ability | 3 times a week for 4 weeks | Stroke (1424) | Able to walk 10m independently | Outpatient (rehabilitation clinic) | 1:3 |
| Harrington et al., 2010) | United Kingdom | Mobility and community integration | 1-hour sessions 2 times a week for 8 weeks (followed by an education session) | Stroke (not reported) | Living in the community | Outpatient (community hall) | Up to 1:9 (carers or family members also present) |
| Marsden et al., 2010 | Australia | Mobility and community integration | 1-hour sessions once a week for 7 weeks (followed by an education session) | Stroke (1159) | Living in the community, have a carer | Outpatient (local hospital) | Not stated |

*personal communication with author

Table 3. Details of published papers describing circuit class therapy provided later after stroke.

*How active are stroke survivors?*

A recent systematic review identified 24 studies conducted between 1980 and 2009 which measured activity levels of stroke survivors in hospital (West & Bernhardt 2011). Fifteen of these studies measured patient activity over the whole day and 10 examined the physical activity of people in therapy sessions specifically. Of the studies that examined physical activity levels over the whole day, all used a technique known as behavioural mapping. Behavioural mapping involves participant being observed for a one minute period every 10 minutes and their activities, where they were and who they were with are recorded on a checklist. From this data, the time spent in various levels of physical activity (none, low, moderate to high) is estimated as minutes. Data can also be expressed as a percentage of observations. Participants were observed for the majority of the working day - for example from 7 or 8am in the morning until 5, 6 or even 7pm at night. The authors found that stroke survivors in hospital spent the majority of the day inactive (median 48.1%, inter-quartile range [IQR] 39.6-69.3), and alone (median 53.7%, IQR 44.2%-60.6%) (West & Bernhardt 2011). When only the category of moderate to high physical activity is examined (this includes activities such as standing, walking and using stairs) only a median of 21%(IQR 12.8 to 27.7%) of the day is spent in these activities (West & Bernhardt 2011).

Of the 10 studies examining physical activity levels in therapy sessions, a variety of techniques were used in including behavioural mapping, video-taping and therapist report (West & Bernhardt 2011). Within these studies, participants were reported to be inactive for between 20% and 58% of therapy sessions. A recent study found that therapists systematically overestimate active time by a mean of 28% and underestimate inactive time by a mean of 36% (Kaur 2011). Therefore the validity of relying on therapists' estimates of how much time their clients spend physically active in therapy sessions is questionable. In studies in which used video-taping of therapy sessions to examine content of therapy sessions, participants are reported to be inactive for between 30 and 40% of their therapy sessions (Kaur 2011). Interestingly, the percentage of time spent active appears similar in individual therapy sessions compared to group circuit class therapy sessions (Elson et al., 2009). In a more recent study, again using video footage of therapy sessions, inactive time was higher in group circuit class therapy sessions (37%) compared to individual therapy sessions (28%) (Kaur 2011). The longer duration of group circuit class therapy sessions meant that participants were active for more time in these sessions (mean of 44 minutes compared with 24 minutes in individual therapy sessions). The long periods of inactive time in therapy sessions, coupled with the finding that therapists systematically overestimate active time and underestimate inactive time (Kaur 2011) suggest that therapists should be mindful of maximising the time participants spend active in group circuit class therapy sessions.

Later after stroke, for those people who regain sufficient function to live in their own home, the picture of activity levels remains poor. Table 4 summarises findings from studies of activity levels of people living in the community after stroke. The majority of these studies used objective measures of activity by using accelerometers or pedometers to estimate the number of steps taken each day. While there are known inaccuracies with these measures, as most devices tend to underestimate step counts in people who walk slowly (Taraldsen et al., 2011), the consistent picture is one of very low levels of physical activity. The studies which measured daily step counts reported they ranged as low as 1400 (Michael & Macko 2007) to 4000 (Fulk et al., 2010) steps per day. The exception was the study by Mudge et al., (2007) which found participants took an average of almost 6000 steps per day. Even so, compared to published norms of 6000 to 8500 steps per day for healthy older adults (Tudor-Locke & Myers 2001), these trials show that people after stroke are generally very inactive.

Therefore, stroke survivors are significantly less active than their non-stroke-affected counterparts. Activity levels are generally poor in the rehabilitation setting and following discharge home, which can have negative flow on effects to general health and well-being. It is known that increasing the amount of activity patients engage in each day is beneficial for their general health, so health care providers should tailor rehabilitation programs to stroke survivors to increase their activity levels. Group circuit class therapy is one method of therapy service delivery which facilitates increased activity.

| Author | Country | Participants | Main findings |
|---|---|---|---|
| Michael & Macko 2007 | USA | n=79<br>Able to walk independently | Low step counts (1389±797 per day) and almost none at high intensity (78 ±168 steps per day at >30 steps per minute) |
| Michael et al., 2005 | USA | n = 50<br>Able to walk independently | Low step counts 2837±1503 per day |
| Rand et al., 2010 | Canada | n=40<br>Able to walk independently | The range of measured daily energy expenditure was vast and suggested that some participants spent most of the day sitting in a chair while others were relatively active. |
| Shaughnessy et al., 2005 | USA | n = 19<br>'mild to moderate hemiplegic gait deficits' | Daily step counts 1536±10 at 2 weeks post-discharge from inpatient rehabilitation to the community and 2765±1677 at 3 months after discharge. |
| Haeuber et al., 2004 | USA | n = 17<br>Able to walk independently | Mean steps per day 3035±1944 (range 400 to 6472). |
| Fulk, et al., 2010 | USA | n = 32<br>Able to walk independently at a speed of 0.4 m/s or greater | Mean steps per day 3838±1963 versus age matched controls 6294±1768 |
| Mudge, et al., 2007 | New Zealand | n= 58<br>Able to walk independently | Mean steps per day 5719±3453 |

Table 4. Physical activity levels of community dwelling stroke survivors.

*Neuroplasticity*

Clinicians working with people after stroke should be aware of the key principles underpinning neuroplasticity in order to assist their patients in maximising motor recovery. Neuroplasticity is a term that refers to the process of the brain remodelling and 'rewiring' new connections in response to experience. It is the mechanism by which the brain learns new behaviours and relearns lost behaviours in response to rehabilitation (Kleim & Jones 2008; Cramer et al., 2011). There is now widespread and robust evidence that the brain is constantly undergoing remodelling in response to experience (Kleim & Jones 2008) and this occurs in the healthy brain (Elbert et al., 1995; Karni et al., 1998; Kolb 2003) as well as the brain affected by neurological damage, including stroke. Several studies have now shown evidence of neuroplastic changes in the brain that correlate with functional recovery after stroke, and more specifically, changes that occur in response to motor therapy (Friel et al., 2000; Johansen-Berg et al., 2002; Green 2003). For a comprehensive overview of neuroplasticity principles as they relate to neurological rehabilitation, see Kleim & Jones (2008).

The fundamental principle that underpins neuroplasticity is 'use it or lose it'. There is now a wealth of evidence that areas of the healthy brain that are used repeatedly, such as the area receiving sensory input from the reading finger of a Braille reader or the fingering hand of a violin player, increase in size and connections in response to the increased sensory input (Pascual-Leone & Torres 1993; Elbert, et al., 1995). Likewise, areas of the healthy brain that are not used, for example when a limb is restrained or amputation occurs, lose function and connections (Cohen et al., 1991; Liepert et al., 1995).

The brain affected by stroke therefore needs to receive repeated, specific sensory input in order to remodel in the most effective way for the stroke survivor's functional recovery. Researchers have now demonstrated that thousands of repetitions of a new task are required to establish changes in the stroke-affected brain (Boyd et al., 2010) and this repetition needs to continue even after functional, behavioural changes are seen for lasting change to occur (Kleim & Jones 2008). Specificity and salience are also important - tasks to be practiced must be specific to the function that is being retrained, and be meaningful and provide reward to the person performing the task. Variety in practice is important to increase skill (Shumway-Cook & Woolacott 2001). When a person performs a task that is too easy, such as repetitive practice of an unskilled movement, it is not likely to provide lasting neuroplastic changes (Kleim et al., 1996; Plautz et al., 2000; Boyd et al., 2010).

Clinically, this means that in order to optimise a stroke survivor's neuroplastic changes and functional recovery, therapeutic exercises should be challenging to the patient, but be tailored so the person can gain reward from performing the exercise. Part-practice and tasks that are as closely related as possible to the new skill being learnt should be prescribed and performed. In addition, patients should practice under slightly different conditions; for example sit to stand using different chairs, reaching using different objects, walking over different surfaces, at different speeds and so on. All tasks should be practiced even after functional and behavioural changes are observed.

New evidence is emerging which suggests that not all practice needs to be active. Mirror neurones in the premotor and parietal cortex have been shown to activate when a person observes a movement as well as when that person performs a movement (Garrison et al.,

2010). In other words, when an action is observed, the brain generates activity that is similar to what occurs when the action is actually performed (Garrison et al., 2010). This is thought to be the mechanism by which observation of movement or actions enhances motor learning. Research into how best to capitalize on this function is in its infancy, but there is evidence that observation of movement in conjunction with actual physical practice improves motor learning (Mattar & Gribble 2005). Observation should be congruent; what is observed should be the same as what is practiced (Garrison et al., 2010). The group nature of group circuit class therapy may be an ideal format to incorporate observation of movement with physical practice.

### 2.4 What is the evidence for the effectiveness of circuit class therapy for stroke survivors?

Several meta-analyses and subsequent research trials have shown that task-oriented exercises are effective in improving balance, transfers, gait and gait-related activities (such as climbing stairs) in stroke survivors, especially when applied within the first six months post stroke. (Kwakkel et al., 2004; Van Peppen et al., 2004; French et al., 2009; Veerbeek et al., 2011). Recently, two independently conducted meta-analyses (English & Hillier 2010; Wevers et al., 2009) studied the effect of circuit class therapy on walking related outcomes. In total nine trials were included in these two reviews (Salbach et al., 2004; Pang et al., 2005; Marigold et al., 2006; Pang et al., 2006; Yang et al., 2006; Mead et al., 2007; Mudge et al., 2009). One trial was reported in two papers (Pang et al., 2005; Pang et al., 2006). Due to slightly different aims and inclusion criteria, three studies were included in both reviews (Dean et al., 2000; Blennerhassett & Dite 2004; Pang et al., 2005; Pang et al., 2006). The review of Wevers et al (2009) only included randomised controlled trials, whereas the review of English & Hillier (2010) also included non-randomised, controlled clinical trials. In addition, the review by English & Hillier (2010) primarily focussed on studies using the 6 minute walking test as an outcome measure and only included studies in which circuit class therapy was provided in a group setting. Table 5 summarises the key evidence relating to group circuit class therapy effectiveness.

*Walking distance*

Both meta-analyses showed that circuit class therapy has a positive effect in terms of improving walking distance as measured by the 6 minute walk test. Mean differences of 42.5 meters (Wevers et al., 2009) and 76.6 meters (English & Hillier 2010) were reported. The study by Blennerhassett & Dite (2004), which was conducted in the sub-acute phase after stroke, showed the largest absolute improvement, with a difference of 116m between the intervention and the control group. This difference is greater than the 13% change which has been reported as the minimum clinically relevant change (Flansbjer et al., 2005; Fulk et al., 2008). Three more studies also showed clinically relevant changes (Salbach et al., 2004; Pang et al., 2005; Mudge et al., 2009). A recent pilot study, published after the meta-analyses, showed that high intensity circuit class therapy resulted in significantly greater walking distances compared to low intensity circuit class therapy, although both groups showed relatively small changes (5-10%) (Outermans et al., 2010).

*Walking speed*

Positive results were reported for walking speed when comparing circuit class therapy with control interventions (English & Hillier 2010; Wevers et al., 2009). Five included studies measured walking speed and only the study of (Dean et al., 2000) was included in both meta-analyses. Mean overall improvements were calculated as 0.07 m/s (Wevers et al., 2009) and 0.12 m/s (English & Hillier 2010) following the circuit class therapy interventions. However, as the minimal clinically significant difference in walking speed for people recovering from stroke has been estimated to be 0.16 m/s (Tilson et al., 2010), neither meta-analysis was able to demonstrate that circuit class therapy led to a clinically meaningful difference. Only the studies of (Yang et al., 2006; Mudge et al., 2009), and the pilot study of (Outermans et al., 2010) showed clinically relevant changes in walking speed.

*Impairments*

Leg muscle strength was reported in three studies. Yang et al (2006) concluded that leg muscle strength (hip, knee and ankle) was significantly improved after individually supervised circuit class therapy compared to no therapy. In line with these results (Pang et al., 2005) reported significantly more gain in paretic leg muscle strength in the circuit class therapy group compared to the control group. In both studies a performance test using a handheld dynamometer was used to determine muscle strength. Mead et al. (2007) also studied leg power but no significant differences between the circuit class therapy group and the control group were found.

Two studies examined oxygen consumption determined by a maximal exercise test ($VO_2$ max) (Pang et al., 2005) or during walking ($VO_2$ mL/kg per metre) (Mead et al., 2007). Both trials found benefits related to group circuit class therapy. In the trial by Mead et al (2007), walking economy improved significantly more in the group receiving circuit class therapy compared to the group receiving relaxation therapy while Pang et al (2005) found that participants receiving group circuit class therapy showed significantly more improvement in cardiorespiratory fitness post-intervention, compared to the control group.

*Activities and Participation*

Evidence regarding the effect of circuit class therapy on improving standing balance is somewhat unclear, which may be in part due to the variety of outcome measures used. The Berg Balance Scale (Salbach et al., 2004; Pang et al., 2005; Marigold et al., 2006; English et al., 2007) and the Timed Up and Go test (Dean et al., 2000; Blennerhassett & Dite 2004; Salbach et al., 2004; Marigold et al., 2006; Yang et al., 2006; Mead et al., 2007; Mudge et al., 2009) were the most common measures used. Both meta-analyses (English & Hillier 2010; Wevers et al., 2009) showed no significant differences for the Berg Balance Scale between the intervention and control groups. In contrast to the meta-analysis by English & Hillier (2010), the meta-analysis of Wevers et al., (2009) showed a significant difference in the Timed Up and Go test between the intervention and the control groups. The patients included in the study of (Blennerhassett & Dite 2004) showed the greatest improvements. The Activities Balance Confidence scale was used in two studies (Marigold et al., 2006; Mudge et al., 2009) included in the meta-analysis by English & Hillier (2010) and showed a significant overall effect. The Step Test was used in two studies (Dean et al., 2000; Blennerhassett & Dite 2004) and showed an overall significant effect in favour of the intervention group (English & Hillier 2010). In the review of Wevers et al., (2009) the study of Yang et al (2006) was included, which also reported step test data. Inclusion of this trial in the meta-analysis led to non-significant findings. One included study used the Functional Reach test and showed no significant differences between the groups (Mead et al., 2007). Results of a more recently published randomised controlled trial showed no significant differences between groups using the Functional Reach test and the Berg Balance Scale (Outermans et al., 2010).

Several different outcome measures were used to determine participants' abilities to perform activities of daily living. However, no published studies reported significant between group differences on any of the measures.

Of the trials that included mobility outcomes such as the Rivermead Mobility Index (Mead et al., 2007; Mudge et al., 2007) or the Elderly Mobility Scale (Mead et al., 2007), none showed significant improvements related to circuit class therapy. In the study of (English et al., 2007) the degree of physical assistance required to walk was measured by the Iowa Level of Assistance Scale. In this trial, significantly more people who received group circuit class therapy were able to walk independently on discharge from inpatient rehabilitation when compared to those receiving traditional one-to-one therapy.

*Service related outcomes*

Length of hospital stay was calculated in two studies (Blennerhassett & Dite 2004; English et al., 2007) and although both studies showed trends in favour of the intervention group, no statistically significant differences were found. However, when data were pooled, the difference between the intervention and control group was statistically significant (English & Hillier 2010) and participants receiving group circuit class therapy had a mean length of hospital stay of almost 20 days less than the control group participants.

Most studies reported on falls occurring during the intervention or the whole study period. In the study of (Marigold et al., 2006) patients kept a falls dairy. No statistically significant differences were reported between the intervention and the control group. Several studies found that more falls were reported among participants receiving circuit class therapy. English et al., (2007) reported four falls occurred during group circuit class therapy sessions compared to two during usual care (individual therapy sessions), Pang et al., (2005) reported five falls in the circuit class therapy group and one in the control group and Salbach et al., (2004) reported four falls occurring during individually supervised circuit class therapy. All authors reported that none of the falls resulted in injury. Two trials (English et al., 2007; Mead et al., 2007) also examined the differences in the rate of falls outside of therapy sessions with neither reporting any between group differences. Overall, no serious adverse events were reported which suggests that group circuit class therapy is a safe intervention.

*Gaps in the evidence*

From the studies described above, it can be concluded that circuit class therapy can be an effective method to provide therapy to people following stroke, particularly in relation to improving their walking speed and walking distance. However, several gaps remain in our knowledge regarding the evidence of this approach.

The aspect which makes group circuit class therapy especially attractive, but which has been studied very little, is the cost-effectiveness of the intervention. Since circuit class therapy is provided in groups and patients are not treated with a staff to patient ratio of 1 to 1, this should reduce the costs of the treatment and make group circuit class therapy more cost effective than other therapy interventions. In the study of English et al (2007), six patients were included in the group circuit class therapy sessions under the supervision of one physical therapist and one assistant, leading to a 1:3 staff to patient ratio. The researchers found that participants receiving the group circuit class therapy received a significantly greater amount of therapy than those stroke survivors receiving traditional one-to-one therapy (129 minutes per day compared to 37 minute per day). Therefore, without changing staffing-associated costs, changing the method of service delivery resulted in substantial increases in therapy time for stroke survivors. In addition to providing more therapy to the patients, group circuit class therapy saved the therapist time - for a therapist to provide group circuit class therapy to six patients it took 129 minutes a day, whereas the amount of therapist time required to provide individual therapy sessions for 6 patients was 222 minutes a day (English et al., 2007).

Economic evaluations include not only the costs of the intervention, but also the costs related to the use of health care facilities (primary and secondary care, community care) social care costs (domestic care, meals on wheels), medication and transport. None of the included studies reported on these sorts of costs. Although not completely comparable, the study of (Harrington et al., 2010) included an economic evaluation when studying the effects of a community-based exercise and education scheme for stroke survivors (Harrington et al., 2010). They concluded that the costs were significantly higher in the intervention group compared to the control group receiving usual care. The reason for the higher costs was unexplained, but the method used to analyse the costs can be implemented in future studies. It is highly important to include cost-effectiveness analysis in future trials, since reduced staff to patient ratios will lead to lower staff-related costs without lowering the amount of therapy, which makes circuit class therapy a very attractive treatment method for the growing stroke population.

When looking at the published papers on group circuit class therapy a few things need to be considered. Most studies included patients who were able to walk and were free of significant cognitive problems or language impairments. It is essential that participants understand the set tasks at each of the workstations, however, with appropriate supervision and instruction, participants with minor cognitive problems and aphasia should be able to participate in group circuit class therapy. Therefore these patients should be included in future research. Likewise, group circuit class therapy can be tailored to suit people with moderate to severe mobility impairments (ie people unable to walk independently), which would allow inclusion of patients earlier after their stroke. In the light of the study by Blennerhassett & Dite (2004), which was conducted in the more acute phase after stoke and showed some of the most positive results from all the trials, it would seem that the early post-stroke period is an appropriate time to provide group circuit class therapy. More research is needed on the effects of group circuit class therapy in the early phase after stroke, for example including patients in acute hospital settings, and especially including patients with mobility, communication and cognitive limitations.

It would be interesting to do further study regarding the content of group circuit class therapy. Most tasks from the published studies were focussed on the leg, but studying the effects of arm tasks would be helpful to further improve arm rehabilitation. More research into the effects of educational or behavioural components would be beneficial to build on the positive results reported by Harrington et al (2010) and Marsden et al (2010) particularly in regard to reducing lifestyle stroke risk factors and increasing activity levels, both in the hospital setting and in the community. And while stroke recovery guidelines and research studies suggest that therapy should be given with a high intensity, the exact optimal dose-response relationship remains unclear and needs further study.

| **Study** | **Design** | **Parameter** | **Outcome measure** | **Weighted mean difference** |
|---|---|---|---|---|
| Wevers et al., 2009 | Systematic review | Walking capacity | 6 minute walk test | 42.5 m* |
| | | Walking speed | Gait speed (comfortable) | 0.07 m/s* |
| | | Mobility | Timed Up and Go test¥ | -3.3 seconds* |
| | | Postural control in standing | Step Test | 2.8 seconds |
| | | | Berg Balance Scale | 0.63 points |
| | | | | |
| English & Hillier 2010 | Systematic review | Walking capacity | 6 minute walk test | 76.6 m* |
| | | Walking speed | Gait speed (comfortable) | 0.12 m/s* |
| | | Mobility | Timed Up and Go test¥ | -3.08 seconds |
| | | Postural control in standing | Step Test | 3.00 seconds* |
| | | | Berg Balance Scale | 0.86 points |
| | | | Activities Balance Confidence Scale | 7.76 points* |
| | | Length of stay | Number of days in rehabilitation hospital | -19.73 days* |
| | | | | Mean difference |
| Outermans et al., 2010 | Randomised controlled trial | Walking capacity | 6 minute walk test | 32.6 m* |
| | | Walking speed | Gait speed (fastest) | 0.3 m/s* |
| | | Postural control in standing | Berg Balance Scale | 0.1 points |
| | | | Functional Reach test | -0.4 cm |

*denotes difference was statistically significant

¥This result does not include data from Mead et al (2007) as the paper did not report baseline values

Table 5. Summary of evidence for the effectiveness of circuit class therapy.

Very few studies to date have measured outcomes related to quality of life or psycho-social outcomes. Future research should include these outcomes to further enhance the understanding of the effect of group circuit class therapy on the holistic wellbeing of stroke survivors. Fatigue and depression are often problems for stroke patients (Hackett et al., 2005; Choi-Kwon & Kim 2011). Several studies have shown that physical training produced a positive effect on non-physical outcomes like fatigue and depression (Sjosten & Kivela 2006; Blake et al., 2009; Mead et al., 2009). Therefore, there is the potential that group circuit class training could be a useful modality to address post-stroke depression and fatigue, and research into this area would be very useful for both clinicians and stroke survivors.

Patient and therapist satisfaction with group circuit class therapy is another under-researched area. English et al., (2007) found there were no significant differences in satisfaction with therapy from people receiving group circuit class therapy only compared to people receiving individual therapy sessions, although people in the circuit class arm were significantly more satisfied with the amount of therapy they received. A small study by Lynch et al (2008) investigating patient satisfaction with group circuit class therapy and individual therapy in a population of stroke survivors who received both models of care found that patients were satisfied with both methods of service delivery. Future research should investigate staff perceptions of implementing group circuit class therapy, and include qualitative outcomes regarding patient and therapist satisfaction.

### 2.5 Implementing group circuit class therapy in the clinical setting

Clinical guidelines for rehabilitation following stroke from countries around the world (van Peppen et al., 2007; Intercollegiate Stroke Working Party 2008; Lindsay et al., 2010; National Stroke Foundation 2010) recommend that stroke survivors should be provided with intense task-specific practice to aid in the recovery of arm function, and restoration of balance and mobility. The Australian guidelines specify that task-specific circuit class therapy should be used to increase the amount of practice in rehabilitation. Therefore, centres providing therapy to stroke survivors should consider providing task-specific group circuit class therapy to this population.

Implementing group circuit class therapy can be difficult as it often involves a change of practice for therapists used to providing therapy individually to their clients. Changing practice in the workplace is a specialty in itself, and the matter of implementing change is worthy of a chapter in its own right, however, an attempt will be made here to summarise the crucial points required to ease the transition when introducing an alternate method of therapy service delivery such as group circuit class therapy.

The barriers to implementing group circuit class therapy are similar to those of every other model involving a workplace change - generally when people are comfortable and confident in their practice, they will only want to change things for a compelling reason. Many people resist change as the process of change itself is challenging and at times confronting. Therefore, the implementation of group circuit class therapy requires a leader who can 'sell' the product of group circuit class therapy to the staff and clients involved. The evidence surrounding the effectiveness of this form of therapy, the anecdotally reported satisfaction levels of staff and participants, along with the relative cost-efficiency when compared to individual therapy should assist in this process. In an inpatient setting, negotiating the time of the group circuit class therapy with the wider multi-disciplinary team is important to minimise impact on other disciplines' treatment planning. At sites where orderlies transport patients to therapy, planning is required if large numbers of patients will be attending therapy at one time. In an outpatient setting, organising a time that fits with participant preference and transport services would be beneficial. It is these authors' experience that when the whole team is supportive of implementing group circuit class therapy, any obstacles that arise can be dealt with in a timely and efficient manner, and the introduction of group sessions runs smoothly.

Once the decision has been made to implement group circuit class therapy, a suitable therapy space must be located. The space must be able to accommodate the extra participants and staff along with the equipment required for circuit class therapy. No specialised equipment is necessary beyond that equipment found in the standard therapy area, and the equipment needs will vary to an extent on the population of the circuit class participants. For instance, in an acute stroke ward, more exercises are likely to be chair or plinth based as participants work towards goals of independent sitting balance or more independence in transfers, whereas in an outpatient setting, more exercises are likely to be standing or walking based. In both these examples, some seated exercises would also be required if arm function is being addressed within the circuit class. In general, chairs of different heights with and without armrests, steps of different heights, a designated walking space, access to stairs, tables at which to exercise the arm, real-world items (such as money, pens and paper, eating and cooking utensils, tools specific to participants' work or hobbies) with which to exercise the arm should all be considered when organising group circuit class therapy equipment. A treadmill is an additional useful, but not essential item. Table 6 provides a summary of all the exercises and activities reported in published trials of circuit class therapy. All trials stated that tasks were individually adapted and progressed as required such that the level of complexity, difficulty and dosage (number of repetitions) matched the individual's ability. In addition, the textbook by Carr & Shepherd (2003) is a valuable resource for activities and equipment that can be used in circuit classes.

The duration and frequency of group circuit class therapy may vary, depending on the population of stroke survivors participating in the therapy (whether they be inpatients or outpatients), and may vary if, in an inpatient setting, group circuit class therapy is used as a sole form of therapy, or used in addition to individual sessions. Recommendations from the literature are at least one hour of therapy, three times a week for four or more weeks regardless of the population of stroke survivors participating (English & Hillier 2010). If group circuit class therapy is the sole form of physiotherapy for inpatients following stroke, up to three hours a day, five days per week has been used in one study (English et al., 2007) and was a sustainable and clinically feasible option.

The structure of the class will also vary, depending whether the group circuit class therapy is the only method of therapy or is used as an adjunct to individual sessions. When group circuit class therapy is the sole method of physiotherapy service delivery, programs must be well tailored to each individual participant, and exercises must be constantly progressed to ensure that rehabilitation goals are met. This means that a flexible structure is required to allow the therapist to spend adequate one-on-one time with individual participants. For example a person requiring physical assistance to stand might receive 5 minutes of standing practice with a therapist while other class participants practice independently. The patient may then rest or do leg exercises in sitting or reaching out of the base of support independently while the therapist spends time with another participant practicing challenging balance activities in standing. In contrast, when group circuit class therapy is provided as an adjunct to individual therapy sessions, the aim of the classes may be more centred on increased practice time of tasks, rather than meeting specific rehabilitation goals. Exercises should still be tailored to individuals and progressed regularly, but one-on-one time with therapists within the class setting may occur less often. In the group-circuit-class-only model of care, the occasional appointment with a therapist outside of circuit classes may also be required to address specific aims such as teaching family members how to assist their relative to get in and out of a car.

The number of staff providing group circuit class therapy needs to be determined, and this may vary depending on the nature of the participant group. For instance, in a community setting, where participants are for the most part mobilising independently, a lower staff to participant ratio will be required than in an acute stroke unit, where some participants may need assistance to sit, stand, transfer and mobilise. Suggested staff to participant ratios from the recent literature are 1:2 in very acute settings ([Zanker et al., 2007), and 1:3 or 1:4 for patients in inpatient rehabilitation (Blennerhassett & Dite 2004; English et al., 2007) and outpatient settings (Dean et al., 2000; Marigold et al., 2006; Mudge et al., 2009). Use of family or carers to provide assistance within sessions can also be considered on an individual basis, as this may ease the burden on staff, while providing valuable training to carers and family members regarding their family members' level of function, along with methods to assist their functional tasks and mobility. At least one member of the staff supervising the group circuit class therapy needs to have formal qualifications, as the supervising staff members need to have specific knowledge and skills to be able to tailor and progress the exercises appropriately for the individual participants.

Exercises within the group circuit classes must be tailored to the participants' current levels of function and their individual goals, and should be as specific as possible to the goal tasks. Therefore, it is necessary to have an assessment session or interview with each participant prior to commencing in the group circuit class. Depending on the referral source, this assessment/interview process may be conducted by the therapists delivering group circuit class therapy, or may in fact be done by the referrer, if they are a therapist knowledgeable about stroke rehabilitation. Generally, it is recommended that each participant have a program outline commenced prior to their first group circuit class session, which can be fine-tuned or progressed by the therapists supervising the sessions. It can be useful to have exercises divided into activities participants can do independently, and exercises requiring assistance so staff can manage their time efficiently, providing assistance to each participant when required, and setting them up with safe and effective exercises at other times. Exercises can be graded into levels which can help the therapist with exercise progression ideas; for example reaching to a bench in front, reaching to a stool down low, reaching to the floor. Judicious use of equipment and participant set-up can allow participants to exercise under supervision without hands-on assistance from a therapist where traditionally, a one-to-one treatment would have been used, such as positioning a participant usually requiring standby assistance next to a raised plinth or rail while performing sit-to-stand exercises. Depending on the individuals within the class, participants may be responsible for choosing and completing their exercises from their personalised programs within each session; other staff and participants may prefer to have the therapists directing how many repetitions or time spent on each specific activity. It is important that therapists progress exercises regularly in line with participants' functional abilities and goals in order for group circuit class therapy to maximise its effectiveness.

Organising participants to work together in pairs or small groups can also be an effective way of exercising as well as providing social interaction; for instance setting up participants so they throw and catch a ball to one another, alternately turn playing cards, count for each other as they perform sit-to-stand exercises, do relay races in teams for walking speed. This allows participants to watch other stroke survivors performing tasks which can enhance motor learning, as discussed in the section on neuroplasticity. It also allows the therapist time to provide one-on-one assistance to the participants as one participant performs the exercises while the other participant watches or rests (van de Port et al., 2009).

A final consideration is infection control. More of the community is being colonised with multi-resistant organisms such as multi-resistant staphylococcus aureus (MRSA) and vancomycin-resistant enterococci (VRE), and different health care sites have different regulations regarding how this population is managed. It is necessary to seek advice from local infection control authorities on how to manage people with multi-resistant organisms in a group setting if additional precautions are required. Conscientious hand-washing of staff and participants, use of personal protective equipment such as gowns and gloves by staff, wiping down of equipment and designated exercise areas should allow everyone to participate in a group session, regardless of their multi-resistant organism status.

| **Category of activity** | **Specific aim of exercise** | **Details of exercise** | **Progressions†** | **Reference/s** |
|---|---|---|---|---|
| Warm up | | Marching on spot | | Salbach et al., 2004; Marsden et al., 2010) |
| | | Walking overground or treadmill | | Blennerhassett & Dite 2004; Marigold et al., 2006 |
| | | Stationary bikes | | Blennerhassett & Dite 2004 |
| | | Stretches | | Salbach et al., 2004; Marigold et al., 2006 |
| Sitting balance | Promote loading of the affected leg and activation of the affected leg muscles | Sitting and reaching in different directions for objects beyond arm's length | | Dean et al., 2000; English et al., 2007; Marsden et al., 2010 |

†The usual progressions such as increasing weights, increasing numbers of repetitions and decreasing therapist support are not mentioned in this table

Table 6. Summary of exercises and activities reported in published trials of circuit class therapy.

| Category of activity | Specific aim of exercise | Details of exercise | Progressions† | Reference/s |
|---|---|---|---|---|
| Sit to stand | Strengthen affected leg extensor muscles and practice task of sit to stand | Repeated standing up from a chair and sitting back down | Start with high chair or perching on the edge of a plinth<br>Reduce seat height<br>No arm-rests<br>Non-affected leg on step<br>Add dual task (eg holding cup of water)<br>Increase speed | Dean et al., 2000; Blennerhassett & Dite 2004; Salbach et al., 2004; Pang et al., 2005; Marigold et al., 2006; Yang et al., 2006; English et al., 2007; Mead et al., 2007; Mudge et al., 2009; van de Port et al., 2009; Marsden et al., 2010; Rose et al., 2010 |
| Standing | Improve postural control in standing | Standing with constrained base of support, with feet in parallel and tandem conditions | Narrow base of support<br>Stand on foam<br>Eyes closed<br>Turning upper body<br>Cross arms<br>Stand on one leg | Dean et al., 2000; Pang et al., 2005; Marigold et al., 2006; English et al., 2007; Mudge et al., 2007 |
| | | Reach for objects, including down to the floor.<br>Trace spiral on a whiteboard | Practice in pairs by passing objects<br>Constrain and narrow base of support (eg stand with feet together, or in tandem) | Dean et al., 2000; Yang et al., 2006; English et al., 2007; van de Port et al., 2009; Marsden et al., 2010 |

†The usual progressions such as increasing weights, increasing numbers of repetitions and decreasing therapist support are not mentioned in this table

Table 6. Summary of exercises and activities reported in published trials of circuit class therapy. (Continuation)

| Category of activity | Specific aim of exercise | Details of exercise | Progressions† | Reference/s |
|---|---|---|---|---|
| | | Self sway in standing near a wall | Progress by increasing amplitude, then doing away from the wall | (Mudge et al., 2009 |
| | | Perturbations from a therapist | | Marigold et al., 2006 |
| | | Stepping grid (participants stand with feet in marked areas, then tap one foot out to touch marks on floor, repeating with the other foot) | | English et al., 2007 |
| | | Alternate toe tapping up onto a step in front | Tap foam cup on step without deforming it Higher step Decrease arm support | Salbach et al., 2004; English et al., 2007 |
| | | Kicking ball | Against wall, then dribbling ball around objects | Salbach et al., 2004; Pang et al., 2005; van de Port et al., 2009 |
| | | Throwing and catching balls in pairs or groups | | English et al., 2007 |
| Walking activities | Promote smooth transition between sit to stand and walking | Standing up from chair, walk short distance, return to chair | Chairs with no arms and reduced seat height | Dean et al., 2000; Salbach et al., 2004 |
| | Endurance/fitness | Walking on a treadmill | Increase incline, speed | Dean et al., 2000 |
| | | Shuttle walks/brisk walking | | Pang et al., 2005; Mead et al., 2007 |

†The usual progressions such as increasing weights, increasing numbers of repetitions and decreasing therapist support are not mentioned in this table

Table 6. Summary of exercises and activities reported in published trials of circuit class therapy. (Continuation)

| Category of activity | Specific aim of exercise | Details of exercise | Progressions† | Reference/s |
|---|---|---|---|---|
| | Improve walking speed | Walk at fastest speed | Running | Salbach et al., 2004; van de Port et al., 2009 |
| | Improve adaptability of walking skills | Obstacle courses (include over low obstacles, steps, ramps, foam surfaces) | Dual tasking (eg carrying tray of objects) Picking up objects from floor | English et al., 2007; Dean et al., 2000; Blennerhassett & Dite 2004; Salbach et al., 2004; Pang et al., 2005; Marigold et al., 2006; Mudge et al., 2009; van de Port et al., 2009; Rose et al., 2010 |
| | | Walking up and down stairs | No handrail | English et al., 2007; Dean et al., 2000; Blennerhassett & Dite 2004; Salbach et al., 2004; Marigold et al., 2006; van de Port et al., 2009; Marsden et al., 2010 |
| | | Sudden stops and turns during walking | | Salbach et al., 2004; Pang et al., 2005 |
| | | Walking different step lengths, walking between parallel lines, , braiding (crossing one foot over in front), figure eight walking | Cross arms walking on a line, heel toe walking carry objects such as shopping bags | Salbach et al., 2004; Pang et al., 2005; Marigold et al., 2006; English et al., 2007; Mudge et al., 2009; Marsden et al., 2010 |

†The usual progressions such as increasing weights, increasing numbers of repetitions and decreasing therapist support are not mentioned in this table

Table 6. Summary of exercises and activities reported in published trials of circuit class therapy. (Continuation)

| Category of activity | Specific aim of exercise | Details of exercise | Progressions† | Reference/s |
|---|---|---|---|---|
| | | Walking backwards, sideways | | Salbach et al., 2004; English et al., 2007 |
| | | Walking relay races | | Dean et al., 2000; English., 2007; Outermans et al., 2010 |
| | | Walking outdoors | | English et al., 2007 |
| Leg strengthening in weight-bearing positions | Strengthen ankle plantarflexor muscles | Heel lifts in standing | Perform on a wedge/ramp<br>Perform in single leg stance<br>Jumps | Dean et al., 2000; Pang et al., 2005; Yang et al., 2006; English et al., 2007; Mudge et al., 2009 |
| | Strengthen lower leg extensors | Stepping forward, backward and sideways onto blocks of various heights (participant places affected foot on a step either to the side or in front and raises him/herself onto the step) | Increase step heights | Dean et al., 2000; Blennerhassett & Dite 2004; Yang et al., 2006; English et al., 2007; Mudge et al., 2009; van de Port et al., 2009 |
| | | Squats | Increase angle of knee flexion | English et al., 2007 |
| | Eccentric quads control | Participant stands on a step and lowers unaffected leg to the ground | Touch foot to a foam cup placed on the ground without crushing it before returning to the start position | English et al., 2007 |

†The usual progressions such as increasing weights, increasing numbers of repetitions and decreasing therapist support are not mentioned in this table

Table 6. Summary of exercises and activities reported in published trials of circuit class therapy. (Continuation)

| Category of activity | Specific aim of exercise | Details of exercise | Progressions† | Reference/s |
|---|---|---|---|---|
| Other leg strengthening exercises | | Reciprocal leg flexion and extension using Kinetron in standing | | Dean et al., 2000 |
| | | Stationary bike riding | | English et al., 2007 |
| | | Active hamstrings in sitting (sitting on chair, flex knee backwards) | Use towel or 'slippery sam' material to reduce friction to make easier, or strap on weights to make it harder | English et al., 2007 |
| | | Standing hams curl | | Mudge et al., 2009 |
| | | Lunges | | Mudge et al., 2009 |
| | | Side leg lifts | | Mudge et al., 2009 |
| | | Marching in place | Marching on a mini-tramp | Marigold et al., 2006; Mudge et al., 2009 |
| Endurance/fitness | | Cycling ergonometry | | Mead et al., 2007 |
| | | Raising and lowering 1.4kg medicine ball | | Mead et al., 2007 |
| Arm strengthening | | Resistance band exercises for shoulder flexion, abduction, extension, external rotation | | Pang et al., 2006; English et al., 2007; Mead et al., 2007 |

†The usual progressions such as increasing weights, increasing numbers of repetitions and decreasing therapist support are not mentioned in this table

Table 6. Summary of exercises and activities reported in published trials of circuit class therapy. (Continuation)

| Category of activity | Specific aim of exercise | Details of exercise | Progressions† | Reference/s |
|---|---|---|---|---|
| | | Cuff weights for elbow flexion and extension, wrist extension and flexion | | Pang et al., 2006; English et al., 2007 |
| | | Hand muscle strength (putty and grippers, pinch, grip, finger extension) | | |
| | | Electrical stimulation for those with no wrist extension | 100 Hz, 150 microseconds, ON, 10 sec OFF time 10 sec, ramp 1 sec, treatment time 10-15 minutes | Pang et al., 2006 |
| | | Upper extremity weight-bearing on a physiotherapy ball or push up on arms of a chair | | Pang et al., 2006 |
| | | Active shoulder girdle movement with arm supported on high table, including protraction, external rotation | | English et al., 2007 |
| Arm range of movement | | Passive or self assisted range of movement exercises to paralyzed joints | | Pang et al., 2006 |

†The usual progressions such as increasing weights, increasing numbers of repetitions and decreasing therapist support are not mentioned in this table

Table 6. Summary of exercises and activities reported in published trials of circuit class therapy. (Continuation)

| Category of activity | Specific aim of exercise | Details of exercise | Progressions† | Reference/s |
|---|---|---|---|---|
| | | Prolonged shoulder positioning in either forward flexion or abduction | Use of circumferential elbow foam splint where elbow contracture or stiffness is an issue | English et al., 2007 |
| | | Active pronation and supination | | English et al., 2007 |
| Functional arm tasks | | Playing cards | Could be done in pairs | Pang et al., 2006; English et al., 2007 |
| | | Picking up objects various sizes and shapes | | Pang et al., 2006; English et al., 2007 |
| | | Taking lids on and off jars | | English et al., 2007 |
| | | Pegging washing on a line | | English et al., 2007 |
| | | Folding washing | | English et al., 2007 |
| | | Scooping coins of the edge of a table | | English et al., 2007 |
| | | Spooning or pouring water from cup to cup | | English et al., 2007 |
| | | Lifting and moving a pen around a marked grid | | English et al., 2007 |

†The usual progressions such as increasing weights, increasing numbers of repetitions and decreasing therapist support are not mentioned in this table

Table 6. Summary of exercises and activities reported in published trials of circuit class therapy. (Continuation)

## 3. References


Blake, H., P. Mo, et al. (2009). How effective are physical activity interventions for alleviating depressive symptoms in older people? A systematic review. Clinical Rehabilitation 23(10): 873-87.

Blennerhassett, J. &W. Dite (2004). Additional task-related practice improves mobility and upper limb function early after stroke: a randomised controlled trial. Australian Journal of Physiotherapy 50(4): 219-24.

Bovier, P. A., D. P. Martin, et al. (2005). Cost-consciousness among Swiss doctors: a cross-sectional survey. BMC Health Services Research 5: 72.

Boyd, L. A., E. D. Vidoni, et al. (2010). Motor learning after stroke: is skill acquisition a prerequisite for contralesional neuroplastic change? Neuroscience Letters 482(1): 21-5.

Carr, J. &R. Shepherd (2003). Stroke rehabilitation. Guidelines for exercise and training to optimize motor skill. Butterworth-Heinemann London.

Choi-Kwon, S. &J. S. Kim (2011). Poststroke fatigue: an emerging, critical issue in stroke medicine. International Journal of Stroke 6(4): 328-36.

Cohen, L., S. Bandinelli, et al. (1991). Motor reorganization after upper limb amputation in man. A study with focal magnetic stimulation. Brain 114: 615-27.

Cramer, S. C., M. Sur, et al. (2011). Harnessing neuroplasticity for clinical applications. Brain 134(Pt 6): 1591-609.

Dean, C. M., C. Richards, et al. (2000). Task-related circuit training improves performance of locomotor tasks in chronic stroke. A randomized controlled pilot trial. Archives of Physical Medicine and Rehabilitation 81(4): 409-17.

Elbert, T., C. Pantev, et al. (1995). Increased cortical representation of the fingers of the left hand in string players. Science 270: 305-7.

Elson, T., C. English, et al. (2009). How much physical activity do people recovering from stroke do during physiotherapy sessions? International Journal of Therapy and Rehabilitation 16(2): 78-83.

English, C. &S. L. Hillier Circuit class therapy for improving mobility after stroke. The Cochrane Database of Systematic Reviews 7: CD007513.

English, C. K., S. L. Hillier, et al. (2007). Circuit class therapy versus individual physiotherapy sessions during inpatient stroke rehabilitation: a controlled trial. Archives of Physical Medicine and Rehabilitation 88(8): 955-63.

Flansbjer, U. B., A. M. Holmback, et al. (2005). Reliability of gait performance tests in men and women with hemiparesis after stroke. Journal of Rehabilitation Medicine 37(2): 75-82.

French, B., L. H. Thomas, et al. (2009). Repetitive task training for improving functional ability after stroke (review). The Cochrane Database of Systematic Reviews(Issue 4, 2007).

Friel, K., A. Heddings, et al. (2000). Effects of postlesion experience on behavioural recovery and neurophysiologic reorganization after cortical injury in primates. Neurorehabilitation and Neural Repair 14: 187-98.

Fulk, G. D., J. L. Echternach, et al. (2008). Clinometric properties of the six-minute walk test in individuals undergoing rehabilitation poststroke. Physiotherapy Theory and Practice 24(3): 195-204.

Fulk, G. D., C. Reynolds, et al. (2010). Predicting home and community walking activity in people with stroke. Archives of Physical Medicine and Rehabilitation 91(10): 1582-6.

Garrison, K. A., C. J. Winstein, et al. (2010). The mirror neuron system: a neural substrate for methods in stroke rehabilitation. Neurorehabilitation and Neural Repair 24(5): 404-12.

Green, J. B. (2003). Brain reorganization after stroke. Topics in Stroke Rehabilitation 10(3): 1-20.

Hackett, M. L., C. Yapa, et al. (2005). Frequency of depression after stroke: a systematic review of observational studies. Stroke 36(6): 1330-40.

Haeuber, E., M. Shaughnessy, et al. (2004). Accelerometer monitoring of home- and community-based ambulatory activity after stroke. Archives of Physical Medicine and Rehabilitation 85(12): 1997-2001.

Harrington, R., G. Taylor, et al. (2010). A community-based exercise and education scheme for stroke survivors: a randomized controlled trial and economic evaluation. Clinical Rehabilitation 24(1): 3-15.

Intercollegiate Stroke Working Party (2008). National Clinical Guidelines for stroke, 3rd Edition. London: Royal College of Physicians.

Johansen-Berg, H., H. Dawes, et al. (2002). Correlation between motor improvements and altered fMRI activity after rehabilitative therapy. Brain 125: 2731-42.

Karni, A., G. Meyer, et al. (1998). The acquisition of skilled motor performance: fast and slow experience-driven changes in primary motor cortex. Proceedings of the National Academy of Sciences USA 95(3): 861-8.

Kaur, G. (2011). How accurate are physiotherapists at estimating therapy time in stroke rehabilitation? School of Health Sciences. Adelaide, University of South Australia. Bachelor of Physiotherapy with Honours Unpublished Thesis

Kleim, J., E. Lussnig, et al. (1996). Synaptogenesis and fos expression in the motor cortex of the adult rate after motor skill learning. The Journal of Neuroscience 16: 4529-35.

Kleim, J. A. &T. A. Jones (2008). Principles of experience-dependent neural plasticity: implications for rehabilitation after brain damage. Journal of Speech, Language and Hearing Research 51(1): S225-39.

Kolb, B. (2003). Overview of cortical plasticity and recovery from brain injury. Physical Medicine and Rehabilitation Clinics of North America 14(1 Suppl): S7-25, viii.

Kwakkel, G., R. van Peppen, et al. (2004). Effects of augmented exercise therapy time after stroke: a meta-analysis. Stroke 35(11): 2529-39.

Lee, C. D., A. R. Folsom, et al. (2003). Physical activity and stroke risk: a meta-analysis. Stroke 34(10): 2475-81.

Liepert, J., M. Tegenthoff, et al. (1995). Changes of cortical motor area size during immobilization. Electroencephalography and Clinical Neurophysiology 97: 382-6.

Lindsay, M. P., G. Gubitz, et al. (2010). Canadian Best Practice Recommendations for Stroke Care (Update 2010). On behalf of the Canadian Stroke Strategy Best Practices and Standards Writing Group. Canadian Stroke Network. Ottawa, Ontario Canada.

Lynch, E., R. Harling, et al. (2008). Patient satisfaction with circuit class therapy and individual physiotherapy. International Journal of Therapy and Rehabilitation 15(4): 167-173.

Marigold, D., J. Eng, et al. (2006). Exercise leads to faster postural reflexes, improved balance and mobility, and fewer falls in older persons with chronic stroke. Journal of the American Geriatric Society 53: 416-23.

Marsden, D., R. Quinn, et al. (2010). A multidisciplinary group programme in rural settings for community-dwelling chronic stroke survivors and their carers: a pilot randomized controlled trial. Clinical Rehabilitation 24(4): 328-41.

Mattar, A. &P. Gribble (2005). Motor learning by observing. Neuron 46: 153-60.

Mead, G. E., C. A. Greig, et al. (2007). Stroke: a randomized trial of exercise or relaxation. Journal of the American Geriatric Society 55(6): 892-9.

Mead, G. E., W. Morley, et al. (2009). Exercise for depression. The Cochrane Database of Systematic Reviews (3): CD004366.

Michael, K. &R. F. Macko (2007). Ambulatory activity intensity profiles, fitness, and fatigue in chronic stroke. Topics in Stroke Rehabilitation 14(2): 5-12.

Michael, K. M., J. K. Allen, et al. (2005). Reduced ambulatory activity after stroke: the role of balance, gait, and cardiovascular fitness. Archives of Physical Medicine and Rehabilitation 86(8): 1552-6.

Mudge, S., P. A. Barber, et al. (2009). Circuit-based rehabilitation improves gait endurance but not usual walking activity in chronic stroke: a randomized controlled trial. Archives of Physical Medicine and Rehabilitation 90(12): 1989-96.

Mudge, S., N. S. Stott, et al. (2007). Criterion validity of the StepWatch Activity Monitor as a measure of walking activity in patients after stroke. Archives of Physical Medicine and Rehabilitation 88(12): 1710-5.

National Stroke Foundation (2010). Clinical Guidelines for Stroke Management. Melbourne, Australia.

Outermans, J. C., R. P. van Peppen, et al. (2010). Effects of a high-intensity task-oriented training on gait performance early after stroke: a pilot study. Clinical Rehabilitation 24(11): 979-87.

Pang, M., J. Eng, et al. (2005). A community-based fitness and mobility exercise program for older adults with chronic stroke. A randomized controlled trial. Journal of the American Geriatric Society 53: 1667-74.

Pang, M., J. Harris, et al. (2006). A community-based upper-extremity group exercise program improves motor function and performance of functional activities in chronic stroke. A randomised controlled trial. Archives of Physical Medicine and Rehabilitation 87: 1-9.

Pascual-Leone, A. &F. Torres (1993). Plasticity of the sensorimotor cortex representation of the reading finger in Braille readers. Brain 116: 39-52.

Plautz, E., G. Milliken, et al. (2000). Effects of repetitive motor training on movement representations on adult squirrel monkeys: role of use versus learning. Neurobiology of Learning and Memory 74: 27-55.

Rand, D., J. J. Eng, et al. (2010). Daily physical activity and its contribution to the health-related quality of life of ambulatory individuals with chronic stroke. Health and Quality of Life Outcomes 8: 80.

Robertson, J., E. J. Walkom, et al. (2011). Health systems and sustainability: doctors and consumers differ on threats and solutions. PLoS One 6(4): e19222.

Rose, D., T. Paris, et al. (2010). Feasibility and effectiveness of circuit training in acute stroke rehabilitation. Neurorehabilitation and Neural Repair 25(2): 140-8.

Rosen, P. &I. Karlberg (2002). Opinions of Swedish citizens, health-care politicians, administrators and doctors on rationing and health-care financing. Health Expectations 5(2): 148-55.

Salbach, N. M., N. E. Mayo, et al. (2004). A task-orientated intervention enhances walking distance and speed in the first year post stroke: a randomized controlled trial. Clinical Rehabilitation 18: 509-19.

Shaughnessy, M., K. M. Michael, et al. (2005). Steps after stroke: capturing ambulatory recovery. Stroke 36(6): 1305-7.

Shumway-Cook, A. &M. Woolacott (2001). Motor control. Theory and practical applications. Lippincott Williams and Wilkins Philadelphia.

Simmons-Mackie, N. &R. J. Elman (2011). Negotiation of identity in group therapy for aphasia: the Aphasia Cafe. International Journal of Language and Communication Disorders 46(3): 312-23.

Sjosten, N. &S. L. Kivela (2006). The effects of physical exercise on depressive symptoms among the aged: a systematic review. International Journal of Geriatric Psychiatry 21(5): 410-8.

Taraldsen, K., T. Askim, et al. (2011). Evaluation of a body-worn sensor system to measure physical activity in older people with impaired function. Physical Therapy 91(2): 277-85.

Tilson, J. K., K. J. Sullivan, et al. (2010). Meaningful gait speed improvement during the first 60 days poststroke: minimal clinically important difference. Physical Therapy 90(2): 196-208.

Tudor-Locke, C. E. &A. M. Myers (2001). Methodological considerations for researchers and practitioners using pedometers to measure physical (ambulatory) activity. Research Quarterly in Exercise and Sports 72(1): 1-12.

van de Port, I. G., L. Wevers, et al. (2009). Cost-effectiveness of a structured progressive task-oriented circuit class training programme to enhance walking competency after stroke: the protocol of the FIT-Stroke trial. BMC Neurology 9: 43.

van Peppen, R. P., H. J. Hendriks, et al. (2007). The development of a clinical practice stroke guideline for physiotherapists in The Netherlands: a systematic review of available evidence. Disability and Rehabilitation 29(10): 767-83.

Van Peppen, R. P., G. Kwakkel, et al. (2004). The impact of physical therapy on functional outcomes after stroke: what's the evidence? Clinical Rehabilitation 18(8): 833-62.

Veerbeek, J. M., E. E. Van Wegen, et al. (2011). Is accurate prediction of gait in nonambulatory stroke patients possible within 72 hours poststroke?: The EPOS study. Neurorehabilitation and Neural Repair 25(3): 268-74.

Wevers, L., I. van de Port, et al. (2009). Effects of task-oriented circuit class training on walking competency after stroke: a systematic review. Stroke 40(7): 2450-9.

Yang, Y. R., R. Y. Wang, et al. (2006). Task-oriented progressive resistance strength training improves muscle strength and functional performance in individuals with stroke. Clinical Rehabilitation 20(10): 860-70.

Zanker, K., C. English, et al. (2007). Interdisciplinary circuit class therapy: increasing therapy time on an acute stroke unit. International Journal of Therapy and Rehabilitation 14(11): 483-8.

West, T. and Bernhardt, J. (2011). Physical Activity in Hospitalised Stroke Patients. Stroke Research and Treatment Volume 2012, Article ID 813765, 13 pages doi:10.1155/2012/813765

# Part 8

# Physical Therapy and New Concepts of Wheelchair Design

# 18

# Reaching for Independence: Challenges for a New Concept of Wheelchair Design

Fausto Orsi Medola[1], Carlos Alberto Fortulan[2],
Carla da Silva Santana[1,3] and Valeria Meirelles Carril Elui[1,3]
[1]*Programme of Post-Graduation Interunits in Bioengineering – EESC/FMRP/IQSC - University of Sao Paulo,*
[2]*Department of Mechanical Engineering, Sao Carlos School of Engineering, University of Sao Paulo,*
[3]*Department of Neuroscience and Behavioural Sciences, Faculty of Medicine of Ribeirao Preto, University of Sao Paulo, Brazil*

## 1. Introduction

The wheelchair has been used for many years as the main means of mobility for many people who have diseases or conditions leading to movement impairment. As an assistive technology device, the wheelchair aims to improve locomotion and promote functional independence, allowing the user to perform his/her activities of daily living (Scherer & Cushman, 2001). Although being one of the most used mobility technology, the wheelchair is still referred by users as the main limiting factor in community participation (Chaves et al., 2004).

Why does the wheelchair, as mobility equipment, fail in providing full independence to its users? To understand this limitation, several factors must be considered. Studies have shown high prevalence of pain among wheelchair users, which negatively affects their quality of life and increases their dependence of caregivers (Desroches et al., 2008; Boninger et al., 2004). The pushrim propulsion has been shown to contribute to the development of upper limb overload injuries, mainly due to its mechanical inefficiency (Van der Woude et al., 2001). Equally important, cost and specific features of the equipment such as weight, size, structure and appearance can also determine the success of its use.

So, what can be done to improve the wheelchair design in terms of performance, comfort, functionality and accessibility in order to provide the user full independence? For doing so, it is crucial that the design of the equipment considers not only the users' age, physical and motor conditions, but also their preferences, lifestyles, work, leisure and sport activities, users' history, and finally, their future objectives (Scherer, 2002; Trefler et al., 2004). Recently, alternative modes of wheelchair ambulation have been proposed. However, pushrim propulsion still remains the main mean of wheelchair ambulation, which exposes users' difficulties in adapting to different systems with changes in wheelchair configuration and dimensions.

Over the last years, bioengineering and ergonomics research have successfully generated evidence base for developing wheeled-technology, ergonomic design and fitting procedures, which has significantly contributed to this purpose. Despite of it, the wheelchair technology remains remarkably old-fashioned compared to other mobility technologies already well-established in daily life. This study was aimed at presenting a critical review of wheelchair research and development, focusing on the analysis of how equipment features can determine the success or failure of its use.

## 2. An overview of the wheelchair evolution

Looking at historical records enables the understanding of the evolution of wheeled mobility devices, as well as the adaptations and innovations in face of individual needs. The first image of a seated mobility device was found in a Chinese sarcophagus dating from 525 A.D (Figure 1a). A wheelchair with footrest was designed for the King Phillip of Spain in the sixteenth century as an adaptation to his throne due to his rheumatic disease (Figure 1b).

(a) (b)

Fig. 1. Wheelchairs: (a) image found in a Chinese sarcouphagus (525 a.C.); (b) King Phillip´s throne adapted with a footrest (XVI century). Source: Sawatzky.

In 1655, Stephen Farfler built a wheelchair propelled by the user himself using the upper limbs. This equipment was the precursor of what we know as hand-cycling. Although different from the current wheelchairs, it represented an important advance as the user could control his/her locomotion, thus enabling the user to be socially reintegrated (Figure 2).

Fig. 2. Self-propelling wheelchair (1655). Source: Sawatzky.

In the eighteenth century, the new wheeled mobility devices began to show a clear concern for the user's comfort. Figure 3a shows the representation of a wheelchair with reclining backrest and footrest-height adjustment. In 1916, a wheelchair made of Indian straw had two notable innovations: its lower weight due to the lighter manufacturing material and its configuration with big rear wheels and small front wheels (Figure 3b).

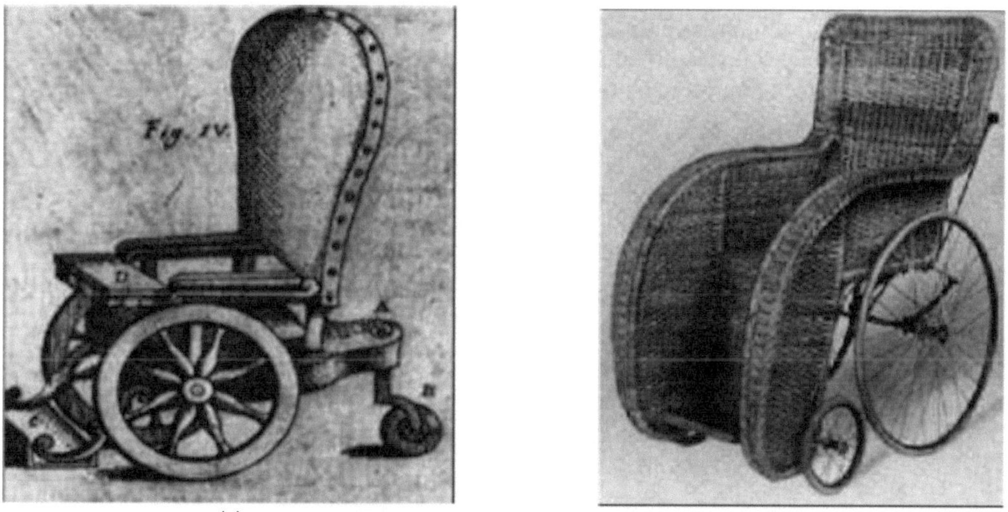


(a) (b)

Fig. 3. Wheelchairs: (a) reclining backrest and footrest adjustment (XVIII century); (b) larger rear wheels and reduced weight due to the use of Indian straw (1916). Source: Sawatzky.

In 1933, Herbert A. Everest, an american who became paraplegic due to a disease, in partnership with the engineer Harry C. Jennings, developed a wheelchair with flexible seat, folding structure made of steel tubes, and pushrim on the rear wheels (Figure 4). This concept of wheelchair remained throughout the last century, still representing the standard model from which other improvements are proposed despite the lack of significant conceptual changes.

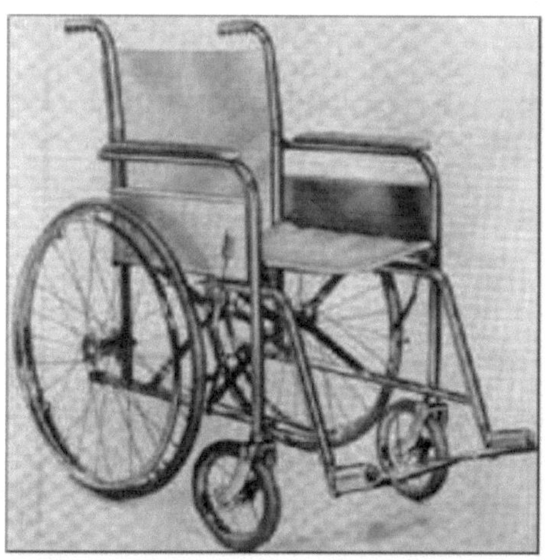

Fig. 4. Mettalic folding structure (1933). Source: Sawatzky.

Through the years, it can be noted some factors determining changes in the design of wheelchairs. Among these factors, one can cite the introduction and popularization of automobiles, which increased the number of car accident victims, most becoming wheelchair-dependent. In addition, the wheelchairs need to be transported in vehicles. The development of rehabilitation programs and improvement in medical services, including the growing number of disabled individuals and the emergence of adapted sports, have favored the development of more sophisticated equipment to better meet the needs of wheelchair users. After the 1948 Paralympic Games in England, the concern was to find lighter materials and meet requirements of versatility and usability for a better sporting performance (Figure 5).

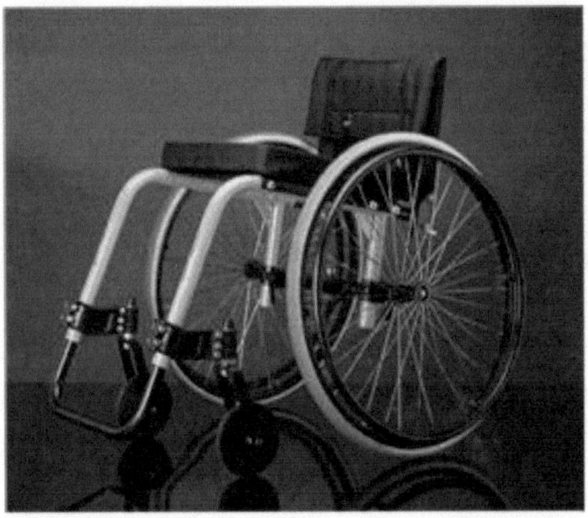

Fig. 5. "*Champion 3000*", sport wheelchair (1986). Source: Carriel, 2007.

## 3. Injuries related to manual wheelchair propulsion

To fully understand the problems involved with prolonged used of manual wheelchairs, it is worth noting that the upper limbs no longer function as before because the loss of motor function in lower limbs causes the upper limbs to perform the task of locomotion. Manual propulsion is the primary means of mobility for wheelchair users, comprising two distinct phases: propulsion phase (or impulse), where there is full contact between hands and push-rims, and recovery phase, where the hands leave the pushrim and swing back to start new contact for another propulsion, thus being characterized as a highly repetitive task (Boninger et al., 2000). Ideally, the recovery phase should be almost entirely done without muscle activity, but differences in wheelchair design and configuration as well as in propulsion technique may contribute to an active recovery phase, adding work load to shoulder muscles. Thus, manual propulsion requires the user to adjust to a particularly stressful work for the upper limb muscles (Wei et al., 2003).

As a result of years of manual wheelchair propulsion, it is believed that the active muscles during the push phase become stronger, while the muscles involved in the recovery phase remain with the same force, creating a muscle imbalance in the shoulder joint (Ambrosio et al. 2005; Mulroy et al., 1996). In addition, the muscles that play an important role in stabilizing the shoulder (rotator cuff, deltoid and long head of biceps) may be changed due to the repetitive nature of wheelchair propulsion (Burnham et al., 1993, Miyahara et al. 1998).

Upper limb pain is a highly prevalent complaint among manual wheelchair users. The study by Sie et. al (1992) found a prevalence of 64% of upper limb pain in persons with paraplegia, with shoulder being the most frequently mentioned site (32%). Curtis et al. (1999) found that 42% of wheelchair users report shoulder pain. Moreover, carpal tunnel syndrome (CTS) has been commonly diagnosed in people who use manual wheelchair. The incidence of CTS in this population ranges from 49% to 63% (Aliure et al. 1985; Gellman, 1988; Tun and Upton, 1988, Davidoff et al. 1991; Steadward and Burnham, 1994; Sie et al. 1992).Furthermore, a correlation was demonstrated between median nerve function and the propulsion rate: higher cadence and larger forces applied to the pushrim are related with reduced median nerve function (Boninger et al., 2004). In addition, ulnar nerve injury has also been reported (Tun and Upton, 1988; Steadward and Burnham, 1994). In consequence, upper limb pain has been associated with poorer quality of life and increased dependence (Boninger et al., 2004, Subbarao et al., 1995).

Several factors may contribute to upper limb injury among wheelchair users, such as body weight and prolonged wheelchair use (Boninger et al., 1999). In addition, it has been suggested that the repetitive and selective activity of muscle groups contributes to the development of a muscle imbalance in the shoulder joint (Myamahara, 1998). Such consequences of the wheelchair propulsion can induce a potentially harmful condition, since wheelchair users rely on their upper limbs for mobility, transfers and most activities of daily living. Understanding the mechanisms involved in this alteration of the upper limbs' mechanics is, therefore, essential to find solutions that minimize or eliminate the risk inherent to the manual use of the wheelchair.

## 4. Critical analysis of the wheelchair as mobility equipment

Conceptually, a manual wheelchair aimed at promoting independent mobility should consider performance, safety, comfort, independence, and transport ease, besides not being harmful to the upper limbs. Despite the diversity of current models and proposed improvements, an equipment covering all these aspects has not been developed yet. In all proposed solutions, improvement in one aspect leads to the impairment of another, which generally limits the acceptance by users. As a result, manual wheelchair propulsion still remains the most widely used form of locomotion among wheeled mobility technologies.

Manual wheelchairs have limitations that make it difficult for the user to reach full independence. Firstly, going uphill is almost impossible due to both the difficulty of propelling and the risk of the wheelchair toppling over, causing the user to fall down. Thus, the user needs the help from another person. Another difficulty is to move around for relatively long distances, because this task requires long-term activity with relatively high frequency use of the upper limbs, causing fatigue and discomfort. The most immediate solution to both problems listed above is the to use a motorized wheelchair. However, although it enables the user to move over long distances and on slopes, the motorized equipment makes the user a "passenger," in a passive condition, resulting in the risk of weight gain and development of cardiovascular disease. In addition, the motorized wheelchairs have higher cost, weight and difficulty in transporting it.

Despite being equipment for promoting mobility, the wheelchair is perceived by the users as the main cause of their limitation at and away from home (Keys et al., 2003). Surprisingly, users find the wheelchair more limiting than their own physical and functional condition. The main complaints are related to weight and higher dimensions of the equipment, making it hard to maneuver, especially in places where space is restricted (Post et al., 1997). In accordance to this statement, Mann et al. (1997) found that 26% of the problems with a wheelchair were related to its weight and size: too heavy to push, too wide to use inside the home.

## 5. Autonomous wheelchairs

The propulsive mechanism of a mobility equipment can be obtained by several resources, including the explosion of fuel, pneumatic system and electric motorization, the latter being the representative of the vast majority of autonomous wheelchairs.

Selecting a wheelchair is a major and complex decision for people with limited mobility. For certain populations with specific functional conditions, there is no clear recommendation for wheelchair prescription regarding mobility mode: manual or powered wheelchair. Thus, the pros and cons of both types of wheelchairs must be considered when choosing the best mode of mobility, depending on the personal lifestyle and preferences, home environment, community accessibility and functional needs (Cooper et al., 2002). Patients with low cervical spinal cord injury typically face this doubt when selecting a wheelchair. The very recent study of Hastings et al. (2011) found that Individuals with C6 and C7 tetraplegia who use manual wheelchairs had significantly better physical function, mobility, and a higher employment rate than those who use power wheelchairs (Hastings et al., 2011). Although the important findings of Hastings et al. (2011) point in the direction to the use of manual wheelchairs, the great diversity among wheelchairs users highlights the need for a customized view of individual's features and needs when prescribing the equipment.

Eletric-powered wheelchairs (EPWs) have been shown to provide independence mobility for children with disabilities (Butler et al, 1982). Understanding the driving behavior of users of electric-powered wheelchairs is critical for designing EPWs, wheelchair components, battery (Cooper et al., 2002). However, when prescribing a powered wheelchair to children, some problem areas can be identified: education of the child about wheelchair usage and drivability, education of the general public about the use of the wheelchair on community spaces and public transportation systems, safety of the users and general public, and establishment of legal status for wheelchair ambulation (Breeed; Ibler, 1982). These factors are critical for a safe and successful usage of EPW and must be taken into account when selecting the equipment.

The advantages of EPW are related to the requirement of very little of the user's strength and endurance which, however, may not be desirable in all instances (Geisbrecht et al., 2009). Physical inactivity seems to contribute to obesity and a cycle of deconditioning and functional decline (Cooper et al., 1999). Furthermore, the weight of the devices (typically 150 lbs or greater) and difficulty in transporting are also limitations of EPW (Geisbrecht et al., 2009; Levy et al., 2004) which, therefore, require expensive vehicle modifications and mechanical lifts (Levy et al., 2010). . In addition, as any electrical equipment, powered wheelchairs have specific issues that can affect overall mobility. Studies have shown a wide disparity in the performance of the batteries and also the performance of the battery chargers of powered wheelchairs (Fisher et al., 1988; Garrett et al, 1990).

Finally, considering that one of the main objectives of an EPW is to provide greater mobility than manual wheelchairs, it is notably surprising that, in terms of daily distance traveled, there is no well-established difference between the two modes of wheeled mobility: while adult manual wheelchair users showed mean daily distance traveled of 1877+1131 meters (Oyster et al., 2011), adult electric-powered wheelchair users were reported to drive an average of 1667 m/day (Cooper et al., 2002). Thus, the motorized equipment does not solve the limitations found in manual wheelchairs related to the amount of mobility, and as a result highlights the need for innovative solutions for wheeled mobility devices.

## 6. Alternative modes of wheelchair propulsion

Alternative modes of wheeled mobility have been proposed in an attempt to enhance the performance, increasing functionality and independence of the users. The hub-crank propulsion system, through which a handle connected to the hub of the rear wheels allows for continuous movement of the hands around the wheel axle, required less effort and showed greater efficiency when compared the pushrim wheelchair propulsion (Van der Woude et al., 1995a; Van der Woude et. al., 1995b; Van der Vlies et al., 1999). To justify the good results, it is believed that the propulsive force exerted by the hands corresponds to only 20% of the cycle, whereas the hub-crank propulsion allows the hands to exert continuous pushing and pulling force through the handle around the hub of the wheel. Thus, both the flexor and extensor group of muscles are involved in the movement cycle, with better distribution of the muscle workload, thereby reducing the amount of work per unit (Van der Woude et al., 2001). However, the use of the hub-crank propulsion wheelchair has been restricted to outdoor environments because of the difficulty in maneuvering it in tight spaces due to its larger width (Van der Woude et al., 2001).

Another proposed solution is the use of lever propulsion systems, in which the arms move cyclically, synchronously or asynchronously. Propulsion systems equipped with a gear on the rear wheels have been recently developed, allowing the wheelchair to be controlled by an activation mechanism located at the top of the lever where the user's hand keeps in contact (Figure 6). Lever propulsion systems have been described as more efficient, requiring less physical effort compared to the pushrim propulsion (Engel et al., 1976; Van der Woude et al., 1993; Van der Woude et al.., 1997). Also, Requejo et al. (2008) found that the use of wheelchair with lever propulsion system reduced and altered the demand for work on the shoulder muscles. Woude et al. (2001) consider the lever propulsion system an interesting alternative for outdoor use, although it can also be used internally, especially for those wheelchair users with lower exercise capacity or those who need to move over greater distances. However, maneuvering and moving with a wheelchair equipped with levers in tight spaces is still a problem that limits its widespread use.

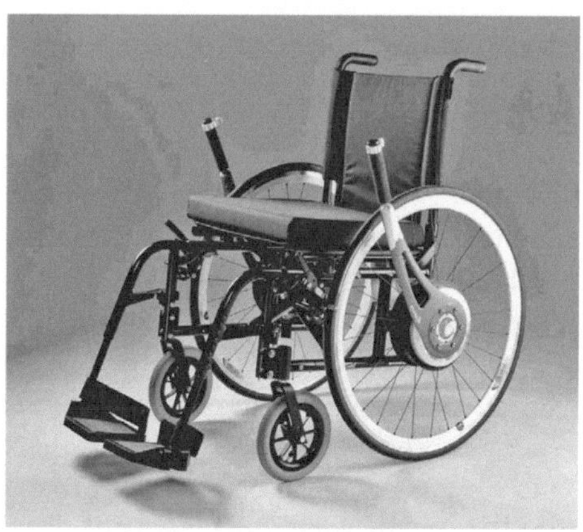

Fig. 6. Lever propulsion system (Requejo et al., 2008)

Stationary arm-crank ergometry has been highly used in exercises for upper body (DiCarlo et al., 1988; Wicks et al., 1983; Sawka et al., 1980), being shown to be a more efficient propulsion mechanism than the hand-rim propulsion (Tropp et al., 1997; Martel et al., 1991). Based on this concept, the arm-crank tricycle propulsion, also called hand-cycling, has become popular in wheeled mobility devices for daily life use and sports (Figure 7). Both synchronously and asynchronously, there is a continuous use of the arm and trunk muscles. However, although hand-cycling wheelchair seems to be the most appropriate mobility system for outdoors, its large dimensions make maneuverability in tight spaces difficult, thus limiting its overall use.

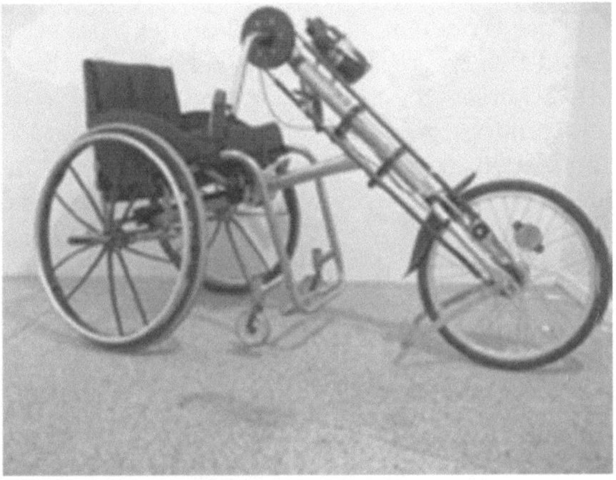

Fig. 7. Hand cycling (Valent et al., 2009).

In the study by Mukherjee et al. (2005), four distinct propulsion systems (pushrim, arm-crank using both arms, arm-crank using one arm and arm lever) were compared in terms of physiological variables (oxygen consumption - VO2 and heart rate). Although no difference had been found, and authors suggested that wheelchair users might have developed certain self-regulatory mechanisms in order to overcome the variation induced by the different propulsion systems.

## 7. Wheelchair engineering: The emergence of an integrative approach

Research and development of wheelchair involves different areas of knowledge so that the perspective of the interaction between body, human movement and equipment's design can be addressed as a whole, with the wheelchair being an extension of the user's body. Woude et. al. (2005) present three important areas in the research and development of wheelchair: mechanics of the equipment, human movement system, and user-wheelchair interface. Thus, it is important not only to adress both disciplines, but also their interaction as this depends on the success use of technology in favor of the user. Because of this multidisciplinary knowledge, the responsibility in conducting research and development of wheelchairs should be shared between health professionals (mainly physical and occupational therapists) and engineers (Mikołajewska; Mikołajewski, 2010). The first evaluate the equipment according to the functional needs of the patient and the latter seek to meet such needs and optimize the functionality of the equipment. Working together, health professionals and engineers can maximize the potential of interaction between humans and equipment.

At this point, it is worth returning to the question that guides this study: Why, compared to the technologically advanced products in a variety of areas, is wheelchair evolution so limited? The technological evolution of the wheelchair has not yet optimally gathered in a single project the three major areas presented by WOUDE et al. (2005): mechanics of the equipment, human movement system, and user-wheelchair interface. Although the use of lighter and stronger materials provided equipment with lower weight, better reliability and durability, the manual locomotor system remains unchanged as high loads are exerted on the upper limbs, which originally were not prepared to develop this function. The advent of the motorized wheelchair seems to solve this problem, but it also involves the imposition of a sedentary life, which increases the risks to the user's health.

The emergence of a special model of care, focused on patients with limited mobility, is essential to reach the advances needed in both healthcare and research and development of mobility equipment. For this new thinking, not only the knowledge of engineering and health should be taken into account, but the frontiers among the disciplines should be trespassed in order to create an open science, with fertile ground for boosting creative thinking. Furthermore, psychological condition, family dynamics in which the patient is inserted, and social relations should also be understood as factors inherent to the patient´s life. Likewise, an insight into work activities and socio-economic conditions favors the understanding of the condition in which the patient lives, thus allowing the equipment to be more appropriately adjusted to this context. Finally, the history of the patient's life is equally important. This includes the patient's expectations, frustrations as well as skills, leisure and sport activities, which make up a range of highly relevant information for determining the ideal equipment capable to promote acceptance and satisfaction. Figure 8 shows a schema of the integrative approach to the wheelchair user.

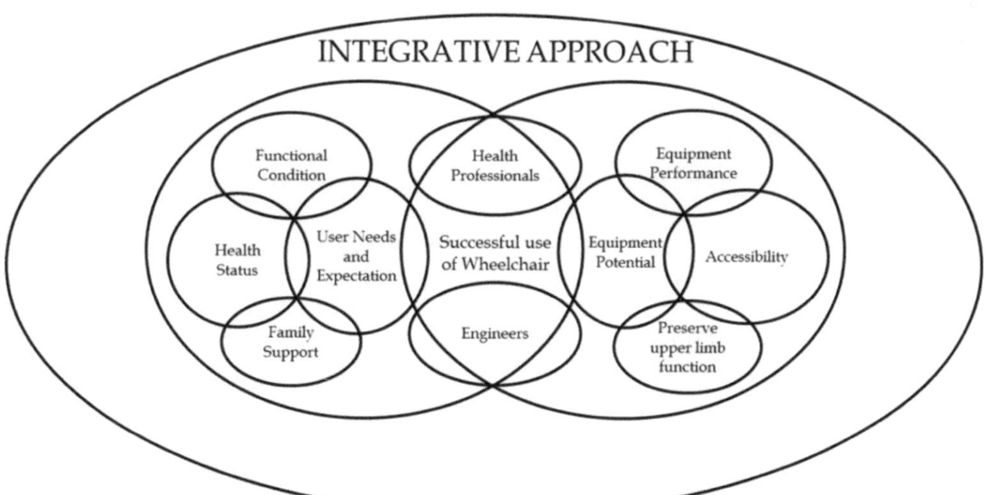


Fig. 8. Model of an integrative approach to the wheelchair user.

## 8. Ergonomics of the wheelchair pushrim design: A case study

As the interface by which the user drives the wheelchair, the pushrim plays a determinant role in the user's ability to control the wheelchair. The conventional pushrim, found in the majority of manual wheelchairs, is made of circular metal tubes located at a distance of 20 mm from the wheel. The tube diameter (20 mm) of these push-rims is too small for adults, whose hand length is approximately 180 mm (5). In consequence, the contact area between hand and pushrim is limited, leading to an increased pressure on the contact points of the delicate structures of the hand (Figure 9). Furthermore, the inability to hold the pushrim with the entire palm and fingers reduces the mechanical efficiency, as more muscle activity is required to stabilize the hand instead of promoting power for propulsion of the wheelchair (Van der Woude et al., 2003).

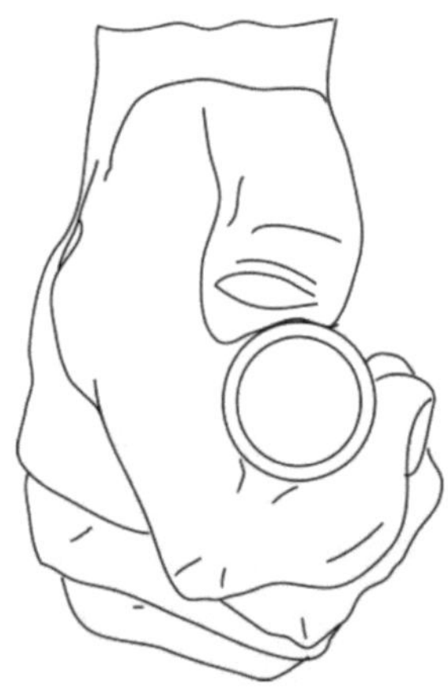

Fig. 9. Cross-sectional view of conventional pushrim (Medola et al., 2011).

It is noteworthy that wheelchair users often report that the design of the pushrim does not fully meet their needs during wheelchair propulsion, and in a survey with manual wheelchair users, only 39% reported using solely the pushrim for propulsion, and the majority (54%) reported holding both the pushrim and tire simultaneously (Perks et al., 1994).

### 8.1 An ergonomic approach

Based on ergonomics concept, a new design of wheelchair pushrim must have, firstly, larger contact surface without increasing wheelchair's dimensions. Also, the shape of the new pushrim must be proper to a comfortable and secure hand grip. In order to reach this goal, the new device features a slightly curved upper surface on which the thumb, the thenar eminence, and the base of the hypothenar eminence can rest; a lateral surface to support the distal half of the palm and proximal phalanges of fingers II, III, IV and V as well as a lower surface to support the medial and distal phalanges of the fingers are also available. Based on these features, a preliminary proposal for a new design of the wheelchair push-rim is shown in Figure 9a. In contrast to the conventional pushrim (20 mm diameter), which provides 68.8 mm of contact surface, the new pushrim was designed with approximately 123 mm of surface for hands with a length of approximately 180 mm (Pheasant & Haslegrave, 2006). This design leads to a better posture of the hands for a proper control of the wheelchair, thus allowing the hands to be fully supported for a stable, firm and functional grip (FIGURE 10)

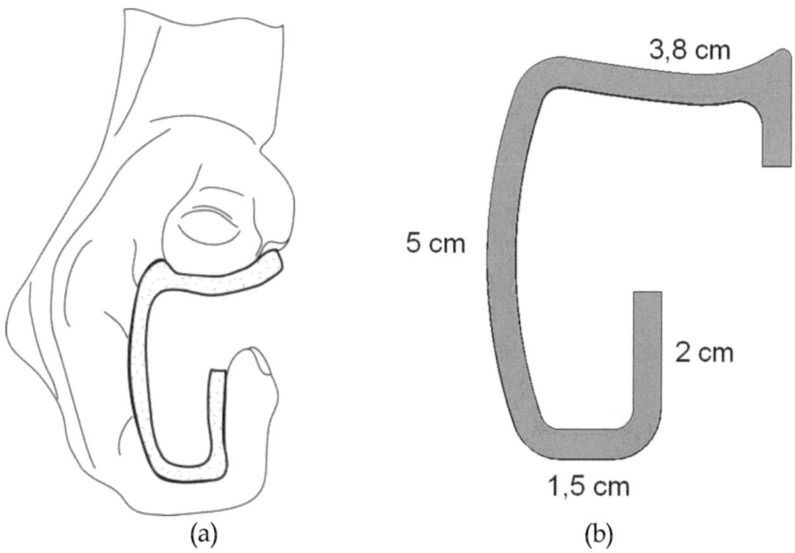


Fig. 10. Applying ergonomics: (a) preliminary proposal for the pushrim design; (b) defined project: shape and dimensions (Medola et al., 2011).

As for any hand operated device, the material used should provide both thermal comfort and adequate friction for the hands. In general, metallic materials should be avoided because they mechanically compress the tissues of the hand and increase the transmission of cold, heat and vibration. Conversely, in addition to the high stiffness and resistance to deformation under load, polymers offer a gain in comfort during wheelchair use by reducing the sensation of heat on the user's hands. For these reasons the polyurethane was used to develop the pushrim prototype (Medola et al., 2011).

### 8.2 Preliminary results of the ergonomic wheelchair pushrim

Some features of the ergonomic pushrim really contributed for an improvement on its design and, therefore, will be briefly described. First, the larger surface contact positioned the fingers with a less flexed posture than the conventional pushrim. Furthermore, the convex shape of the lateral surface provided adequate support for the entire palm, requiring less effort of the fingers to hold the pushrim. Figure 11 shows the positioning of the hand in the pushrim, without the excessively flexor posture of the fingers as observed for the conventional pushrim. It can also be noted that the thumb has a proper support in the upper surface of the new pushrim. By using the space between push-rim and wheel, the new device was able to provide adequate support to the thumb without, however, increasing the width of the wheelchair, which could make it difficult to reach tight spaces.

Thus, with an innovative design, the ergonomic approach showed to be potentially beneficial for the old concept of wheelchair pushrim. The use of anthropometrics in the pushrim design allowed the development of a prototype suitable for a firm and stable hold, by providing a larger contact area between the hand and the device, thus reducing the effort of the fingers to hold the pushrim (Medola et al., 2011).

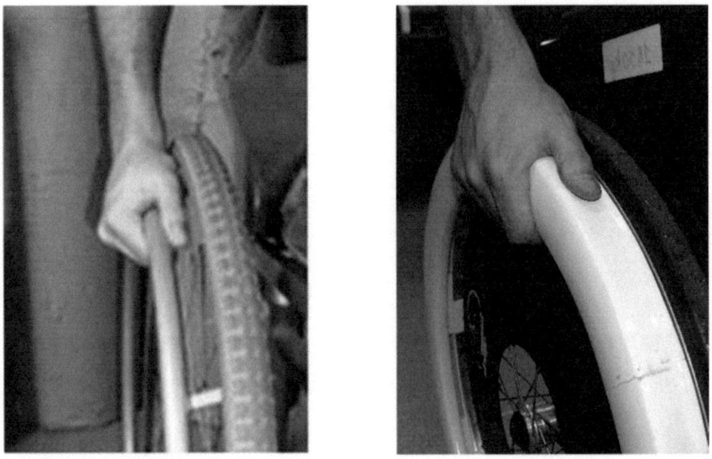

(a) (b)

Fig. 11. Hand coupling to: (a) conventional pushrim; (b) ergonomic pushrim. (Medola et al., 2011).

## 9. Conclusion

By observing the evolution of the wheelchair as mobility equipment over time, it can be noted that the user's individual needs have led to the creation of equipment with characteristics based on a concept of a wheelchair that, for nearly a century, has been widely used as the technology available for people with mobility problems. However, the manual wheelchair propulsion imposes a condition potentially harmful in long term use, which can cause the upper limbs to fail in promoting independent mobility. In an attempt to minimize or even eliminate the adverse effects of using the manual wheelchair, several devices have been created with important improvements in design and pattern of muscle work of the upper limbs. Because the equipment's size and weight affect the user's mobility in tight spaces and make it difficult to adapt to new techniques of upper limb movement, these devices have their overall acceptance limited and thus the pushrim propelled wheelchair remains as the main wheeled mobility device. Finally, we present a model for a new approach to rehabilitation engineering, based on a holistic view, that integrates patient, equipment and environment. The holistic framework for the integrated work between health professionals and engineers favors the emergence of creative solutions to major problems found in the current concept of manual wheelchair. The emergence of the integrative concept will also contribute to research, development and, consequently, production of scientific knowledge that goes beyond the very narrow limits of the disciplines, creating a unique and integrated science between the different knowledge areas.

## 10. Acknowledgements

This research was supported by CAPES (Coordination for the Improvement of Higher Level -or Education- Personnel) - Brazil. The authors would also like to thank the financial support provided by the Fund for the Support of Teaching, Research and Assistance (FAEPA) of the Clinical Hospital of the Faculty of Medicine of Ribeirao Preto - University of Sao Paulo.

## 11. References


Aliure J, Eltorai I, Bradley WE, Lin JE. Johnson B (1985). Carpal tunnel syndrome in paraplegic patients. *Paraplegia*, v.23, pp. 182-186.

Ambrosio F, Boninger ML, Souza AL, Fitzgerald SG, Koontz AM, Cooper RA (2005). Biomechanics and Strength of Manual Wheelchair Users. *Journal of Spinal Cord Medicine*, v.28, pp. 407-414.

Boninger ML, Impink BG, Cooper RA, Koontz AM (2004). Relation between median and ulnar nerve function and wrist kinematics during wheelchair propulsion. *Archives of Physical Medicine Rehabilitation*, v.85, pp.1141-1145.

Boninger ML, Baldwin MA, Cooper RA, Koontz AM, Chan LC (2000). Manual wheelchair pushrim biomechanics and axle position. *Archives of Physical Medicine Rehabilitation*, v.81, pp.608-613.

Boninger ML, Cooper RA, Baldwin MA, Shimada SD, Koontz AM (1999). Wheelchair pushrim kinetics: body weight and median nerve function. *Archives of Physical Medicine and Rehabilitation*, v. 80, pp.910–915.

Breed AL, Ibler I (1982). The motorized wheelchair: new freedom, new responsibility and new problems. *Developmental Medicine and Child Neurology*, v.24, pp. 366-371.

Burnham RS, Steadward RD (1994). Upper extremity peripheral nerve entrapments among wheelchair athletes: prevalence, location, and risk factors. Archives of Physical Medicine and Rehabilitation, v.75, pp;519–524.

Butler C, Okamotot GA, McKay TM (1983). Powered mobility for very young disabled children. *Developmental Medicine and Child Neurology*, v.25, pp. 472-474.

Carriel IRR (2007). Ergonomic recommendations for the design of wheelchair: considering the physiological and cognitive aspects of the elderly. Dissertation (Master of Industrial Design) - Faculty of Architecture, Arts and Communication, Sao Paulo State University, Bauru, Sao Paulo, Brazil, 244p.

Chaves ES, Boninger ML, Cooper R, Fitzgerald SG, Gray DB, Cooper RA (2004). Assessing the influence of wheelchair technology on perception of participation in spinal cord injury. *Archives of Physical Medicine Rehabilitation*, v.85, pp.1854-1858.

Cooper RA, Thorman T, Cooper R, Dvorznak MJ, Fitzgerald SG, Ammer W, Song-Feng G, Boninger ML (2002). Driving characteristics of electric-powered wheelchair users: How far, fast, and often do people drive? *Archives of Physical Medicine and Rehabililitation*, v.83, pp.250–255.

Cooper RA, Quatrano LA, Axelson PW, Harlan W, Stineman M, Franklin B, Krause JS, Bach J, Chambers H, Chao EY, Alexander M, Painter P (1999). Research on physical activity and health among people with disabilities: A consensus statement. *Journal of Rehabilitation Research and Development*, v.36, pp. 142–54.

Curtis KA, Tyner M, Zachary L (1999). Effect of a standard exercise protocol on shoulder pain in long-term wheelchair users. *Spinal Cord* 1999;37:421–9.

Davidoff G, Werner R, Waring W (1991). Compressive mononeuropathies of the upper extremity in chronic paraplegia. *Paraplegia*, v. 29, pp.17-24.

Desroches G, Aissaoui R, Bourbonnais D (2008). Relationship between resultant force at the pushrim and the net shoulder joint moments during manual wheelchair propulsion in elderly persons. *Archives of Physical Medicine Rehabilitation*, v. 89, pp. 1155-1161.

Engel P, Neikes M, Bennedik K, Hildebrandt G, Rode FW (1976). Work physiological studies performed to optimate the lever propulsion and the seat position of a lever propelled wheelchair. *Rehabilitation (Stuttg)*, v.15, n.4, pp. 217–228.

Fisher WE, Garrett RE, Seeger BR (1988). Testing of gel-electrolyte batteries for wheelchairs. *Journal of Rehabilitation Research and Development*, v.25, pp.27-32.

Garrett RE, Hartridge M, Seeger BR (1990). Comparative evaluation of chargers for wheelchair gel cell batteries. *Australian Physical & Engineering Sciences in Medicine*, v.13, pp.148-56.

Gellman H, Chandler DR, Petrasek J, Sie I; Adkins R, Waters RL (1988). Carpal tunnel syndrome in paraplegic patients. *Journal of Bone and Joint Surgery American*, v.60, p.517-519, 1988.

Giesbrecht EM, Ripat JD, Quanbury AO, Cooper JE (2009). Participation in community-based activities of daily living: Comparison of a pushrim-activated powerassisted wheelchair and a power wheelchair. *Disability and Rehabilitation: Assistive Technology*, v.4, pp.198–207.

Hastings J, Robins H, Griffiths Y, Hamilton C. The differences in self-esteem, function, and participation between adults with low cervical motor tetraplegia who use power or manual wheelchairs (2011). *Archives of Physical Medicine and Rehabilitation*, v. 92, pp. 1785-1788.

Levy CE, Chow JW, Tillman MD, Hanson C, Donohue T, Mann WC (2004). Variable ratio power assist wheelchair eases wheeling over a variety of terrains for elders. *Archives of Physical Medicine and Rehabilitation*, v. 85, pp. 104–112.

Levy CE, Buman MP, Chow JW, Tillman MD, Fournier KA, Giacobbi P (2010). Use of power assist wheels results in increased distance traveled compared with conventional manual wheeling. *American Journal of Physical Medicine and Rehabilitation*, v.89, pp. 625–634.

Mann WC, Hurren D, Charvat B (1996). Problems with wheelchair experienced by frail elders. *Technology and Disability*, v. 5, pp. 101-111.

Martel GM, Noreau L, Jobin J (1991). Physiological responses to maximal exercise on arm cranking and wheelchair ergometer with paraplegics. *Paraplegia*, v. 29, pp. 447 – 456.

Medola FO, Fortulan CA, Purquerio BM, Elui VMC (2011). A new design for an old concepto f wheelchair pushrim. *Disability and Rehabilitation: Assistive Technology*, Epub ahead of print.

Mikołajewska E, Mikołajewski D (2010). Wheelchair Development from the Perspective of Physical Therapists and Biomedical Engineers. *Advances in Clinical Experimental Medicine*, v. 19, n. 6, pp. 771-776.

Miyahara M, Sleivert GG, Gerrard DF (1998). The relationship of strength and muscle balance to shoulder pain and impingement syndrome in elite quadriplegic wheelchair rugby players. *International Journal of Sports Medicine*, v.19, n.3, pp.210–214.

Mulroy SJ, Gronley JK, Newsam CJ, Perry J (1996). Electromyographic activity of the shoulder muscles during wheelchair propulsion by paraplegic persons. *Archives of Physical Medicine Rehabilitation*, v.77, n.2, pp.187–193.

Oyster ML, Karmarkar AM, Patrick M, Read MS, Nicolini L, Boninger ML (2011). Investigation of factors associated with manual wheelchair mobility in persons with spinal cord injury. *Archives of Physical Medicine and Rehabilitation,* v.92, n.3, pp.484-490.

Perks BA, Mackintosh R, Stewart CP, Bardsley GI (1994). A survey of marginal wheelchair users. *Journal of Rehabilitation Research & Development, v.*31, pp.297–302.

Pheasant S, Haslegrave CM (2006). *Bodyspace:* anthropometry, ergonomics, and the design of work. Taylor & Francis, Philadelphia, 332 pp.

Post MW, Asbeck F, Dijk AJ, Schrijvers AJ (1997). Services for spinal cord injured: availability and satisfaction. *Spinal Cord,* v. 35, pp. 109-15.

Requejo PS, Lee SE, Mulroy SJ, Haubert LL, Bontrager, EL, Gronley JK, Perry J (2008). Shoulder Muscular Demand During Lever-Activated Vs Pushrim Wheelchair Propulsion in Persons With Spinal Cord Injury. *Journal of Spinal Cord Medicine,* v.31, pp. 568–577.

Sawatzky B. Wheeling in the new millenium: The history of the wheelchair and the drivng forces in the wheelchair design today. Available in: <http://www.wheelchairnet.org/WCN_WCU/SlideLectures/Sawatzki/WC_history.htm> Acess in: September, 2010.

Sawka MN, Glaser RM, Wilde S (1980). Metabolic and circulatory responses to wheelchair and arm crank exercise. *Journal of Applied Physiology Environmental Exercise Physiology,* v. 49, pp. 784–788.

Scherer MJ (2002). *Assistive technology:* matching device and consumer for successful rehabilitation. Washington, DC: American Psychological Association, 325pp.

Scherer M, Cushman L (2001). Measuring subjective quality of life following spinal cord injury: a validation study of assistive technology device predisposition assessment. *Disability and Rehabilitation,* v.23, pp. 387-393.

Sie IH, Waters RL, Adkins RH, Gellman H (1992). Upper extremity pain in the postrehabilitation spinal cord injured patient. *Archive of Physical Medicine Rehabilitation,* v.73, pp. 44-48.

Subbarao JV, Klopfstein J (1995). Prevalence and impact of wrist and shoulder pain in patients with spinal cord injury. *Journal of Spinal Cord Medicine,* v.18, n. 1, pp. 9-13.

Trefler E, Fitzgerald SG, Hobson DA, Bursick T, Joseph R (2004). Outcomes of wheelchair systems intervention with residents of long-term care facilities. *Assistive Technology,* v.16, n.1, pp.18-27.

Tropp H, Samuelsson K, Jorfeldt L (1997). Power output for wheelchair driving on a treadmill compared with arm crank ergometry. *British Journal of Sports Medicine,* v. 31, pp. 41-44.

Tun CG, Upton J (1988). The paraplegic hand: electrodiagnostic studies and clinical findings. *Journal of Hand Surgery American,* v.13, pp. 716-719.

Linda J.M. Valent, Annet J. Dallmeijer, Han Houdijk, Hans J. Slootman, Thomas W. Janssen, Marcel W.M. Post, Lucas H. van der Woude (2009). Effects of Hand Cycle Training on Physical Capacity in Individuals With Tetraplegia: A Clinical Trial. *Physical Therapy,* v. 89, pp. 1051–1060.

Van der Vlies FWJ, Gerritsma CJ, Veeger HEJ (1999). Physiological responses in hubcrank and hand rim wheelchair propulsion using a computer controlled wheelchair ergometer, in van der Woude LHV, et al. (eds): *Biomedical Aspects of Manual Wheelchair Propulsion: State of the Art II.* Amsterdam, IOS Press, pp. 190–194.

Van der Woude LHV, Groot S (2005). Wheelchair propulsion: a straining form of ambulation *Indian Journal of Medical Research,* v.121, pp. 719-22.

Van Der Woude LHV, Formanoy M, Groot S (2003). Hand rim configuration: effects on physical strain and technique in unimpaired subjects? *Medical Engineering & Physics,* v. 25, pp. 765–774.

Van der woude LHV, Dallmeijer AJ, Janssen TWJ, Veeger D (2001). Alternative modes of manual wheelchair ambulation: An overview. *American Journal of Physical Medicine and Rehabilitation,* v. 80, pp. 765–777.

Van der Woude LHV, Botden E, Vriend I, Veeger D (1997). Mechanical advantage in wheelchair lever propulsion: effect on physical strain and efficiency. *Journal of Rehabilitation Research and Development,* v. 34, n.3, pp. 286–294.

Van der Woude LHV, Kranen E, Ariens G (1995). Physical strain and mechanical efficiency in hubcrank and handrim wheelchair propulsion. *Jounal of Medical Engineering Technology,* v. 19, pp.123-131a.

Van der Woude LHV, Maas K, Rozendal RH (1995). Physiological responses during hub crank and hand rim wheelchair propulsion. Journal of Rehabilitation Science, v.8, pp.13–19b.

Van der Woude LHV, Veeger HE, de Boer Y, Rozendal RH (1993). Physiological evaluation of a newly designed lever mechanism for wheelchairs. *Jounal of Medical Engineering Technology,* v. 17, n.6, pp.232–240.

Veeger HE, van der Woude LHV, Rozendal RH (1991). Load on the upper extremity in manual wheelchair propulsion. *Journal of Electromyography Kinesiology,* v.1, n.4, pp. 270–280.

Wei SH, Huang SL, Jiang CJ, Chiu JC (2003). Wrist kinematic characterization of wheelchair propulsion various seating positions: implication to wrist pain. *Clinical Biomechanics,* v.18, pp.46-52.

Wicks JR, Oldridge NB, Cameron BJ (1983). Arm cranking and wheelchair ergometry in elite spinal cord injured athletes. *Medicine Science and Sports Exercise,* v. 3, pp. 224–231.